Molecular Physical Chemistry for Engineering Applications

Florin Emilian Daneş • Silvia Daneş
Valeria Petrescu • Eleonora-Mihaela Ungureanu

Molecular Physical Chemistry for Engineering Applications

Springer

Florin Emilian Daneş
École Polytechnique
de l'Université de Nantes
Institut des Sciences de l'Ingénieur en
Thermique-Énergétique et Matériaux
Nantes, France

Silvia Daneş
Department of Inorganic Chemistry,
Physical Chemistry and Electrochemistry
University POLITEHNICA of Bucharest
Bucharest, Romania

Valeria Petrescu
Department of Inorganic Chemistry,
Physical Chemistry and Electrochemistry
University POLITEHNICA of Bucharest
Bucharest, Romania

Eleonora-Mihaela Ungureanu
Department of Inorganic Chemistry,
Physical Chemistry and Electrochemistry,
University POLITEHNICA of Bucharest
Bucharest, Romania

ISBN 978-3-030-63898-6 ISBN 978-3-030-63896-2 (eBook)
https://doi.org/10.1007/978-3-030-63896-2

This Springer imprint is published by the registered company Springer Nature Switzerland AG
The registered company address is: Gewerbestrasse 11, 6330 Cham, Switzerland

The words are insufficient to express the great pain we feel today, on 6th of June 2021. The main author, Florin-Emilian DANEŞ, went to heaven. He went to the stars! We will miss Florin very much. He was a learned man who knew how to answer your every question. I always mentioned "this, only Florin knows."

He has suffered a lot the last year. He was a good man, he helped everyone. Where did all that erudition, all that knowledge, all that culture that he had go?

Goodbye, dear friend of a lifetime! We will miss you enormously. God bless you!

The other authors

To learn a trade, in ancient times, an apprentice was given by his family to a master to live and train with him. Professor Florin Danes, like the craftsmen of times past, trained his students in physical chemistry, in how to do research, how to write a thesis, how to start a teaching career, how to play bridge, and how to lead a life of continuous learning. He taught us how and when to start sharing with our "disciples" what we learnt from him. Florin was always ready to listen, to help in a way only family is supposed to help, and was "there" whenever his former students needed him. Now we must continue in a world where our Teacher is no more. Let's think about everything he gave us and be grateful that we met him and grew into our own looking up at him.

Mihaela LEONIDA, Professor, Dept.
Chemistry, Biochemistry, and Physics
Fairleigh Dickinson Univ. Teaneck, NJ, USA

Professor Florin Emilian DANEȘ was a "Primus inter pares" among our eminent professors formed after the Second World War, influencing the Romanian School of Chemistry and in particular the Applied Sciences—Chemical Engineering one. Now this wide influence can be noticed and admired, among the chemistry professors all over the world (students or coworkers). His dimension is difficult to be described, since his work is still ongoing with several generations of chemists, acting and building a future with a profound European accent. His education of German and French origin influenced and offered a higher "valence" to its Romanian gifted talent for teaching. His origin on the beautiful Olt Valley Romania was not by chance, since he belongs to a family with a fine tradition in teaching and clergy. Later he fulfilled the call of destiny and mentored generations of researchers or professors in France and in Romania. France was also the adoptive mother of our professor and offered him the life, difficult to achieve at homeland in a certain period. Like an arch between worlds and time, he always initiated the right "thermodynamic system" for anyone, fulfilling his profound and universal beliefs in science and peoples. A light is guiding us forever in the endless space of universal knowledge.

Mihai Cosmin COROBEA, PhD Eng, Senior Researcher I, INCDCP-ICECHIM Bucharest, Romania

Authors

All four authors have experience in teaching thermodynamics and other topics of physical chemistry to students studying chemical engineering, metallurgy and electronics at the Polytechnic Institute of Bucharest (IPB) – today the University POLITEHNICA of Bucharest (UPB). They conducted fundamental and applied research in the university's chemistry lab.

- Chemical engineering graduate from IPB; Doctor of Science from the University of Halle, Germany; HDR Dr. Habil from the Polytechnic Institute of Grenoble; and professor at the University of Nantes, France, **Florin Emilian DANEŞ** is specialised in foreseeing the thermodynamic characteristics of compounds and modelling the coupled transfer of mass/electricity/heat at the interfaces for industrial separations as well as for producing thermal-conducting electro-insulating polymers.
- Chemical engineer and Doctor of *Processes and Devices* from the Faculty of Industrial Chemistry at IPB, **Silvia DANEŞ** is specialised in thermodynamics of the separation process of complex compounds of soluble mineral substances – through extraction, ionic exchange and precipitation – as well as in the thermodynamics of solutions of electrolytes, power sources and corrosion.
- Graduate from the Faculty of Chemistry at the University of Bucharest, then Doctor from the Centre of Physical Chemistry of Bucharest of the Romanian Academy, **Valeria PETRESCU** is specialised in the electrochemistry of saline compounds, thermodynamics of melted or watery electrolytes, technology of combustion files and the polarisation of porous electrodes.
- Chemical engineer and then Doctor of Engineering from the Faculty of Chemical Engineering at IPB, **Eleonora-Mihaela UNGUREANU** took postdoctoral perfecting courses in France and Denmark. Specialised in the electrochemistry of organic compounds, she is professor and doctoral studies manager in engineering sciences at the University POLITEHNICA of Bucharest, with annual research in the fields of the electrochemical processes in organic solvents and modified electrodes, and molecular recognition and functionalisation of carbon nanotubes.

From the Same Authors

Daneş F., Daneş S.: Echilibrul chimic şi calcularea lui, Ed. Tehnică, Bucureşti 1982.

Daneş F., Daneş S., Petrescu V., Ungureanu E.-M.: Termodinamica chimică – o termodinamică a materiei, Ed. AGIR, Bucureşti 2013.

Daneş F., Garnier B.: "Maîtrise de l'utilisation de l'énergie– bilans et utilisation efficiente et rationnelle, illustrés par des exemples et exercices corrigés. Coll. Technosup/énergie et procédés", Ed. Ellipses, Paris 2012.

Daneş F., Ungureanu E.-M.: "Cinetica transformărilor fizico-chimice", Ed. AGIR, Bucureşti 2009.

Petrescu S., Petrescu V.: "Principiile termodinamicii – evoluţie, fundamentări, aplicaţii", Ed. Tehnică, Bucureşti 1983.

Petrescu S., Petrescu V.: "Metode şi modele în termodinamica tehnică. Seria Termo-Frig", Ed. Tehnică, Bucureşti 1988.

Sternberg S., Daneş F.: "Termodinamica chimică aplicată", Ed. Academiei, Bucureşti 1978.

Ungureanu E.-M.: "Introducere în electrochimia organică", Ed. Tehnică, Bucureşti 1999.

Ungureanu E.-M.: "Experimente de electrochimie organică", Ed. Politehnica, Bucureşti 2006.

Ungureanu E.-M.: "Electrochimia organică, de la fundamente la aplicaţii", Ed. Politehnica, Bucureşti 2010.

Preface

Macroscopic and Microscopic in Matter Sciences

This is a textbook of chemical thermodynamics dedicated to physical molecular chemistry.

The concept of a "molecule" is auxiliary to the practical usage and to the understanding of the axiomatic construction of thermodynamics. Indeed, the passing from the characteristics of energy exchange processes to the properties of matter is attainable just as well not only through reporting it to the system's number of moles (proportionally to its number of molecules), but also through dividing at the system's mass, just as it is often done in technical energetics and almost always in the so called "technical thermodynamics" by the thermic energy specialists.

However, for the physical chemist, the concepts of classical theoretical thermodynamics, a phenomenological discipline – consisting of the generalisation of measurement data through the four principles that stand at the base of a rigorous axiomatic construction – prove to be insufficient: thermodynamics allows for validating one through the other, the values obtained for the different characteristics of substances, but it does not provide any help in foreseeing these values, which obviously depends on the chemical compositions of the matter. Using modelling at a molecular scale is the way to obtain this predicting goal.

As compared to the physical-chemical treatment of the classical thermodynamics, with its formal nature at a phenomenologically, empirically and macroscopically limited scale, and with rigorous, predominantly deductive reasonings, the molecular stand point of this book is a lot more intuitive, and its scope a lot larger, the theory being constructed on a structural base at a microscopical scale, the modelling mechanisms and the inductive phases playing an important role (to be noted here, microscopic means "that cannot be seen even using a microscope"!). There is, of course, only one price for these advantages of the molecular treatment over the phenomenological one, a drop in the rigorous nature of the conclusions, the results being a bit uncertain, insufficiently accurate, and less exact.

The Document Subdivision

The book is organised into three parts, as follows:

Part I – Molecular Structure of Matter
Part II – Molecular Models in Thermodynamics
Part III – Transportation: Phenomenon and Mechanism

In the first part, the characterisation methods of the nature of the composing molecules of substances from a given system are introduced, thoroughly described, justified, and exemplified. In this introductory part, the results of the three subfields of molecular physics are exposed in a simplified manner: the quantic theory, statistical physics, and the kinetic-molecular theory of heat. The following two parts use the methods of molecular physics, exposed in the first part, for studying the states of material systems and of processes that may happen at the level of these systems.

The second part of the book is dedicated to the molecular modelling of the matter's properties in a state of equilibrium, and the third part describes the application of the molecular concepts in a state of disequilibrium of the matter, especially in modelling some of the most important processes that may occur in a situation of disequilibrium: the transportation phenomena.

The macroscopic phenomenological laws are treated differently in the two parts dedicated to applications.

Indeed, the basic phenomenological aspects given by thermodynamics are assumed to be known. They are discussed in two textbooks (Daneş et al. 2013 [1], 2016 [2]) to which this book refers numerous times. For kinetics, i.e., the science of phenomena rates, the macroscopic laws are not commonly familiar to students of technical disciplines. These laws are introduced at the beginning of the textbook dedicated to physical kinetics (Daneş and Ungureanu 2009 [3]). This textbook treats more extensively the basic laws of chemical kinetics as well as their applications in industrial processes. Each three big topics of the present textbook is subdivided into chapters. The logic behind is described in the introductory text of each part.

The authors hope that these explanations will ease the reader into the text and into understanding it.

We would like to acknowledge our collaborators in the elaboration of this book: Dr.Ing. Rodica-Magdalena BUJDUVEANU, the PhD students Ovidiu-Teodor MATICA, Alina-Alexandra VASILE, Alina BROTEA, Cornelia MUŞINĂ, Laura-Bianca ENACHE, Veronica ANĂSTĂSOAIE, and Alexandru IVANOV, Dr.Ing. Cristina ŞERBĂNESCU (who corrected most of the typing errors).

Grenoble and Bucharest
June 2020

F. E. Daneş
S. Daneş
V. Petrescu
E. -M. Ungureanu

Contents

Notations (Symbols)

Latin Symbols

A	electron affinity
	* atomic mass
B	coefficient of virial
c	rate (of movement)
	* concentration (mol/m^3)
	* specific heat
C	caloric capacity (molar)
d	intermolecular distance = molecular diameter
D	diffusivity
	* inter-nuclear distance
E	molar energy of the level
f	number of degrees of freedom
	* Boltzman factor of the level
	* force on the unit of surface
	* fugacity
F	standard integral distribution function
	* molar free energy
g	multiplicity of a (energetic) level
	* conservative extensive unit (transported)
G	molar free enthalpy
	* function of integral distribution
	* shear mode
G'	conservative intensive property

H	molar enthalpy
i	(electric) ionising potential
I	moment of inertia
I, J	definite integral
j	flow density
K	material constant
l	length
ℓ	free path
L	variance (=number of macroscopic degrees of freedom)
	* code of network coordination
m	mass (extensive)
M	molar mass
n	amount of substance (in moles)
p	momentum
	* pressure
P	probability (mathematical)
Q	heat
r	distance (to the axis or point)
s	surface (= area)
s	entropy (extensive)
S	molar entropy
t	time
T	temperature
u	internal energy (extensive)
	* rate of oscillations' propagation (including of sound)
	* intermolecular potential (molecule's internal energy)
U	molar internal energy
v	speed (of movement)
	* rate (of reaction or process)
	* volume (extensive)
$\mathbf{v_f}$	free volume (available for the molecule)
V	molar volume
w	rate of movement
	* weight (to a mediation)
W	thermodynamic probability
X	molar fraction
z	compression factor P·V/(R·T)
	* frequency of molecule's collisions with other molecules,
Z	density of collisions, $m^{-3} \cdot s^{-1}$
	* amount of state = partition function
$\mathbf{Z_s}$	collision density molecule/surface, $m^{-2} \cdot s^{-1}$

Greek Symbols

Γ differential distribution function
Δ difference operator
$\varepsilon\varepsilon$ energy of one particle
ε_c particle's kinetic energy
ε_p particle's potential energy
ξ Einstein function
η viscosity
θ characteristic temperature

* fraction of voids (within the network)

λ thermal conductivity

* wavelength
* latent heat

Λ quantum number of molecular orbital

* number of neighbours

μ molecules density, m^{-3}

* atomicity = no. of atoms from molecule
* moment of electrical dipole

μ_i chemical potential of the i-th component
ν kinematic viscosity

* frequency

$\bar{\nu}$ wave number
Π product operator
ρ density (in fact), in kg/m^3
Σ amount operator
σ symmetry code

* equilibrium intermolecular distance

Φ standard differential distribution function
φ fugacity coefficient
χ compressibility
ψ wave function

Indices

B Boyle
c critic
cr crystal

D	Debye
E	Einstein
e	equilibrium
ef	actual
el	electronic
ext	extensive
f	boiling (vaporisation at 1 atm)
	* physical
fr	friction
g	geometric
G	gas
int	internal
K	Knudsen
L	liquid
m	multiplicity of spectral term)
med	mean (arithmetic) of distribution
mg	magnetic
n	nuclear
p	quadratic (q. mean)
P	constant pressure (isobar)
pc	pseudo-critical (pc. unit)
r	relative
r	reduced by division with the critical parameter
red	reduced by division with the reference value
rot	rotation to Z
rt	of network
s	constant entropy (isentropy)
S	solid
sb	sublimation
spc	absolute value =spectroscopic
T	constant temperature (isothermal)
tch	practical value, thermochemical
tr	translation
v	constant volume (isochore)
	* speed
	* vaporisation
VdW	Van der Waals
vib	vibration to Z
w	the most probable value

Exponents

X^*	reference value of the X *propriety*
X^0	reference value of the X unit
X°	standard X value

Prefixes

abs	absolute value
d	simple derivation; differential
exp	exponential
f(X)	(or F, φ) function of X
lg	decimal logarithm
lim	limit
ln	Neperian logarithm
log	logarithm of any base
max	maximum
min	minimum
mod	$mod_k\, n$ = the rest of division of n by k
Prob	probability
δ	infinitesimal variation
Δ	difference (finite)
$\Delta^f X$	(unit) X of forming
$\Delta^r X$	(unit) X of reaction
∇	gradient $\partial/\partial X + \partial/\partial Y + \partial/\partial Z$
∇^2	nabla $\partial^2/\partial X^2 + \partial^2/\partial Y^2 + \partial^2/\partial Z^2$
∂	partial derivation
δ	any infinitesimal variation
∀	for any
§	chapter or sub-chapter

Binary Operators

·	multiplication
÷	from to
→	implication
⇒	conclusion
≡	identity or definition
∈	belonging
↔	intersection: a↔b
≈	reunion

Other Operators

$\overline{X}$	arithmetic mean of X
$\overline{X}$	unit X partially-molar
D/D	*D*(. . .)/*D*(. . .) Jacobian
$\langle X \rangle_k$	mean of k order of x

Abbreviations

cap.	chapter
ES	equation of state
fig.	figure
pag.	page
S.I.	International System of Units
tab.	table
VdW	Van der Waals
[Pr]	Prandtl number
[Re]	Reynolds number
[Sc]	Schmidt number
&	and
§	paragraph (= subchapter)
$\in$	belonging
$\leftrightarrow$	intersection: $a \leftrightarrow b$
$\approx$	reunion

Constants and Units of Measure

Universal Physical Constants

Notation	Name and value
c_0	Speed of light in vacuum: $2997.92 \cdot 10^8$ m/s
e_o	Elementary electric charge: $1602.18 \cdot 10^{-19}$ C
F	Faraday constant: $N_A \cdot e_o = 96.485$ C/mol
g	Conventional gravitational acceleration 9806.65 m/s^2
k_B	Boltzmann constant: $R/N_A = 1380.65 \cdot 10^{-23}$ J/K
N_A	Avogadro number: 6022.14 mol^{-1}
R	Regnault constant (of perfect gases): 8314.46 J/(mol·K)≅ 0,08206 L·atm/(mol·K) = 19864 cal/(mol·K)

Six Fundamental Quantities of the S.I.

Notation	Name	Physical quantity
kg	Kilogram	Mass
m	Metre	Length
s	Second	Time
K	Kelvin	Temperature
mol	Mol	Amount of substance
A	Ampere	Electric current intensity

Other Quantities (of measure)

Notation	Name	Physical quantity	Equivalence
°C	Degree Celsius	Temperature	t/°C = T/K – 273,15
at	Technical atmosphere	Pressure	0981 bar
atm	Atmosphere physical	Pressure	1.01325 bar= 760 Torr
bar	Bar	Pressure	10^5 Pa
C	Coulomb	Electric charge	1 A·s
cal	Calorie	Energy	4184 J
g	Gram	Mass	0001 kg
J	Joule	Energy	1 kg·m^2/s^2
L	Litre	Volume	0001 m^3
L·atm	Litre-atmosphere	Energy	101,325 J
M	Molar	Molarity	1 mol/L
N	Newton	Force	1 kg·m/s^2
Pa	Pascal	Pressure	1 kg/(m·s^2)
Torr	mm, Hg	Pressure	133.322 Pa
V	Volt	Electric voltage	1 J/A

Part I
Molecular Structure of Matter

The systems taken into account in this first part, which has an introductory role, are especially the simplest ones, in which interactions between molecules are as small as possible: namely, the ideal gases. Real gasses, as well as liquids or solids, show strong molecular interactions and will be treated later, in parts II and III of this book.

Part 1, entitled "Molecular Structure of Matter", presents the calculation procedure of the probability of a macrostate occurrence – defined as the global state of a common material system (containing 10^{15}–10^{30} particles) depending on the mean or probable features of each *microstate*, understood as the state of an isolated particle – namely, the molecule.

Statistical physics is greatly reduced to *statistical mechanics*, as the main molecular features taken into account are the mechanical ones, related to molecules motion: energy and speed of motion, collisions – with system walls or intermolecular.

Part 2, entitled "Molecular models in thermodynamics", is focused on the first of these mechanical characteristics – the energy – in its general understanding, that is without being limited to the kinetic energy of the molecule motion as a whole (translation) or the molecule parts motion (rotation and vibration). The energy of a microstate encompasses the potential energies corresponding to changes of the quantum states of electrons from the molecule or of subatomic particles from the atoms nuclei which constitute the molecule.

The exposed methods and procedures allow the calculation of particle distribution on energy levels – the *partition functions*, synonym with *sum-over-states* – and then the transition from the sum-over-states to the numerical equilibrium values characterizing macrostates, mainly from the thermodynamic point of view.

In Part 3, "Transport phenomena and their mechanism", methods of calculating the mean values and distributions of some molecular displacement characteristics are introduced, such as: kinetic energy, velocity or its projections on one or more directions, or the duration between two collisions and the travelled distance.

On this basis, several relations are introduced, which allow that, starting from experimental, microscopic or macroscopic methods, to determine the apparent geometric parameters of the molecule.

The Atom and the Molecule

The belief in the existence of an *atomic* entity – the idea according to which matter is discontinuously structured – has emerged, at least as a hypothesis, ever since antiquity. But the concept of *molecule*, as the smallest features carrier of a substance, is an innovation of the XIXth century: between 1811 (when Amedeo Avogadro introduces the notion) and 1911 (when Jean Perrin manages to explain the Brownian motion by molecule shocks), the majority of scientists could not distinguish clearly between atoms – the last particles in which matter can decompose under ordinary conditions – and molecules, which are the smallest particles still possessing most of substance's macroscopic features from which molecules originate.

At the boundary between physics and chemistry, physical chemistry could only appear as a distinct science once the definitive acceptance of the molecule idea following the contributions from J. Perrin. The difference between the fields of study of physics and physical chemistry coincides with the distinction between the phenomenological laws having an *atomic* basis and those related to the *molecular* nature of matter properties:

- the structure of the atom is studied by atomic physics, while
- the manner by which atoms are bound to form molecules forms the study object of a discipline intermediate between physics and theoretical chemistry: chemical physics (Anglo-Saxon denomination which has been inappropriately translated by "matter structure". and
- the transition from the isolated molecule features and interactions to the macroscopic properties is realized by a branch of physical chemistry: the molecular physical chemistry, the bases of which are introduced in this chapter.

Molecule, mer, mole. The molecule definition in physical chemistry is broader than that from physics: the physicochemical molecule can be:

- an ion (with electrons in addition to or less than the molecule itself),
- a monoatomic molecule – comprised of a single "free atom",
- a molecule somewhat different from the other molecules of the same substance (which is a polymer); or
- a group of atoms or ions which repeats in space more or less periodically, as in the atomic, ionic or metallic crystalline networks.

In the last two cases, one mole of matter consists of $6.02 \cdot 10^{23}$ *mers* and not as many *molecules*. One mer is the repeating unitary unit: one Cl^- ion plus one Na^+ ion on the salt lattice, one carbon atom from graphite or diamond, or the $-CH_2-$ group in polyethylene $(C_2H_4)_n$.

Chapter 1
Molecular Physics

1.1 Structure of Matter and Molecular Physics

Classical physics, fully consolidated until the 1890s, proved to have a series of deficiencies, by allowing only a limited comprehension of phenomena. Thus, in classical physics different laws apply for the case of matter having invariant mass (laws of mechanics) and for the case of radiations (laws of thermodynamics, electromagnetism, and optics).

Classical physics assumes that both matter and energy have a continuous structure. Material constants involved in the laws of classical physics can be obtained only experimentally, as classical physics cannot establish connections between the macroscopic laws and the behavior of matter at an atomic and a molecular level.

In classical physics all substances have identical behavior, only being differentiated by the numerical value of several empirical constants. However, the numerical values of these constants strongly depend on the properties of atoms and molecules, properties that are being studied by a branch of physics (as well as by one of theoretical chemistry) called the *structure of matter*.

Molecular physics as a branch of science has the purpose of settling connections between the macroscopic and microscopic laws of physics, which demands the use of three fields of theoretical physics: statistical physics, quantum theory, and the theory of relativity. Only the first two of these three development directions are important for *molecular physical chemistry*.

Quantum Physics

Matter quantum theory studies the specific properties of the particles that constitute matter, starting from a series of hypotheses confirmed by the validity of the obtained results, namely:

F. E. Daneş et al., *Molecular Physical Chemistry for Engineering Applications*,
https://doi.org/10.1007/978-3-030-63896-2_1

- Particles of the same kind are indiscernible (in classical physics only the objects formed by molecules are studied, which can be distinguished one from the other).
- The position of a particle x and its momentum p cannot be simultaneously known with whatsoever degree of accuracy; between their "indeterminacy" (characterized by the precisions by which Δx and Δp can be known) exists the correlation called the *relation of uncertainty* of Heisenberg:

$$\Delta \mathrm{x} \cdot \Delta \mathrm{p} = \mathrm{h}/(2 \cdot \pi) \tag{1.1}$$

where h is the Planck's constant.

- The behavior of matter is described by a wave function, for which *Schrödinger equation* is valid. For the stationary case (invariable behavior in time) in the case of a single particle, this equation has the form:

$$\nabla^2 \Psi + \left(8 \cdot \pi^2 \cdot m/h^2\right) \cdot H\Psi = 0 \tag{1.2}$$

In the relation (1.2), ∇^2 and H are operators applied to function Ψ, indicating: the first one, summation of second-order derivatives $\nabla^2 = \frac{\partial^2}{\partial x^2} + \frac{\partial^2}{\partial y^2} + \frac{\partial^2}{\partial z^2}$ over the three axes of coordinates, and the second one, multiplication by ε_c, the kinetic energy of the particle.

Whereas the total energy ε is the sum of the kinetic energy and of the potential energy ε_p (which depends on the spatial coordinates x, y, z), Schrödinger equation will become:

$$\partial^2\Psi/\partial x^2 + \partial^2\Psi/\partial y^2 + \partial^2\Psi/\partial z^2 + 8 \cdot \pi \cdot m \cdot \left(\varepsilon - \varepsilon_p\right)/h^2 = 0$$

When solving Schrödinger equation, for the wave function ψ (x, y, z), one or more solutions are obtained—solutions that are referred as *eigenfunctions*. To each eigenfunction corresponds an *eigenvalue* of the total energy ε of the particle. The spectrum of eigenvalues is discrete. This is what differentiates the quantum behavior from the classical one, where energy can vary continuously.

Quantum theory allows unitary treatment of both radiation and matter that possesses momentum. This allows the obtaining of concrete values of energies for different motion types of electrons, atoms, and molecules.

Statistical Physics

Statistical Physics Studies of matter's properties at a microscopic scale, where matter is treated as a set of really small particles (photons, electrons, or molecules) that establishes the dependence on its macroscopic properties—both the static ones and the dynamic ones—on the characteristics of the constituent particles.

Statistical Mechanics To establish the connection between thermodynamic properties (macroscopic) and molecules' properties, statistical thermodynamics roots from the results obtained from statistical mechanics. The latter is a branch of statistical physics, having as object of study the distribution and the median values of the positions, speeds, momenta, and energy levels of particles.

Statistical Thermodynamics Another branch of statistical physics, studies on the one hand, the connection between the results of statistical mechanics about the *mean values* of molecules' properties or of other particles, and on the other hand, the values of macroscopic thermodynamic characteristics of the assemblies formed by these particles.

The results obtained by statistical thermodynamics allow the overcoming of several limits of the phenomenological thermodynamics, object of study of the first book [1] of the series "Physical chemistry for engineers" (AGIR editions, Bucharest, 2013), as phenomenological thermodynamics can establish only the *relations* between state properties, while statistical thermodynamics provides the *values* of these properties.

Molecular Kinetics

While only the *mean values* of the molecule's mechanical characteristics are used in statistical thermodynamics, *the distribution* according to the values of these properties allows for the development of a quantitative kinetic–molecular theory of matter, which can explain and quantify a series of non-equilibrium phenomena, such as:

- Transport phenomena (diffusion, heat transfer, flow) studied by the *physical kinetics*, which is treated in the third part of this book, and in [2].
- Chemical reaction rates—object of *chemical kinetics*, science branch presented in the book of Daneş and Ungureanu [3], already published in 2009 at AGIR editions.

1.2 Statistical Physics of Particles

Microstate: An Elementary Configuration of Particles

In the field of statistical mechanics only the more typical characteristics of particles are considered, like the possibility to find a particle in one of the system's parts, for example in its left or right side. Any distinct arrangement of particles within the different parts of a system is called "elementary configuration" or *microstate* of the system.

Therefore, a box with two compartments of equal volume is considered. Between each compartment an amount of N number of particles, namely $N = 4$, is distributed.

Table 1.1 The 16 manners of particle arrangements

	1	2	3	4	5	6	7	8	9	10	11	12	13	14	15	16
I	A															
	B	B	A	A	A											
	C	C	C	B	B	C	B	B	A	A	A					
	D	D	D	D	C	D	D	C	D	C	B	A	B	C	D	0
II	0	A	B	C	D	A	A	A	B	B	C	B	A	A	A	A
						B	C	D	C	D	D	C	C	B	B	B
												D	D	D	C	C
																D

The distinct particles are denoted by A, B, C, and D. There are 16 distribution possibilities of the particles between the two compartments, denoted I and II, and given in Table 1.1.

A fundamental theorem of statistical mechanics is the *theorem of equiprobability of microstates*, formulated as follows:

any microstate has the same chances of achievement.

According to the theorem of equiprobabilities, the possibility for the system to be found in any of the 16 microstates is the same: $P^{(1)} = P^{(2)} = ... = P^{(16)}$, where the indices 1, 2, ..., 16 are the order numbers of microstates from Table 1.1.

Since the summation of probabilities of all microstates leads to value 1 (the system is obligatorily found in one or another of the microstates), each microstate is achieved with a probability equal to $1/16 = 6.25\%$.

Microstates and Macrostates

Statistical mechanics also uses the notion of *macrostate*, which is differentiated from the one of microstate and opposed to it.

Macrostate designates any state of the system that can be distinguished from a macroscopic point of view by any measurement. Usually, the number of macrostates is smaller than the one of microstates, due to the fact that at least one microstate corresponds to any considered macrostate.

Within a given system, the macrostates can be differentiated by the total numbers of particles from each compartment. Within this system, five macrostates are possible, as it can be seen in Table 1.2.

Macrostates' achievement conditions more than those of microstates are of practical interest, because only the system's macrostates—and not his microstates—are differentiated through a measurable characteristic.

Table 1.2 Characteristics of macrostates

Order number of macrostate	5	4	3	2	1
Number of particles in compartment I	4	3	2	1	0
Number of particles in compartment II	0	1	2	3	4
Order numbers of the microstates	1	2–5	6–11	12–15	16
Mathematical probability of the macrostate – Number of microstates, *W*	1	4	6	4	1
Mathematical probability of macrostate *P* (%)	6.25	25	37.5	25	6.25

Thus, for example, within the considered system, the listed macrostates (1–5) are differentiated by the *density* within each of the two compartments.

Thermodynamic Probability

The probability of achieving a macrostate is equal to the sum of the probabilities of achieving the corresponding microstates; for example, the probability of the macrostate no. 2 will be: $P_2 = p^{(12)} + p^{(13)} + p^{(14)} + p^{(15)} = 4{\cdot}(1/16) = 25\%$.

As microstates have the same chance of occurrence, it is possible to determine the chances of achieving a macrostate by its corresponding number of microstates.

This number is referred to as *thermodynamic* probability, *W*, of the given macrostate. On the last line of Table 1.2, the actual mathematical probability of the respective macrostate is presented.

The use of the thermodynamic probability, *W*, is equivalent to the use of the mathematical probability, *P*, due to their proportionality:

$$\mathrm{P_j} = \mathrm{W_j}/\mathrm{W_T} \tag{1.3}$$

where P_j and W_j are the mathematical and thermodynamic probabilities of the *j* macrostate and W_T is the sum of thermodynamic probabilities of all macrostates:

$$W_T = \sum_{j=1}^{j=M} W_j \tag{1.4}$$

where *M* represents the *number of macrostates*. According to Eq. (1.4), W_T will be equal to *m* – the *total number of microstates* from the system. Thus, in the given system, characterized by $m = 16$ and $M = 5$: $W_1 = W_5 = 1$, $W_2 = W_4 = 4$, and $W_3 = 6$, while $W_T = 1 + 4 + 6 + 4 + 1 = 16$.

For each of the five microstates, the mathematical probabilities can be obtained from Eq. (1.3): $P_1 = P_5 = 1/16 = 6.25\%$; $P_2 = P_4 = 4/16 = 25\%$; $P_3 = 6/16 = 37.5\%$.

Mathematical and Thermodynamic Probabilities

Mathematical and thermodynamic probabilities are proportional, one to the other, meaning that the ratio between the thermodynamic probabilities W' and W'' is equal, as seen above, to the ratio of the mathematical probabilities P' and P'':

$$W'/W'' = P'/P''$$

Both ratios, $P_3/P_2 = 0.375/0.25$ and $W_3/W_2 = 6/4$, have the value 1.5. The use of W instead of P thus presents the advantage of allowing a direct comparison of occurrence chances for two macrostates, without the necessity of prior calculation of the total number of microstates.

Therefore, in the given example, the macrostates 2 and 3 have been compared knowing the values of W_2 and W_3, and without prior knowing of the values for W_1, W_4, W_5, or M.

The simplification brought is even more important as the number M of system's macrostates is more elevated.

It can be observed that W is always a positive integer, while P a positive real subunitary integer. An impossibility corresponds to $W = P = 0$.

In the absence of any specification, the term "probability" without the indication of its type refers to the *mathematical* probability.

A physical interpretation can be made for the thermodynamic probabilities. Thus, among the system's macrostates, the macrostate no. 3 has the highest value for W, so that the system will preferably be found in the macrostate no. 3—the density state being equal for the two compartments.

This result corresponds from the physical point of view to a kinetic–molecular theory premise, according to which the gas is homogenous. This means that at equal volumes of a given gas, at equilibrium state, an equal number of molecules will be correlated.

It can be observed that generally the macrostates have different occurrence chances as opposed to microstates, which are equiprobable according to the aforementioned theorem.

Combinatorial Analysis Calculation

Calculation of the numbers of microstates, macrostates, and thermodynamic probabilities is based on the combinatorial analysis formulae.

Therefore, *for a system consisting of two equal compartments*, containing a number N of distinct particles, the total number of microstates is $m = 2^N$, and the number of macrostates is $M = N + 1$.

Let j be the number of particles from compartment I; j can take the values 0, 1, . . ., N and serves as an index of the respective macrostate. The thermodynamic probability W_j is in this case (two equal compartments, distinct particles, no energetic bonding) given by the simple combinations formula:

$$W_j = C_N^j = \frac{N!}{j!(N-j)!} \tag{1.5}$$

where C_N^j is the combinations number of N elements taken j at a given time. For example, for a system of 10 particles, the thermodynamic probability of the state with 7 particles in compartment I and with 3 particles in compartment II is:

$$W_7 = C_{10}^7 = \frac{10!}{7! \cdot 3!} = 120$$

This probability is higher than the probability of the state with 10 particles in the same compartment, $W_{10} = \frac{10!}{10! \cdot 0!} = 1$, but lower than the probability of the macrostate with an equal number of particles in the two compartments, $W_5 = \frac{10!}{(5!)^2} = 252$. Thus, the homogeneity of molecules' spatial distribution is true for any number of particles.

Stirling's Approximations

In the case of elevated values for N and j, W calculation using factorials is difficult, so that a Stirling approximation for the factorials is used:

$$N! \cong \left(\frac{N}{e}\right)^N \cdot \sqrt{2 \cdot \pi \cdot \left(N + \frac{1}{12}\right)} \tag{1.6a}$$

For the combinatorial analysis calculations, a simpler form is used, namely:

$$ln\,(\mathrm{N!}) = \mathrm{N} \cdot (ln\ \mathrm{N}) - \mathrm{N} + 0.5 \cdot ln\,(2 \cdot \pi \cdot \mathrm{N}) \tag{1.6b}$$

where the last term from the second member of the equation is often omitted:

$$ln\,(\mathrm{N!}) = \mathrm{N} \cdot (ln\ \mathrm{N}) - \mathrm{N} \tag{1.6c}$$

Thus, for the case of 100 particles, W value for the state with 48 particles in compartment I is: $W_{48} = 100!/[(48!)\cdot(52!)]$.

By applying the logarithm, the following expression is obtained: $lnW_{48} = ln\ 100! - ln\ 48! - ln\ 52!$. Or by using the third approximation of Stirling—Eq. (1.6c):

$$lnW_{48} = (100 \cdot ln\,100 - 100) - (48 \cdot ln\,48 - 48) - (52 \cdot ln\,52 - 52)$$
$$= 100 \cdot ln\,100 - 48 \cdot ln\,48 - 52 \cdot ln\,52 = 70.6548$$

from where it results that $W_{48} = 1.171 \cdot 10^{30}$.

If the particle number, N, is distributed between K compartments ($K > 2$), each macrostate is characterized by a set of K indices, corresponding to the order of N_k particle numbers from compartments 1, 2, ..., k, ..., K; therefore, the macrostate is denoted by N_k. These N_k numbers (N_1, N_2, ..., N_K) satisfy the equation of particle number balance: $N_1 + N_2 + \ldots + N_K = N$. W is calculated in this case by the repeating combinations formula:

$$W(N_1, N_2, \ldots, N_k) = \frac{N!}{N_1! \cdot N_2! \cdot \ldots \cdot N_k!} \tag{1.7a}$$

Using the third approximation of Stirling, it results:

$$lnW(N_1, N_2, \ldots, N_k) = N \cdot lnN - \sum_{k=1}^{k=K} N_k \cdot lnN_k \tag{1.7b}$$

Equations (1.6) and (1.7) allow the thermodynamic probability calculation for the discussed cases.

Energy Conservation for a Set of Particles

The requirement of respecting the law of *conservation of energy* (*the system's total energy is constant*, being defined as the sum of the energies of all particles) is a supplemental limitation not considered during W calculation using Eqs. (1.6) and (1.7).

Therefore, besides the law of conservation of the *total* number of particles, N:

$$\sum_{k=0}^{k=K} N_k = N = const. \tag{1.8}$$

the law of conservation of *total* energy, ε, is also considered:

$$\sum_{k=0}^{k=K} N_k \cdot \varepsilon_k = \varepsilon = const. \tag{1.9}$$

where ε_1, ε_2, ..., ε_K are the corresponding energies of compartments 1, 2, ... K. These energies are usually different, the compartments being consequently also called "energy levels" of the system.

In "classical" thermodynamics, which is phenomenological and macroscopic, to the conservation law (1.9) of the *total* energy corresponds the invariance of the *internal* energy u within isolated systems, or the achievement of constant and minimal values of several thermodynamic potentials, for systems kept in different conditions.

In general case, for thermodynamic probability calculation, the type of particles must be additionally considered, as it is going to be discussed forward.

Discernibility of Particles and Limitation of their Number

Boltzons, Bosons, and Fermions The particles can be considered *discernible*, that is, that can be distinguished one from another, or, on the contrary, indiscernible. The above *Boltzmann statistics*, which considers that particles—named *boltzons*—are discernible, is valid in the case of classical physics, but not in the case of quantum physics (where particles of the same nature cannot be distinguished one from another).

The number of *indiscernible* particles in a certain compartment can be either unlimited, as it is the case with the discernible particles, or limited.

- *Fermi–Dirac statistics* corresponds to the situation where there is a limited number of particles. For example, there are cases in which each compartment contains at least one particle, as in the case of occupying the energy levels of an atom (according to the Pauli principle, at most one electron can be found in each quantum state).
- In *Bose–Einstein statistics* particles are supposed to be indiscernible, but the number of particles in a compartment is not submitted to limitations.

In quantum mechanics, it is shown that:

- The elemental-integer spin particles (so-called *bosons*, like the photon or some nuclei) follow the Bose–Einstein statistics.
- But the half-integer spin particles (called *fermions*, like the electron, the proton, the neutron, or certain atomic nuclei) follow the additional restriction of Fermi–Dirac statistics.

1.3 Distribution of Particles on Energy Levels

System's Discrete Energy Values

As quantum mechanics proves it, the particles of a given system cannot have any energy, yet they can take certain energy values.

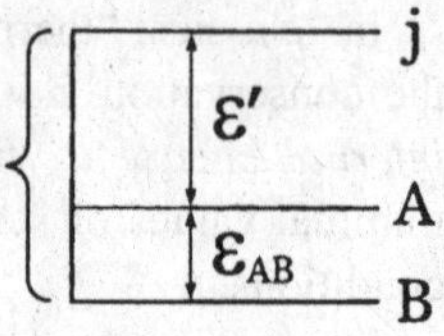

Fig. 1.1 Zero energy level

These values depend on the system's nature: each system is therefore characterized by a series of energy levels, which represent the permitted energy values of system's particles (the "eigenvalues" mentioned in 1.1.).

As benchmark (point zero) for energy measurement, an arbitrary value of energy can be adopted.

Indeed, making the notations as in Fig. 1.1, that is, ε_j' is the energy value of level j relative to the reference level (benchmark) A and ε_j'', to benchmark B, the following relation will exist between ε_j' and ε_j'':

$$\varepsilon_j'' = \varepsilon_j' + \varepsilon_{AB}$$

where ε_{AB} is the difference between energy benchmarks A and B.

Let N_j be the number of particles on the energy level j. The total energy relative to benchmark B will be, according to Eq. (1.9):

$$E'' = \sum_j N_j \varepsilon_j^N = \sum_j N_j(\varepsilon_j + \varepsilon_{AB}) = \varepsilon_{AB} \sum_j N_j + \sum_j N_j \varepsilon_j'$$

where $\sum_j N_j = N$ and $\sum_j N_j \cdot \varepsilon_j = E$ are the total energy relative to benchmark A. Therefore, we obtain $E'' = E' + N \cdot \varepsilon_{AB}$ so that, if the system's energy is constant compared to benchmark A, it will be also constant compared to benchmark B.

The lowest energy level is called *ground state*, and all the other levels are called *excited* states.

Therefore, the energy balance is valid in any of its measurement scales. It is convenient to consider, as a benchmark, the *ground* state energy, noted with a zero index:

$$\varepsilon_0 = 0.$$

In this measurement scale, the energy ε_j values for any excited state ($j = 1, 2, ...$) represent the difference between the excited state and the ground state energies.

Distribution on Non-degenerate Energy Levels

The illustration of particle distribution in a closed system can be made considering the conservation of particle number and total energy, while applying the rules of combinatorial analysis.

Therefore, a system with four equidistant energy levels $\varepsilon_0 = 0$, $\varepsilon_1 = 1$, $\varepsilon_2 = 2$, $\varepsilon_3 = 3$, (energy measurement is in arbitrary units of measure) is considered.

If the system contains a total number of particles of $N = 4$ and its total energy is $E = 3$ (measured in the same units), then the following system microstates will be possible:

$$\mathrm{A}: \mathrm{N_0} = 3, \mathrm{N_1} = 0, \mathrm{N_2} = 0, \mathrm{N_3} = 1$$
$$\mathrm{B}: \mathrm{N_0} = 2, \mathrm{N_1} = 1, \mathrm{N_2} = 1, \mathrm{N_3} = 0$$
$$\mathrm{C}: \mathrm{N_0} = 1, \mathrm{N_1} = 3, \mathrm{N_2} = 0, \mathrm{N_3} = 0$$

It can be proved that the following conditions are met for each macrostate (A, B, C): $\sum N_j = 4$ and $\sum N_j \varepsilon_j = 3$. It also can be proved that there exists no *other* macrostate that simultaneously fulfils these conditions.

Assuming the discernible particles (Boltzmann statistics), the thermodynamic probability of each macrostate can be calculated through the formula of repeating combinations (1.7a): $W = \frac{4!}{N_0! \cdot N_1! \cdot N_2! \cdot N_3!}$, from where:

$$W_A = \frac{4!}{3! \cdot 0! \cdot 0! \cdot 1!} = 4; W_B = \frac{4!}{2! \cdot 1! \cdot 1! \cdot 0!} = 12; W_C = \frac{4!}{1! \cdot 3! \cdot 0! \cdot 0!} = 4.$$

It results that the ratios between thermodynamic probabilities of A, B, C macrostates are 1: 3: 1, with macrostate B being the most favorable one.

Degeneracy of Energy Levels

Due to the multiplicity of energy levels, a change occurs in thermodynamic probabilities calculation method.

Quantum mechanics shows that each of its *eigenfunctions*—the wave functions Ψ (x, y, z) from Schrödinger's equation—corresponds to a single "eigenvalue" of energy ε. But to a given value of energy may correspond several eigenfunctions.

Yet the eigenfunction is the quantity describing particle's state—the probability for the particle to exist in a certain element of volume is proportional to the wave function square value.

Each eigenfunction therefore corresponds to a statistical "compartment," so that any energy level is characterized by the *multiplicity* of the level, g_j, which represents the number of wave functions corresponding to the same energy level j.

The energy levels having $g_j = 1$ are called *simple* or *non-degenerate* states, while the ones having $g_j > 1$ are called *multiple* (double, triple, ...) or *degenerate* states.

The existence of degenerate states increases macrostates' thermodynamic probabilities: apart from the different distribution modes of particles between levels, additional combinatorial possibilities occur due to particles' distribution on different such level compartments.

For example, if there are two particles (a and b) on a given level, the macrostates' number increases nine times in the case of a triple degenerate level (whose sublevels are referred to as I, II, and III). Its particles can be disposed in the following nine modalities, numbered from 1 to 9:

	1	2	3	4	5	6	7	8	9
I	Ab	a	A	b	0	0	b	0	0
II	0	b	0	a	Ab	a	0	b	0
III	0	0	b	0	0	b	a	a	Ab

For the discernible particle case, the degeneracy of an energy level increases by $(g_j)^{N_j}$ both the number of microstates and the thermodynamic probability. Considering the possible degeneracy of all energy levels, the equation of comparisons with repetitions of *Boltzmann statistics* changes by its multiplying with the coefficients (g_j) of each level, raised to power N_j:

$$W = N! \cdot \prod_j \left[(g_j)^{N_j} / (N_j!) \right] \tag{1.10}$$

Distribution on Degenerate Energy Levels

In the case of degeneracy, macrostates' probabilities change not only in absolute value but also in relative value. If, for example, in the previous system the multiplicities of the four levels are $g_0 = 2$, $g_1 = 4$, $g_2 = 1$, and $g_3 = 1$ as in Fig. 1.2, then the thermodynamic probabilities change, according to Eq. (1.10); the following is obtained:

$$W_A = 4! \cdot \frac{2^3 \cdot 4^0 \cdot 1^0 \cdot 1^1}{3! \cdot 0! \cdot 0! \cdot 1!} = 32,\ W_B = 4! \cdot \frac{2^2 \cdot 4^1 \cdot 1^1 \cdot 1^0}{2! \cdot 1! \cdot 1! \cdot 0!} = 192 \text{ and } W_C$$
$$= 4! \cdot \frac{2^1 \cdot 4^3 \cdot 1^0 \cdot 1^0}{1! \cdot 3! \cdot 0! \cdot 0!} = 512.$$

Thus, all thermodynamic probabilities have increased with different increasing factors (W_A, 8 times; W_B, 16 times; and W_C, 128 times), so that the proportion of

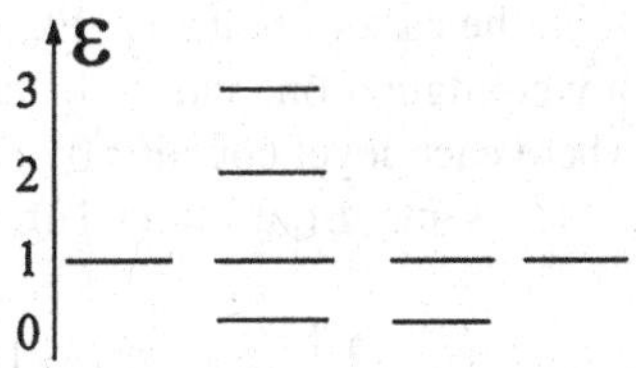

Fig. 1.2 A system with four different multiplicity levels

thermodynamic probabilities was modified from 1:3:1 to 1:6:16, C macrostate becoming the most probable state.

For the case of *indiscernible particles*, combinatory analysis can be used to prove the validity of Eq. (1.11) for bosons and the validity of Eq. (1.12) for fermions, namely:

$$W = \prod_j (g_j + N_j - 1)!/[(g_j - 1)! \cdot N_j!] \tag{1.11}$$

$$W = \prod_j g_j!/[(g_j - N_j)! \cdot N_j!] \tag{1.12}$$

Highly Degenerated Systems

In systems having technical interest, the degeneration degree of each energy level is highly elevated, $g_j >> 1$, while the *occupancy* rate of each level is extremely low, $N_j/g_j << 1$. This simplifies the distribution equations for bosons and fermions. Therefore, Eq. (1.11) can be written in the following form:

$$W = \prod_j \frac{1}{N_j!} [(g_j) \cdot (g_j + 1) \cdot \ldots \cdot (g_j + N_j - 2) \cdot (g_j + N_j - 1)]$$

Yet, since g_j is higher than any of the numbers 1, 2, ..., (N_j—1), each factor from the square bracket can be replaced by g_j; this bracket having N_j factors, the following is obtained:

$$W = \prod_j \left[(g_j)^{N_j} / N_j! \right] \tag{1.13}$$

Equations (1.11), (1.12), and (1.13) for the thermodynamic probability of degenerated configurations can be customized to the case of simple states, by putting $g_j = 1\ \forall j$.

In the case of boltzons, Eq. (1.7a) is found, in the case of bosons W will be 1 for any configuration, and in the case of fermions W is 1 only for those configurations where each level contains 0 or 1 particle.

The same, from Eq. (1.12), that is valid for fermions and that can be written as:

$$W = \prod_j [(g_j - N_j + 1) \cdot (g_j - N_j + 2) \cdot \ldots \cdot (g_j - 1) \cdot (g_j)]/N_j!$$

It results in the same Eq. (1.13), after prior replacement with g_j of each factor from the square brackets. This equation differs only by the absence of $N!$ factor from the distribution Eq. (1.10) of Boltzmann's statistics for the discernible particles, meaning that the N particles of the system cannot be discerned one from another.

For many of the situations encountered in statistical mechanics, the distribution (1.13) is used (2.4.). It is referred to as the *corrected Boltzmann distribution* and can be obtained by $N!$ division of the simple Boltzmann distribution.

Since $N >> 1$, the factorials are approximated by Stirling equation, which leads to

$$ln\ \mathrm{W} = \mathrm{N} \cdot ln\ \mathrm{N} + \sum_j \mathrm{N_j}\left(ln\ \mathrm{g_j} - ln\ \mathrm{N_j}\right) \tag{1.14a}$$

for the case of Boltzmann statistics, and to

$$ln\ \mathrm{W} = \mathrm{N} + \sum_j \mathrm{N_j}\left(ln\ \mathrm{g_j} - ln\ \mathrm{N_j}\right) \tag{1.14b}$$

for the case of modified Boltzmann statistics.

From all cases results that:

$$ln\ \mathrm{W} = f(\mathrm{N}) + \sum_j \mathrm{N_j}\left(ln\ \mathrm{g_j} - ln\ \mathrm{N_j}\right) \tag{1.15}$$

where f(N) represents $N \cdot lnN$ or N, for simple and respectively modified Boltzmann statistics.

1.4 Boltzmann's Relationship Between Entropy and Probability

Both entropy, which is studied in thermodynamics, and (thermodynamic) probability, which is studied in statistical mechanics, have many common features, including:

- In an isolated system, both entropy s and thermodynamic probability W are maximal in the equilibrium state.

- When a non-equilibrium system evolves, both s and W increase in their value.
- When increasing temperature and shifting from the more orderly states of matter to the less orderly ones (on a scale solid → liquid → gas), both s and W increase in their value.

Therefore, Boltzmann assumed the existence of a functional connection between these two units, namely: $s = F(W)$.

Parallelism of Thermodynamic Probability with Entropy

In order to find the form of $s(W)$ dependence, an arbitrary system consisting of two distinct parts is considered. Let s, s_A, and s_B be the system entropy and respectively the component part entropies, and let W, W_A, and W_B be the corresponding thermodynamic probabilities.

According to the composition law of probabilities: $W = W_A \cdot W_B$. On the other hand, the system entropy is the sum of composing entropies, thus: $s = s_A + s_B$. The dependency $s = F(W)$ satisfies therefore the functional equation:

$$F(W_A \cdot W_B) = F(W_A) + F5(W_B)$$

The only possible function satisfying this functional equation is $F(W) = C \cdot \ln W$, where C is an arbitrary constant, so that:

$$s = C \cdot \ln W \tag{1.16}$$

Entropy of Mixing

A comparison is made in order to determine the constant C, between the classical and statistical thermodynamic expressions for the mixing process case.

Let it be two distinct compartments found in the initial state, the first containing an N' number of type A molecules and the second an N'' number of type B molecules. The probability increases between W_0 and W_a that is due to mixing are expressed through the simple combinations formula:

$$W_a/W_0 = (N' + N'')!/[(N')! \cdot (N'')!]$$

$$W_a/W_0 = (N' + N'')!/[(N')! \cdot (N'')!]$$

After factorials approximation through $ln(Y!) \cong Y \cdot ln\ Y - Y$ from Stirling relation, it results that:

$$ln\,(W_a/W_0) = (N' + N'') \cdot ln\,(N' + N'') - N' \cdot ln\ N' - N'' \cdot ln\ N''$$

By introducing the notation $N = N' + N''$ for the total number of molecules and the molar fraction ratios $x_A = N'/N$, $x_B = N''/N$, it results that:

$$ln\,(W_a/W_0) = -N \cdot (x_A \cdot ln\ x_A + x_B \cdot ln\ x_B) \quad (1.17)$$

Besides, the total entropy of mixing thermodynamic expression is:

$$s_a = -R \cdot (n_A \cdot ln\ x_A + n_B \cdot ln\ x_B)$$

where n_A and n_B designate the number of moles of components. By expressing n_A and n_B according to the total number of n moles from the mixture ($n_i = n\ x_i$), it is obtained that:

$$s_a = -R \cdot n \cdot (x_A \cdot ln\ x_A + x_B \cdot ln\ x_B) \quad (1.18)$$

According to Eq. (1.16), s_a corresponds to the probability increase at mixing, $\Delta ln\ W$, defined as the difference of thermodynamic probability logarithms ($ln\ W_A$ and $ln\ W_O$) of final (after mixing) and initial state:

$$s_a = C \cdot \Delta\ ln\ W \quad (1.19)$$

By replacing in Eq. (1.19) s_a from Eq. (1.18) and $\Delta ln\ W$ from Eq. (1.17), it results:

$$C = R \cdot n/N$$

But $R = k_B \cdot N_A$ (where N_A is the Avogadro number while k_B the Boltzmann's constant) and $n = N/N_A$, therefore: $C = k_B$.

By replacing this value of C within Eq. (1.16), the following equality results:

$$s = k_B \cdot ln\ W \quad (1.20)$$

This equality is known as *Boltzmann relationship between entropy and probability.*

Equation (1.20) offers the possibility of establishing the correlation between the microscopic properties (reflected in W) and the thermodynamic macroscopic properties, such as entropy. This relationship also represents the fundament for the informational interpretation of entropy.

Thermodynamic Interpretation of Boltzmann Equation

From Physics' point of view, Eq. (1.20) reflects the parallelism between entropy increase and the increase in spatial arrangement degree of *disorder* and in material particles' motion.

Thus, in the ideal crystal at 0 K, the order is at the maximum level while the disorder at the minimum; the particles can be ordered within a single possible configuration, namely the perfectly ordered crystalline lattice, so that $W = 1$ and $s = 0$. The Third Law of Thermodynamics thus receives a statistical justification.

As the temperature rises, the degree of disorder increases in the crystal by:

- The creation of some defects (disordered spatial disposition).
- The occurrence of the increasingly intense oscillatory motions of particles from the lattice nodes, and therefore by the emergence of more possibilities for particle motion, which is a second way to increase a crystal's disorder and entropy.

During melting, the degree of disorder suddenly increases through disappearance of molecules' spatial arrangement and through the increase in the possible number of molecules' motions—in liquid state, molecules are capable not only of vibrational motions but also of rotational motions, as well as of translation (braked) displacements.

By further heating, when vaporization temperature is exceeded, both molecules' arrangement disorder (that will be spatially distributed in a completely chaotic manner) and thermodynamic probability of distinct types of motions within the molecule (because in gas state, molecules can move absolutely independently) will sharply increase.

1.5 Distribution of Particles on Energetic Levels

Equilibrium and Evolution in Statistical Mechanics

Statistical mechanics may predict a certain general characteristic of equilibrium states found in the maximal thermodynamic probability:

$$\mathrm{W} = \max \tag{1.21}$$

Equation (1.21) is a double criterion, that is, both of equilibrium and evolution:

- In any non-equilibrium state, W values are lower than the maximum value corresponding to the (unique) equilibrium state.
- During the *evolution toward the equilibrium state*, system's thermodynamic probability increases.

The total particle number, N, of a *closed* system is constant. If the system has no interactions with the exterior (*isolated system*), the system's total energy, E, is also a constant. Therefore, mass and energy balances become:

$$\sum_{j=0}^{j=K} N_j = N = const. \tag{1.22}$$

$$\sum_{j=0}^{j=K} N_j \cdot \varepsilon_j = E = const. \tag{1.23}$$

From a mathematical point of view, equilibrium distribution is an extremum problem in the presence of restrictions: W is the function to be maximized, and depends on the independent variables $N_1, N_2, \ldots, N_K$, while the restrictions are given by Eqs (1.22) and (1.23).

In these two relationships, the energies of levels, ε_j, are the system's characteristic constants, as is the case also for g_j multiplicities from Eqs. (1.14a, 1.14b) and (1.15). But the number of particles N_j on the j energy level is variable, differing from one (macro-) state of the system to another.

Boltzmann solved the problem of determining the number of particles N_j on each energy level of an isolated steady-state system. For this purpose, Eqs. (1.21), (1.22), and (1.23) are considered.

Equation (1.21) is equivalent to ln W = max, from which, by replacing $ln\ W$ from Eqs. (1.14a) and (1.14b), the following equations result for the simple and respectively the corrected Boltzmann statistics:

$$\mathrm{N} \cdot ln\ \mathrm{N} + \sum\nolimits_{\mathrm{j}} \mathrm{N_j} \cdot \left(ln\ \mathrm{g_j} - ln\ \mathrm{N_j} \right) = \max ; \mathrm{N} + \sum\nolimits_{\mathrm{j}} \mathrm{N_j} \cdot \left(ln\ \mathrm{g_j} - ln\ \mathrm{N_j} \right) = \max .$$

$$\mathrm{N} \cdot ln\ \mathrm{N} + \sum\nolimits_{\mathrm{j}} \mathrm{N_j} \cdot \left(ln\ \mathrm{g_j} - ln\ \mathrm{N_j} \right) = \max ; \mathrm{N} + \sum\nolimits_{\mathrm{j}} \mathrm{N_j} \cdot \left(ln\ \mathrm{g_j} - ln\ \mathrm{N_j} \right) = \max .$$

Maximization of Thermodynamic Probability

Since N = const., we can omit from the previous expressions of functions to be maximized the terms $N \cdot ln$ N and N, with both equations being reduced to:

$$\sum_{j=0}^{j=K} N_j \cdot (lng_j - lnN_j) = max. \tag{1.24}$$

For determining the N_j units, Eqs. (1.22), (1.23), and (1.24) are differentiated:

$$\sum_{j=0}^{j=K} dN_j = 0 \tag{1.25}$$

$$\sum_{j=0}^{j=K} \varepsilon_j dN_j = 0 \tag{1.26}$$

$$\sum_{j=0}^{j=K} (lng_j - lng_j - 1)dN_j = 0 \tag{1.27}$$

By applying the method of Lagrange multipliers, a member-by-member addition is performed between: Eq. (1.25) multiplied by constant ($C' + 1$), Eq. (1.26) multiplied by constant C'', and Eq. (1.27) multiplied by 1. The following expression results:

$$\sum_{j=0}^{j=K} (C' + C'' \cdot \varepsilon_j + ln\ \varepsilon_j - ln\ N_j) \cdot dN_j = 0 \tag{1.28}$$

Since the $d\,N_j$ variations are independent, the expression in the left member of Eq. (1.28) is identically null only when the differential coefficient of each variable is zero, namely when, for any j, $C' + C'' \cdot \varepsilon_j + ln\ g_j - ln\ N_j = 0$, from which the following is obtained: $N_j = g_j \cdot \exp(C' + C'' \cdot \varepsilon_j)$ or, by noting with $C' = -ln\ \beta$ and $C'' = -\alpha$:

$$Nj = (g_j/\beta) \cdot exp(-\alpha \cdot \varepsilon_j) \tag{1.29}$$

with α and β yet unknown.

Equation (1.29) is summed up for all the possible j values, in order to determine β as a function of α. The following expression results:

$\beta \cdot \sum_j N_j = \sum_j g_j \cdot exp(\alpha \cdot \varepsilon_j)$; or, because $\sum N_j = N$:

$$\beta \cdot N = \sum_j g_j \cdot exp(\alpha \cdot \varepsilon_j) \tag{1.30}$$

By substituting β from Eq. (1.30) into Eq. (1.29), we obtain N_j as a dependency only of α:

$$N_j = N \cdot g_j \cdot e^{-\alpha \cdot \varepsilon_j} / \sum_{j=0}^{j=K} g_j \cdot e^{-\alpha \cdot \varepsilon_j} \tag{1.31}$$

Partition Functions

The expression found at Eq. (1.31) denominator is a *global* attribute of the system, because on the one hand it depends on all the system's features (all g_j and ε_j values) and on the other hand it does not depend on the total number of particles nor on the particle numbers on j level. This expression is referred to as *sum-over-states* or *partition function* of the system and is denoted by Z:

$$Z = \sum_{j=0}^{j=K} g_j \cdot e^{-\alpha \cdot \varepsilon_j} \tag{1.32}$$

The sum-over-states calculation, when this property corresponds to the different possible system's particle motion types, within systems with different structures, represents the main objective of the statistical mechanics to thermodynamics application.

By deriving Eq. (1.32) in relation to its single independent value, namely α, the following results:

$$\frac{dZ}{d\alpha} = -\alpha \cdot \sum_{j=0}^{j=K} g_j \cdot e^{-\alpha \cdot \varepsilon_j} \tag{1.33}$$

Besides, by replacing Z from Eq. (1.32) into Eq. (1.31), the particle number on j level is obtained, namely:

$$N_j = N \cdot g_j \cdot e^{-\alpha \cdot \varepsilon_j}/Z \tag{1.34}$$

To obtain the α constant value, a thermodynamic reasoning is used, by expressing in advance W, s, and the system's energy as a function of Z. By replacing therefore N_j from Eq. (1.34) into Eq. (1.15), it is found that:

$$lnW = f(N) + \sum_j \frac{N}{Z} \cdot g_j \cdot e^{-\alpha \cdot \varepsilon_j} \cdot ln\left(\frac{Z}{N \cdot e^{-\alpha \cdot \varepsilon_j}}\right)^{-}$$

where function $f(N)$ depends only on the total number of particles and their discernibility (see 1.3 above). By regrouping the terms, the following expression is obtained:

$$lnW = f(N) + \left[\frac{N}{Z} \cdot \left(ln\frac{N}{Z}\right) \cdot \sum_j g_j \cdot e^{-\alpha \cdot \varepsilon_j}\right] + \left[\alpha \cdot \frac{N}{Z} \cdot \sum_j g_j \cdot \varepsilon_j \cdot e^{-\alpha \cdot \varepsilon_j}\right]$$

This expression becomes, after entering the sum $\sum_j g_j \cdot e^{-\alpha \cdot \varepsilon_j}$ from Eq. (1.32) and the sum $\sum_j g_j \cdot \varepsilon_j e^{-\alpha \cdot \varepsilon_j}$ from Eq. (1.33),

$$lnW = [f(N) - N \cdot lnN] + N \cdot lnZ + (N \cdot \alpha/Z) \cdot dZ/d\alpha \tag{1.35}$$

Or, shifting from variable Z to lnZ into the last term, it becomes:

$$ln\,W = \varphi(N) + N \cdot ln\,Z + N \cdot \alpha \cdot d(lnZ)/d\alpha \tag{1.36a}$$

Function $\varphi(N)$ from Eq. (1.36a) is a new function, which depends only on the total number of particles N and is defined by

$$\varphi(\mathrm{N}) \equiv f(\mathrm{N}) - \mathrm{N} \cdot ln\,\mathrm{N} \tag{1.36b}$$

so that

$\varphi(N) = 0$ for the simple Boltzmann statistics—distribution (1.14a)—and $\varphi(N) = N - N \cdot ln\, N$ for the modified one, corresponding to the distribution (1.14b).

Partition Function and Thermodynamic Properties

One can obtain from the sum-over-states system's internal energy expression. It is known from thermodynamics that for the isolated system case the *internal* energy $u = \text{const.}$; in the case of Boltzmann's statistical treatment it was considered that at the equilibrium state of an isolated system, the *total* energy $E = \text{const.}$

But both E and u are defined in relation to an arbitrary benchmark level: see 1.3 for E, while for u it is known that only differences Δu exist in the case of phenomenological thermodynamics, differences that are identical for any zero level of the internal energy. Consequently, the same benchmark for E and u is chosen, so that Eq. (1.23) is equivalent—up to an arbitrary supplemental constant—to:

$$u = \sum_j N_j \cdot \varepsilon_j$$

where N_j is introduced from Eq. (1.34). It results that:

$$u = \frac{N}{Z} \cdot \sum_{j=0}^{j=K} g_j \cdot \varepsilon_j \cdot e^{-\alpha \cdot \varepsilon_j}$$

By replacing the sum from the above equality with the expression $dZ/d(ln\ \alpha)$, according to Eq. (1.33), it is obtained that:

$$\mathrm{u} = -(\mathrm{N}/\mathrm{Z}) \cdot [d(\,ln\,\mathrm{Z})/\mathrm{d}\alpha] \tag{1.37}$$

System entropy s can also be expressed in relation to the sum-over-states, using Eqs. (1.20) and (1.35):

$$s = k_B \cdot \varphi(N) + k_B \cdot N \cdot lnZ - k_B \cdot N \cdot \alpha \cdot \frac{dlnZ}{d\alpha} \tag{1.38}$$

The thermodynamic quantities u and s in Eqs. (1.37) and (1.38) depend on α, both directly and through the state function Z. By deriving these two equations in relation to α, the following Eqs. (1.a) and (1.b) are obtained:

$$-\frac{du}{d\alpha} = -N \cdot \frac{d^2 lnZ}{d\alpha^2} \quad (1.a)$$

$$-\frac{ds}{d\alpha} = -k_B \cdot N \cdot \alpha \frac{d^2 lnZ}{d\alpha^2} \quad (1.b)$$

whose term-by-term division is leading to:

$$du/ds = 1/(k_B \cdot \alpha) \quad (1.39)$$

Maxwell-Boltzmann Distribution of Energies

It is known from phenomenological thermodynamics that in an isolated system volume v remains constant, so that equation $du = T{\cdot}ds - p{\cdot}dv$ becomes:

$$du/ds = T \quad (1.40)$$

When comparing Eqs. (1.39) and (1.40), the following is found:

$$\alpha = 1/(k_B \cdot T) \quad (1.41)$$

Therefore, α constant expression was determined as a temperature dependency.

The final expressions of N_j and Z are obtained by replacing $\alpha(T)$ from Eq. (1.41) into Eqs. (1.31) and (1.32). The following expression results:

$$N_j = N \cdot g_j \cdot e^{-\varepsilon_j/(k_B \cdot T)}/Z \quad (1.42)$$

$$Z = \sum_{j=0}^{j=K} g_j \cdot e^{-\varepsilon_j/(k_B \cdot T)} \quad (1.43)$$

These equations ensure a mechanical–statistical significance for temperature, while phenomenological thermodynamics considers temperature as a macroscopic quantity having an empirical nature.

Equation (1.42) allows the number of particles' calculation for each energy level found in the steady state of a given isolated system and it is called *Maxwell-Boltzmann distribution of energies.*

1.6 Factors Influencing the Equilibrium Distribution

From Eqs. (1.42) and (1.43) it can be observed that particles' distribution according to their energy (N_j values spectrum), as well as the sum-over-states Z, depends both on system's characteristics (g_j and ε_i values for any j level) on the one hand and on an external system quantity—temperature—on the other.

Boltzmann Factor of the Energy Level

Temperature is encountered in Z and N_j expressions in the form of the so-called *Boltzmann factor*, which is defined for the j energy level as:

$$f_j = e^{-\theta_j/T} \tag{1.44}$$

The *characteristic temperature* θ_j of j level is the ratio $\theta_j = \varepsilon_j/k_B$, where ε_j is the energy of the level and k_B the Boltzmann's constant.

The Boltzmann factor stands for the ratio between the occupancy degree of an energetic level at a given temperature and its maximal possible occupational degree. It appears in the expressions of the most different physical and chemical phenomena: viscosity in liquids, diffusion in condensed phases, reaction speed, etc.

The Boltzmann factor f_j increases with temperature, according to Fig. 1.3:

At low T temperatures compared to the characteristic temperature θ, the increase is rapid—exponential—while at elevated temperatures the increase is much slower, f_j asymptotically tending to value 1.

Energy Level Multiplicity and System Size Effects

From Eq. (1.42) it can be observed that particles' number N_j increases with the level multiplicity and decreases with the increase in level energy.

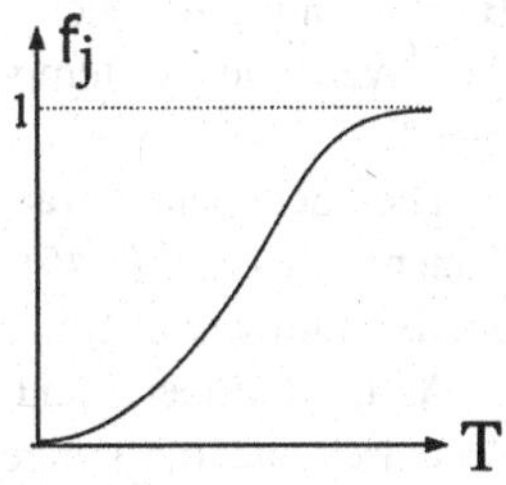

Fig. 1.3 Boltzmann factor dependency on temperature

Fig. 1.4 Occupancy of energy levels

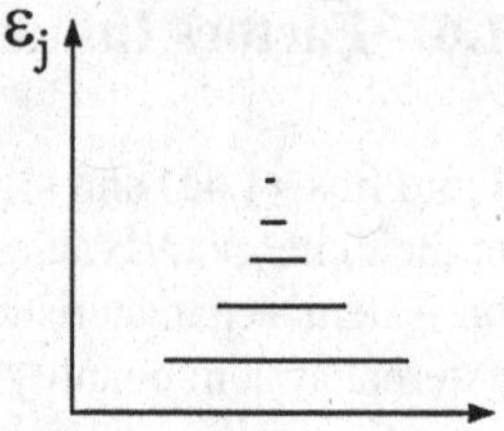

The *occupancy rate* of an energy level—a ratio between the number N_j of occupied places and the number g_j of available places—decreases exponentially with the increase in level's energy:

$$N_j = \left(N \cdot g_j / Z\right) \cdot \exp\left[-\varepsilon_j / (k_B \cdot T)\right] \tag{1.45}$$

which is maximal for the ground state, the one with energy $\varepsilon_0 = 0$, and from where it results that:

$$N_0 = N \cdot g_0 / Z$$

The occupancy rate also depends on the system complexity, which increases with its size: the higher the *K* number of system levels or system multiplicity, the higher will be the sum-over-states *Z* while the occupancy rates N_j/g_j will have lower values, according to Eqs. (1.43) and (1.45).

The occupancy rate will be lower for the excited states having increased energies, as can be seen in Fig. 1.4. The occupation of levels with very high energy levels is zero in the first approximation.

Temperature, *T*, is the main intensive quantity influencing Boltzmann distribution. It can be observed from Eq. (1.43) that *Z* increases with temperature increase.

Therefore, at 0 K all particles are found on the ground state, while at $T \to \infty$ the exponential factor of Eqs. (1.42) and (1.45) becomes equal to 1, so that the occupancy rates are equal for any energy level.

Influence of Temperature on Distribution

By illustrating the occupancy rates through the length of segments corresponding to the given energy, temperature variation of the occupancy rates is as shown in Fig. 1.5.

The occupancy rate of the ground state decreases continuously when *T* increases—Eq. (1.45)—while the occupancy rate of excited states goes through a maximum when temperature increases, as shown in Fig. 1.6.

As the number *j* and therefore the level energy increase, the maximum height decreases and the temperature value T_j at which the maximum appears will also

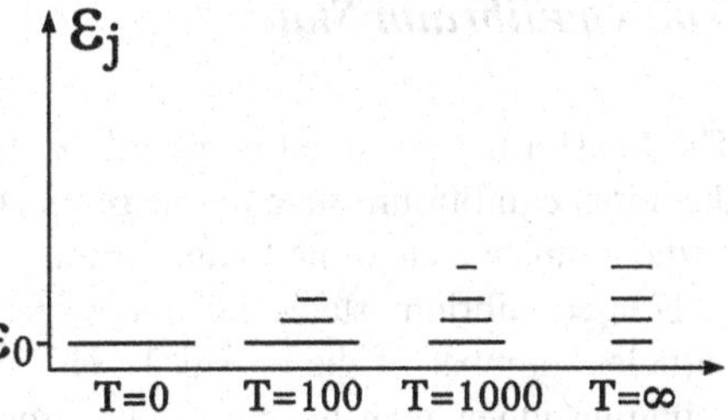

Fig. 1.5 Occupancy rates of energy levels

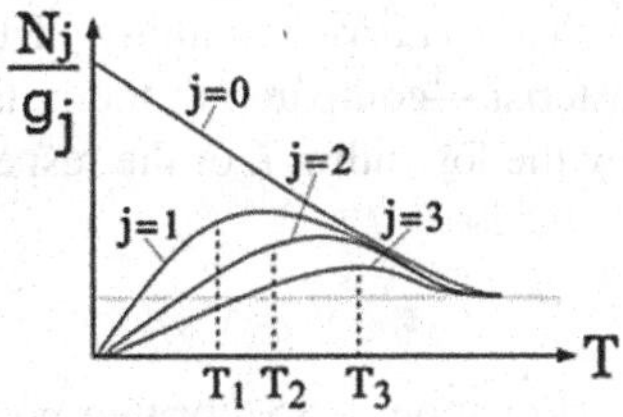

Fig. 1.6 Dependency on temperature of the occupancy rate of energy levels

increase. This temperature T_j, at which the occupancy rate of the j level reaches the maximum value, is the temperature at which the average energy per particle, $\overline{\varepsilon}$, becomes equal to the energy ε_j of the j level: $\overline{\varepsilon}(T) = \varepsilon_j$.

The average energy $\overline{\varepsilon}$ can be calculated by dividing the total energy E to the total number of particles, N:

$$\overline{\varepsilon} = E/N,$$

where the total energy is previously obtained by the sum: $E = \sum N_j \cdot \varepsilon_j$. By replacing N_j from Boltzmann distribution (1.42), the average energy is found:

$$\overline{\varepsilon} = \frac{\sum \varepsilon_j \cdot g_j \cdot e^{-\varepsilon_j/(k_B \cdot T)}}{\sum g_j \cdot e^{-\varepsilon_j/(k_B \cdot T)}} \tag{1.46}$$

1.7 Deviations from Equilibrium Distribution

A *fluctuation* is, in statistical thermodynamics, a temporary deviation from the equilibrium state of an *isolated* system, the system passing during the fluctuation through a series of non-equilibrium states.

In classic macroscopic thermodynamics case, the fluctuation phenomenon is not possible, because once the system reaches the equilibrium state, it cannot evolve until system isolation is ended, by changing the conditions of its interaction with the medium.

Non-equilibrium States

The transformation called *non-equilibrium* designates system's displacement from the initial equilibrium state to one of its possible non-equilibrium states, considered as the ultimate state of non-equilibrium.

Non-equilibrium states *do not* respect Boltzmann distribution (1.42) for the particles' number of the energy levels, hereinafter referred to as W^*, which is by definition lower than the maximum probability corresponding to the equilibrium state, which will be referred to as W without a distinctive index.

The *relative* instability of the final state—of non-equilibrium, marked with an asterisk—compared to the initial unmarked equilibrium state, can be characterized by the logarithm L of the respective probability ratio, that is, through the property:

$$L \equiv ln\, W* - ln\, W \tag{1.47}$$

Equation (1.15) for thermodynamic probability is valid both for the equilibrium and the non-equilibrium states, from where the following is obtained:

$$ln\, W = f(N) + \sum\nolimits_j N_j \cdot ln\left(g_j/N_j\right);\; ln\, W* = f(N*) + \sum\nolimits_j N*_j \cdot ln\left(g*_j/N*_j\right)$$

where $f(N)$ is a dependency function on the total particles' number N, N_j is the particles' number of the level marked by j index, g_j is the degree of degeneracy of this level in the case of equilibrium distribution, while N^*, N^*_j, and g_j are the respective properties for the non-equilibrium distribution. The following expression is obtained:

$$L = f(N*) - f(N) + \sum\nolimits_j \left\{\left[N*_j \cdot ln\left(g*_j/N*_j\right)\right] - N_j \cdot ln\left(g_j/N_j\right)\right]\right\},$$

The equation is simplified by the replacement of $g^*_j = g_j$ for any j level (since both states refer to the same system), and $N^* = N$, since system non-equilibrium does not modify its total number of particles:

$$L = \sum\nolimits_j \left[\left(N*_j - N*_j\right) \cdot \left(ln\, g_j\right) + N_j \cdot \left(ln\, N_j\right) - N*_j \cdot \left(ln\, N*_j\right)\right] \tag{1.48}$$

Simplest Change of State

In thermodynamic–statistical analysis, the system is assumed to be *isolated*—a term to be understood in the sense of absence of any interactions with the medium. This implies not only the system's total particles' number and total energy conservation according to Eqs. (1.8) and (1.9), but also temperature T invariance within a non-equilibrium state.

$$\mathrm{T} = \mathrm{T}* = \text{const.}$$

The "total" energy is the sum of particles' energies ε_j on *any* j level:

$$\mathrm{E} = \sum\nolimits_{j} \mathrm{N_j} \cdot \varepsilon_j \tag{1.a}$$

The non-equilibrium state involves changes in particles' numbers from at least three energy levels: changing the particles' number of a single level does not conserve the total number of particles, and the change at two levels (consisting of—in order to leave N unchanged—the passage of several particles from an energy level to a second level, whose energy is by definition different from the first one's) cannot respect the energy conservation condition.

Therefore, a system with a certain number of energy levels is considered, in which the state change does not modify the number of particles except on three levels, referred to as a, b, and c in the ascending order of energy:

$$\varepsilon_a < \varepsilon_b < \varepsilon_c$$

$$\forall \mathrm{j} \in \{\mathrm{a,b,c}\} : \mathrm{N}*_\mathrm{j} \neq \mathrm{N_j}$$

where $N*_j$ and N_j are particles' number of the j level before and respectively after non-equilibrium state. On all the other levels, particles' number remains unchanged, that is, $\forall \mathrm{j} \not\subset \{\mathrm{a,b,c}\} : N*_j = N_j$, therefore the terms corresponding to other levels than a, b, or c can be excluded from the sum found in the second member of Eq. (1.48). The *relative* stability of the final state of non-equilibrium thus becomes:

$$\mathrm{L} = \sum\nolimits_{\mathrm{j}\in\{\mathrm{a,b,c}\}} \left[(\mathrm{N}*_\mathrm{j} - \mathrm{N_j}) \cdot \left(\mathit{ln}\ \mathrm{g_j} \right) + \mathrm{N_j} \cdot (\mathit{ln}\ \mathrm{N_j}) - \mathrm{N}*_\mathrm{j} \cdot (\mathit{ln}\ \mathrm{N}*_\mathrm{j}) \right] \tag{1.49}$$

The value of L given by Eq. (1.49) depends on too many quantities, namely for 9: 3 numbers N_j, 3 of type $N*_j$, and 3 of g_j. As a first simplification, it can be admitted that the degree of degeneracy is the same—it will be noted by «g» without index—for all the three involved levels:

$$\forall \mathrm{j} \in \{\mathrm{a,b,c}\} : \mathrm{g_j} = \mathrm{g} \tag{1.50}$$

By using simplification (1.50), Eq. (1.49) becomes:

$$\mathrm{L} = \sum\nolimits_{\mathrm{j}\in\{\mathrm{a,b,c}\}} [(\mathrm{N}*_\mathrm{j} - \mathrm{N_j}) \cdot (\mathit{ln}\ \mathrm{g}) + \mathrm{N_j} \cdot (\mathit{ln}\ \mathrm{N_j}) - \mathrm{N}*_\mathrm{j} \cdot (\mathit{ln}\ \mathrm{N}*_\mathrm{j})]$$

where the coefficient of (*ln* g) is 0, since the sum of particles' numbers on the a, b, and c levels has the same value for the initial and the final states:

$$N{*}_a + N{*}_b + N{*}_c = N_a + N_b + N_c \tag{1.b}$$

so that the term proportional to *ln*g will disappear, the stability becoming thus:

$$L = \sum\nolimits_{j \in \{a,b,c\}} N_j \cdot (ln\ N_j) - N^*{}_j \cdot (ln\ N{*}_j)] \tag{1.51}$$

Relative Stability of a Non-Equilibrium State

In the initial state, of equilibrium, Eq. (1.42) is applicable on each level:

$$N_a = N \cdot g_a \cdot e^{-\frac{\varepsilon_a}{k_B \cdot T}}/Z; N_b = N \cdot g_b \cdot e^{-\frac{\varepsilon_b}{k_B \cdot T}}/ZN_c = N \cdot g_c \cdot e^{-\frac{\varepsilon_c}{k_B \cdot T}}/Z$$

By dividing to N_b each of the units N_a and N_c and when considering according to Eq. (1.50) that g_a, g_b, and g_c are equal to each other, the following expressions are obtained:

$$N_a/N_b = exp\,[(\varepsilon_b - \varepsilon_a)/(k_B \cdot T)]; N_c/N_b = exp\,[(\varepsilon_b - \varepsilon_c)/(k_B \cdot T)] \tag{1.c}$$

In order to further reduce the calculation complexity, it is considered that the three energy levels are situated at the same distance *q* one from the other:

$$\varepsilon_a = \varepsilon_b - q; \varepsilon_c = \varepsilon_b + q \tag{1.52}$$

so that, (c) equations will become:

$$N_a = N_b \cdot e^r; N_c = N_b \cdot e^{-r} \tag{1.53}$$

where by *r* was noted the positive ratio:

$$r \equiv q/(k_B\ T) \tag{1.54}$$

Besides, by denoting with V_k the non-equilibrium variation of particle numbers found on the k level, the final numbers of particles $N*_k$ will be:

$$N{*}_a = N_a + V_a; N{*}_b = N_b + V_b; N{*}_c = N_c + V_c \tag{1.d}$$

By introducing the final particle numbers $N*_a$, $N*_b$, and $N*_c$ from Eq. (1.d) within the conservation conditions, on all three levels, particles' number—Eq. (1.b)—and of total energy given by Eq. (1.a), namely:

$$N*_a \cdot \varepsilon_a + N*_b \cdot \varepsilon_b + N*_c \cdot \varepsilon_c = N_a \cdot \varepsilon_a + N_b \cdot \varepsilon_b + N_c \cdot \varepsilon_c.$$

two linear-homogenous correlations are obtained between variations V_a, V_b, and V_c of the number of particles at the change of state, namely:

$$V_a + V_b + V_c = 0; V_a \cdot \varepsilon_a + V_b \cdot \varepsilon_b + V_c \cdot \varepsilon_c = 0 \quad (1.e)$$

By using ε_a and ε_c from Eq. (1.52), the second Eq. (1.e) becomes:

$$\varepsilon_b \cdot (V_a + V_b + V_c) + \varphi \cdot (V_c - V_a) = 0$$

from which we obtain, considering that $\varphi \neq 0$ and also considering the first Eq. (1.e):

$$V_c = V_a; V_b = -2 \cdot V_a \quad (1.f)$$

Role of the Non-Equilibrium Extent

By denoting with x the *relative* value of the non-equilibrium extent:

$$x \equiv -V_b/(2 \cdot N_b) \quad (1.55)$$

we obtain from Eq. (1.f) that particles' number variations at transformation are all proportional to the product $x \cdot N_b$, namely:

$$V_a = x \cdot N_b; V_b = -2 \cdot x \cdot N_b; V_c = x \cdot N_b \quad (1.g)$$

By introducing into Eq. (1.d) the expressions for V_a, V_b, and V_c from Eq. (1.g) as well as the equations for N_a and N_c from the equality (1.53), it is obtained that:

$$N_a* = N_b \cdot (e^r + x); Nb* = N_b \cdot (1 - 2 \cdot x); N_c* = N_b \cdot (e^{-r} + x) \quad (1.56)$$

By replacing into Eq. (1.51) the values of N_a and N_c from Eq. (1.53) and the values $N*_a$, $N*_b$, $N*_c$ from Eq. (1.56), the relative instability of the final state of non-equilibrium becomes:

$$L = -N_b \cdot [(e^r + x) \cdot ln\,(e^r + x) + (e^{-r} + x) \cdot ln\,(e^{-r} + x) + (1 - 2 \cdot x) \cdot ln\,(1 - 2 \cdot x)]$$

or, after terms reduction:

$$L = -N_b \cdot [(e^r + x) \cdot ln\,(1 + x \cdot e^{-r}) + (e^{-r} + x) \cdot ln\,(1 + x \cdot e^r) + (1 - 2 \cdot x) \cdot ln\,(1 - 2 \cdot x)] \quad (1.h)$$

The three logarithms are decomposed according to Taylor series in the neighborhood of 1, namely:

$$ln\,(1 + y) = y - y^2/2 + y^3/3 - y^4/4 + ..$$

Under these conditions, Eq. (1.h) becomes the following decomposed expression of the relative instability *L* of positive integer powers of the non-equilibrium degree *x*:

$$L = -N_b \cdot \sum_{k=2}^{k=\infty} \left[2^k + (-1)^k \cdot \left(e^{k-k \cdot r} + e^{k \cdot r-k}\right)\right] \cdot x^k/k! \quad (1.i)$$

For small-enough non-equilibriums ($x << 1$), the (i) series are well approximated by its first term, as follows:

$$L = -N_b \cdot x^2 \cdot (4 + e^r + e^{-r})/2 \quad (1.57)$$

Relative Probability of a Non-Equilibrium State

Equation (1.57) represents the expression sought for the *relative* probability, W^*/W (in relation to the equilibrium state), of the discussed non-equilibrium state. As expected, the relative probability is subunitary, since $ln\,(W^*/W) < 0$.

Three general consequences of the equation are the following:

- *W** decreases when *x* increases, that is, a thermodynamic state is even more unlikely to occur as it is situated more further from equilibrium. The dependency (ln W*) on the dimension *x* of non-equilibrium is quadratic, so that small non-equilibriums can be still achievable, but the ample ones are practically impossible: when *x* increases 10 times, the ratio W^*/W decreases e^{100} times, namely 10^{44} times.
- *W** decreases when *N* increases, namely the non-equilibrium probability is even lower as the system is more elevated.

- By deriving W^* from Eq. (1.57) in relation to r it results that W^* increases when r decreases. But according to definition (1.54) it results that $dr/dT < 0$, therefore the non-equilibrium probability will be even higher at elevated temperatures.

Fluctuation Errors

The expression $y_{med} = \int_{t=t_{ini}}^{t=t_{fin}} y(t) \cdot dt/(t_{fin} - t_{ini})$ is the mean value of the physical property y, where t_{ini} and t_{fin} are the initial and final moments of the evaluation period.

The appreciation criterion for the y fluctuation of this period is the *relative standard deviation (RSD)*, $\Psi_y \equiv \sqrt{\int_{t=t_{ini}}^{t=t_{fin}} [(y/y_{med}) - 1]^2 \cdot dt/(t_{fin} - t_{ini})}$.

As the duration of existence of each state is proportional to its physical probability W, it can be deduced that the relative standard deviation of a magnitude from its mean value is measurable by the standard deviation of the thermodynamic probability logarithm, referred to as L in this paragraph.

Certainly, the mean of a thermodynamic magnitude that fluctuates around an equilibrium value is assimilated to this equilibrium value, while for small deviations $|u| << 1$, it can be approximated that $ln(1 + u) \cong u$.

An appropriate measure of a thermodynamic y property fluctuations is given by Ψ_y:

$$\Psi_y \cong \sqrt{\sum_{k=1}^{k=K} (y_k/y_{ech} - 1)^2 e^{-L_k} / \sum_{k=1}^{k=K} e^{-L_k}} \tag{1.j}$$

namely Ψ_y is approximated by the quadratic relative standard deviation of values y_k taken by y in each of the K possible states of the system, each of these values being measured by the thermodynamic probability L_k of the respective state, calculated for the k state using Eq. (1.57). As a consequence of Eq. (1.j), the relative magnitude of fluctuations has the order of magnitude of the inverse of the quadratic mean of the N number of system particles:

$$\Psi_y = K/\sqrt{N} \tag{1.k}$$

The K factor, having the order of unity, varies with the nature of y and the existing conditions in the system.

The fluctuation error is usually negligible as compared to the experimental error of measurement of any thermodynamic property, due to the considerable number of atoms practically encountered in the systems.

Even an ultra-miniaturized nano-device, such as the communication nano-relay at frequencies of terahertz range of dimensions (2015) of 1 × 2 × 6 microns, contains 3 billion atoms (because their mean diameter is 0.15 nm) from where—by (k) equation—a fluctuation error is of the order of 0.001%.

1.8 Statistics of Thermodynamic Properties

The results obtained within statistical mechanics allow molar thermodynamic properties of matter determination as dependency functions of the sum-over-states, *Z*.

For this purpose, in the thermodynamic–statistical expressions from the previous paragraphs, the number of *N* particles is replaced by the number of molecules in a mole, that is, the number of Avogadro, N_A.

Simple Boltzmann statistics apply, because the molecules in the macroscopic system are discernible particles.

Internal Energy

With $N = N_A$ and replacing α by $1/(k_B \cdot T)$ from Eq. (1.41) into Eq. (1.37), the molar internal energy results to be: $U = N_A \cdot d\, lnZ/\, d\,[1/(k_B \cdot T)]$ or, as the product $(k_B \cdot N_A)$·is equal to: $U = R \cdot T^2 \cdot d\,(ln\, Z)/\, d\, T$.

The above derivative is in fact a partial derivative, because the system is supposed to be isolated (v = const). The exact form of the equation is therefore:

$$U = R \cdot T^2 \cdot [\partial(ln\, Z)/\partial T]_v \tag{1.58}$$

If we add the value, denoted by U_0, of the energy at 0 K, it results that:

$$U = U_0 + R \cdot T^2 \cdot [\partial(ln\, Z)/\partial T]_v \tag{1.a}$$

where unit U_0 corresponds to the ground state energy ε_0:

$$U_0 = N_A \cdot \varepsilon_0 \tag{1.b}$$

Entropy

Similarly, by replacing *N* with N_A and *α* with $1/(k_B \cdot T)$ into Eq. (1.38), the *molar* entropy equation will be obtained:

$$S = k_B \cdot \varphi(N_A) + k_B \cdot N_A \cdot ln\ Z + k_B \cdot N_A \cdot [d\ ln Z / d\ ln (k_B \cdot T)]$$

with $\varphi(N)$ defined by Eq. (1.37) and zero for the case of simple Boltzmann statistics. We obtain, therefore, after derivation (at constant volume) and putting $k_B \cdot N_A = R$:

$$S = R \cdot [ln\ Z + T \cdot (\partial\ ln Z / \partial T)_V] \qquad (1.59)$$

Free Energy

The free energy F expression is obtained from the thermodynamic equality $F = U - T \cdot S$ by replacing U and S from Eqs. (1.58) and (1.59):

$$F = -R \cdot T \cdot ln Z \qquad (1.60)$$

From Eq. (1.60) it results that the sum-over-states has a direct thermodynamic significance, *ln* Z being directly proportional to the system free energy, measured relative to a benchmark corresponding to 0° K temperature.

Caloric Capacity

The equation of caloric capacity at constant volume, C_V, is calculated using the thermodynamic equation $C_v = T \cdot (\partial S / \partial T)_V$, from which it results that:

$$Cv = (R/T^2) \cdot \left[\partial^2 \ln Z / \partial(1/T)^2\right]_V \qquad (1.61)$$

Properties Depending on Pressure

As $P = -\cdot(\partial S / \partial V)_T$, it is found that:

$$P = (R \cdot T) \cdot (\partial\ ln\ Z / \partial V)_T \qquad (1.62)$$

The *Z(V)* dependency is yet difficult to establish, because volume variation modifies both statistical weights (multiplicities) g_j and levels' energies ε_j. From relation (1.62), *P(Z)* can be further calculated, as dependency function on the sum-over-states: C_p, H = U + P·V, G = F + P·V; etc.

1.9 Five Worked Examples

Example 1.1
Compare the thermodynamic probabilities of boltzons', bosons', and fermions' distribution for the system with degeneracy and energy çonservation from Fig. 1.2 (in 1.3.), where $N = 4$, the total energy $E = 3$, and for energy levels having the order number (0, 1, 2, 3) and multiplicities (2, 4, 1, 1).

The three macrostates to be compared, referred to as A, B, and C, are the ones from the example in Sect. 1.4.2, "Distribution on Non-Degenerate Energy Levels," characterized by the following particle numbers N_0, N_1, N_2, N_3: 3, 0, 0, 1 (A); 2, 1, 1, 0 (B); and 1, 3, 0, 0 (C).

Solution

- Boltzmann statistics led to the thermodynamic probabilities already calculated in an earlier section: $W_{dB}(A) = 32$, $W_{dB}(B) = 192$, and $W_{dB}(C) = 512$.
- For a same system, Bose–Einstein statistics implies the use of Eq. (1.11): $w_{dBE} = \prod_{j=0}^{j=3}\left(g_j + N_j - 1\right)!/\left[\left(g_j - 1\right)! \cdot N!\right]$; or after the replacement of g_j values: $w_{dBE} = \left[\frac{(N_0+1)!}{1! \cdot N_0!}\right] \cdot \left[\frac{(N_1+3)!}{3! \cdot N_1!}\right] \cdot \left[\frac{N_2!}{0! \cdot N_2!}\right] \cdot \left[\frac{N_3!}{0! \cdot N_3!}\right]$.
  And after simplifying:

$$w_{dBE} = (N_0 + 1) \cdot (N_1 + 1) \cdot (N_1 + 2) \cdot (N_1 + 3)/6 \tag{1.a}$$

where the number of particles on a level differs for the three A, B, C macrostates.
Thus, in the case of macrostate A, for which $N_0 = 3$ and $N_1 = 0$, Eq. (1.a) is obtained: $w_{dBE}(A) = \frac{4 \cdot 1 \cdot 2 \cdot 3}{6} = 4$, similarly for macrostates B and C it results that:

$$w_{dBE}(B) = (3 \cdot 2 \cdot 3 \cdot 4)/6 = 12, w_{dBE}(C) = (2 \cdot 4 \cdot 5 \cdot 6)/6 = 40.$$

- Equation (1.12) of Fermi–Dirac statistics becomes, for the studied degeneracy system, $W_{dFD} = \prod_{j=0}^{3} \frac{g_j!}{(g_j - N_j)! \cdot N_j!}$, from where, by replacing g_j:

$$W_{dFD} = \frac{2!}{(2 - N_0)! \cdot N_0!} \cdot \frac{4!}{(4 - N_1)! \cdot N_1!} \cdot \frac{1!}{(1 - N_2)! \cdot N_2!} \frac{1!}{(1 - N_3)! \cdot N_3!} \tag{1.b}$$

- FD statistics requires that, $\forall j$: $N_j \le g_j$, therefore macrostate A, for which $N_0 = 3 > g_0 = 2$, is not achievable for fermions, and $W_{dFD}(A) = 0$.
- By replacing into Eq. (1.b) the values $N_0 = 2$, $N_1 = N_2 = 1$, and $N_3 = 0$ of macrostate B, we obtain: $W_{dFD}(B) = 4$.

– Similarly, for macrostate C, for which N_j (j = 0, 1, 2, 3) takes the values 1, 3, 0, and 0, Eq. (1.b) leads to: $W_{dFD}(C) = 8$.

The results for the three statistics are compared in the following final table:

Statistics type		Boltzmann	Bose–Einstein	Fermi–Dirac
Macrostate Thermodynamic probability	A	32	4	0
	B	192	12	4
	C	512	40	8

Example 1.2 By using the **s** indices for the simple system, **d** for the degenerate one, and further indexes **B**, **BE**, and **FD** for the particles subjected to Boltzmann, Bose–Einstein, and Fermi–Dirac statistics, it is required to:

I. Calculate the thermodynamic probabilities in a simple system of the three macrostates referred to as (A), (B), or (C) at Example 1.1, for the three types of particles.
II. Compare the mathematical probabilities of the three macrostates for B, BE, and FD statistics, between the degeneracy systems at Example 1.1 and the systems without degeneracy at point **I** of this example.
III. Discuss qualitatively the obtained results.

Solution

I. For a simple system, the value of multiplicity is 1 for any energy level

- Equation (1.7a) is valid for the case of indiscernible particles. Thermodynamic probabilities have been already calculated in the example at 1.3 above: $W_{sB}(A) = 4$, $W_{sB}(B) = 12$ and $W_{sB}(C) = 4$.
- Equation (1.11) of Bose–Einstein statistics takes the following form when putting $g_0 = g_1 = g_2 = g_3 = 1$: $W_{sBE} = \prod_{j=0}^{j=k}(1 + N_j - 1)!/[(1-1)! \cdot (N_j!)] = 1$ for any macrostate. It results that $W_{sB}(A) = W_{sB}(B) = W_{sB}(C) = 1$.
- By putting $g_j = 1\ \forall j$, equation (1.12) of Fermi–Dirac statistics becomes $W_{sFD} = \prod_{j=0}^{j=k} 1!/[(1-N_j)! \cdot (N_j!)]$ with two possible values: $W_{sFD} = 1$ for the macrostates where there is at most one particle on any energy level and $W_{sFD} = 0$ if at least one of the particle number of an energy level is superior to 1. It results that: $W_{sFD}(A) = W_{sFD}(B) = W_{sFD}(C) = 0$, respectively because $N_0(A) = 3$, $N_0(B) = 2$, and $N_1(C) = 2$. Therefore, none of the macrostates can be achieved for the case of fermions.

Table 1.3 Percent probabilities of macrostates at Example 1.2

Macrosta		A	B	C
System				
Simple	Indiscernible	20	60	20
	Bosons	33.3	33.3	33.3
	Fermions	Impossible		
With degeneracy	Indiscernible	4.3	26.1	69.6
	Bosons	7.1	21.4	71.5
	Fermions	0	33.3	66.7

II. The mathematical probability of achieving the macrostates is obtained by dividing the macrostate thermodynamic probability to the sum of thermodynamic probabilities of the three macrostates—Eq. (1.3) (Table 1.3).

III. It can be observed that:

- Degeneracy increases the number of microstates of a macrostate (its thermodynamic probability W): in some cases (macrostates B, C) only the appearance of degeneracy makes possible fermions' distribution.
- For a given system, W decreases in the following order: indiscernible particles > bosons > fermions.
- Degeneracy changes the ratio of thermodynamic probabilities of any two macrostates: for the simple system, macrostate B is the most probable—the one with particles as evenly distributed as possible (at least for boltzons)—while for the degenerate system, macrostate C becomes the most plausible for all types of particles.

The reason for favoring macrostate C is that in its case of configuration, particles are distributed on the energy levels approximately proportionally to the degree of degeneracy of each level.

Example 1.3
Let ε be the relative calculation error of unit *ln* W for fermions by using modified Boltzmann statistics, for a system with K energy levels each one having the degeneracy multiplicity B and θ occupancy rate, if $\theta \ll 1$ and B multiplicity and N_j particles' number on the level are both much higher than 1.

The values of occupancy rate, for which $\varepsilon < 1\%$, are required.

Solution
By adopting the indices **FD** and **Bm** for the Fermi–Dirac and modified Boltzmann statistics, the above error definition becomes:

$$\varepsilon = [ln\,(W_{FD}/W_{Bm})]/\,ln\;W_{Bm} \tag{1.3a}$$

Equation (1.15) is used for W_{Bm} from the denominator of expression (1.3a), where for any level $j \in [1, J]$: $g_j = B$ while $N_j = B \cdot \theta$.

The total number of particles is therefore $N = K \cdot B \cdot \theta$. It results that:

$$ln\ W_{Bm} = K \cdot B \cdot \theta + \sum_{j=1}^{j=K} \{B \cdot \theta \cdot [ln\,(B) - ln\,(B \cdot \theta)]\}$$

or, because the sum Σ_j includes K identical terms:

$$ln\ W_{Bm} = K \cdot B \cdot \theta \cdot (1 - ln\ \theta) \quad (1.3b)$$

For the ratio W_{FD}/W_{Bm} from the numerator of expression (1.3a), the following products are introduced, as taken from Eqs. (1.12) and (1.13):

$W_{FD} = \prod_{j=1}^{j=K} g_j!/\left[\left(g_j - N_j\right)! \cdot (N_1)!\right]$, $W_{Bm} = \prod_{j=1}^{j=K} \left(g_j\right)^{N_j}/(N_j)!$, from where, by dividing and by applying the logarithm, the identical terms' sum K is obtained:

$$ln\,(W_{FD}/W_{Bm}) = \sum_j \left\{ ln\left(g_j!\right) - ln\left[\left(g_j - N_j\right)!\right] - N_j \cdot ln\ g_j \right\}.$$

By replacing $g_j = B$, $N_j = B \cdot \theta$ it is obtained that:

$$ln\,(W_{FD}/W_{Bm}) = K \cdot \{ ln\,(B!) - ln\,[(B - B\theta)!] - B\theta \cdot ln\ B\} \quad (1.3c)$$

In order to evaluate the logarithms of the two factors from expression (1.3c), the third Stirling approximation is used, (1.6c):

$$ln\,(B!) = B \ln B - ln\ B \text{ and } ln\,[(B - B\theta)!] = (B - B\theta) \cdot ln\,(B - B\theta)$$
$$= (B - B\theta).$$

After replacement into (1.3c) the following is obtained:

$ln(W_{FD}/W_{Bm}) = -K \cdot [(B - B \cdot \theta) \cdot ln\,B - (B - B \cdot \theta) \cdot ln\,(B - B \cdot \theta) - B \cdot \theta]$ or, after regrouping the terms:

$$ln\,(W_{FD}/W_{Bm}) = -K \cdot B \cdot [(1 - \theta) \cdot ln\,(1 - \theta) + \theta] \quad (1.3d)$$

By replacing in (1.3a) $ln\ W_{Bm}$ from (1.3b) and $ln(W_{FD}/\ W_{Bm})$ from (1.3d), it results that:

$$\varepsilon = -[(1 - \theta) \cdot ln\,(1 - \theta) + \theta]/[\theta \cdot (1 - ln\ \theta)] \quad (1.3e)$$

It can be observed that ε depends only on θ, but not on g_j, N_j, or K.

For $\theta \ll 1$, by decomposition according to the McLaurin series $ln\,(1-\theta)$ from the numerator of expression (1.3e) it results that: $ln(1-\theta) = -\theta - \theta^2/2 - \theta^3/3 - \cdots$.

The square bracket from the right member of the numerator from (e) therefore becomes:

$$(1-\theta)\cdot\left(\theta+\theta^2/2+\theta^3/3+\cdots-\theta\right)\ \text{or}\ \theta^2/2+\theta^3/6+\cdots-\theta$$

which, after simplifying by θ between denominator and numerator, leads to:

$$\varepsilon = \left(\theta/2+\theta^2/6+\cdots-\theta\right)/(1-ln\ \theta) \tag{1.3f}$$

For a relative error of 1% (expressed in module) and keeping only the first two terms of the series from the numerator of Eq. (1.3f), it is obtained that:

$$50\cdot\theta\cdot(1+\theta/3)+ln\ \theta-1=0 \tag{1.3g}$$

The transcendental Eq. (1.3g) has only one positive solution: $\theta = 7.12\%$. Therefore, the modified Boltzmann's statistic approximates the Fermi–Dirac statistics with an accuracy better than 1% in the range: $\mathbf{0 < \theta < 0.071}$.

Example 1.4

A system with an infinite number of energy levels, numbered by the index j starting at 0 is considered. The energy of the j level and its multiplicity are respectively $\varepsilon_j = A\cdot j$ and $g_j = 2j+1$, where A is a constant. Calculate:

I. Sum-over-states, Z, as a dependency function on temperature T.
II. Temperature T_j where the occupancy rate of **j** level is maximum.

Solution

I. In the given case, Eq. (1.43) becomes:

$Z=\sum_{j=0}^{\infty}(2\cdot j+1)\cdot e^{-j\cdot A/(k_B\cdot T)}$ or $Z=\sum_{j=0}^{\infty}(2\cdot j+1)\cdot a^j$, where $a = e^{-A/(k_B\cdot T)}$. But $\sum_{j=0}^{\infty} a^j = \frac{1}{(1-a)}$ and $\sum_{j=0}^{\infty} ja^j = \frac{a}{(1-a)^2}$, from where

$$\Rightarrow Z=\{1+exp\,[-A/(k_B\cdot T)]\}/\{1-exp\,[-A/(k_B\cdot T)]\}^2$$

II. By noting with θ_j the occupancy rate of j level, Eq. (1.45a) becomes: $\theta_j = \left(\frac{1}{z}\right)\cdot exp\left(-\frac{\varepsilon_j}{k_B\cdot T}\right)$; by replacing, as in Example 1.3, $\varepsilon_j = A\cdot j$ and $a = exp[-A/(k_B\cdot T)]$, the following is obtained:

$$\theta_j = a^j\cdot(1-a)^2/(1+a) \tag{1.4a}$$

By differentiating equality (1.4a) with respect to variable a, it is found that:

$$d\theta_j/da = (1-a)\cdot a^{j-1}\cdot\left[-(j+1)\cdot a^2 - 3\cdot a + j\right]/(1+a)^2 \quad (1.4b)$$

The maximal occupancy rate corresponds to equation $d\theta_j/\,dT = 0$ or, as T is a continuous function of a: $d\,\theta_j/\,d\,a = 0$.

The required solution is therefore found among the values of a that cancel the numerator from the left member of equality (b), without canceling its denominator. These values, four by number, and which will be furtherly denominated by A, B, C, and D, are the following:

IV. : $1-a=0$ or $a=1$ corresponds to a minimum $\theta_j = 0$ situated at $T=\infty$.
V. : $a=0$ corresponds to the *minimum* $\theta_j = 0$ (for $j \neq 0$) placed at $T=0$.
VI. : the negative solution of the algebraic equation of second order in a:

$$(j+1)\cdot a^2 + 3\cdot a - j = 0 \quad (1.4c)$$

does not make sense here (a, as exponential, cannot be negative).

VII. : Only the positive solution of the equation remains, that is:

$a = \frac{\left[-3+\sqrt{4j^2+4j+9}\right]}{2\cdot(j+1)}$, which will correspond to the maximal value θ.

Returning from variable a to T, it is obtained that:

$$T_j = (A/k_B)\cdot ln\left[2\cdot(j+1)/\left(-3+\sqrt{4\cdot j^2+4\cdot j+9}\right)\right]$$

Example 1.5 Calculate the temperature T_j at which the energy levels having the order numbers (j – 1) and j contain the same number of particles, for an equilibrium system, characterized by $g_j = (j+1)^2$ and $\varepsilon_j = A - A/\,(j+1)^2$, where A is a constant.

Solution

At temperature equal to T_j, the logarithmic form of Eq. (1.42) is:

$$ln\left(\frac{N_j\cdot Z}{N}\right) = -\frac{\varepsilon_j}{k_B\cdot T_j} + ln\ g_j \quad (1.5a)$$

By replacing for the same temperature the j level with the (j – 1) level in terms of energy and multiplicity, it results that:

$$ln\left(N_{j-1}\cdot\frac{Z}{N}\right) = -\frac{\varepsilon_{j-1}}{k_B\cdot T_j} + ln\ g_{j-1} \quad (1.5b)$$

By subtracting member by member the equality (1.5b) from equality (1.5a), the following is obtained:

$$ln\left(N_j/N_{j-1}\right) = \left(\varepsilon_{j-1} - \varepsilon_j\right)/\left(k_B \cdot T_j\right) + ln\left(g_j/g_{j-1}\right)$$

from where, for $N_j = N_{j-1}$:

$$\left(\varepsilon_j - \varepsilon_{j-1}\right)/\left(k_B \cdot T_j\right) = ln\left(g_j/g_{j-1}\right) \tag{1.5c}$$

By introducing ε_j, similarly to the statement: $\varepsilon_j = A - A/(j+1)^2$, $g_j = (j+1)^2$, and analogue:
$\varepsilon_{j-1} = A - A/j^2$, $g_{j-1} = j^2$, Eq. (1.5c) becomes:

$$A \cdot \left[1/j^2 - 1/(j+1)^2\right]/\left(k_B \cdot T_j\right) = ln\,[(j+1)/j]^2$$

so that:

$$T_j = \frac{A}{k_B} \cdot \frac{(j+1/2)}{j^2 \cdot (j+1)^2 \cdot ln\,(1+1/j)}$$

Chapter 2
Statistical Thermodynamics of Ideal Gas

Partition Functions Calculation

The partition function Z, also designated as *sum-over-states*, is defined by the relation (1.43) from Sect. 1.5, namely $Z = \sum_{j=0}^{j=K} g_j \cdot e^{-\varepsilon_j/(k_B \cdot T)}$, where the addition is made for all K energy levels of a certain possible motion type for a given particle. In relation (1.43), g_j and ε_j are the multiplicity and the energy of j level, respectively, while k_B represents Boltzmann's constant and T the temperature.

The thermodynamic probability of the state of equilibrium for a certain substance and its thermodynamic characteristics in different conditions are calculated by statistical mechanics procedures starting from the sum-over-states, using the methods applied in this chapter to the simplest case—of ideal gas—for the different possible motions in the different types of molecules.

2.1 Components of the Partition Function

Composition of the Sum-Over-States

For the case of 1 mole, Eq. (1.20) becomes: $S = R \cdot \ln W$, where the thermodynamic probability W refers to 1 mole, as the molar entropy S. The introduction of this expression for S in Eq. (1.59) leads to:

$$\ln W = N_A \left[\ln Z + T \cdot (\partial \ln Z / \partial T)_V \right] \quad \text{(a)}$$

One of the consequences of Eq. (a) is the law of composition of sum-over-states for independent types of motions: to the independence of two phenomena denoted by A and B corresponds a multiplication of probabilities, according to the combinatory analysis, that is, $W_C = W_A \cdot W_B$, where W_C is the probability of achievement of a

F. E. Daneş et al., *Molecular Physical Chemistry for Engineering Applications*,
https://doi.org/10.1007/978-3-030-63896-2_2

"composed" event C, consisting in simultaneous achievement of A and B events. By applying the logarithm, the following expression is obtained:

$$\ln W_C = \ln W_A + \ln W_B \tag{b}$$

Probabilities (W_A, W_B, W_C) can be expressed as a dependency function of the corresponding sum-over-states (Z_A, Z_B, Z_C) through Eq. (a); by further substitution of (W_A, W_B, W_C) into Eq. (b), by simplifying with N_A and regrouping the terms, it results:

$$\{d[T \cdot \ln (Z_A \cdot Z_B/Z_C)]/dT\}_V = 0 \tag{c}$$

Equation (c) being true at any temperature, it results:

$$Z_C = Z_A \cdot Z_B \tag{d}$$

Therefore, to the law of composition by multiplication of probabilities of independent events corresponds a law of composition by sum-over-states multiplication of different types of motions—Eq. (d); for a given number M of motion types the following law of *sum-over-states composition* is obtained by generalizing the following equation:

$$Z = \prod_{i=1}^{i=M} Z_i \tag{2.1}$$

where Z is the total sum-over-states, and Z_i ($i = 1, 2, \ldots.., M$) are the sum-over-states of the different independent motion types. In its logarithmic form, Eq. (2.1) becomes a composition by addition:

$$\ln Z = \prod_{i=1}^{i=M} Z_i \tag{e}$$

In Sect. 1.8, it was shown that thermodynamic functions have a linear dependence on ($\ln Z$) and of derivatives of $\ln Z$ with respect to T or V. Consequently, to the additive composition (*e*) of the sum-over-states corresponds a series of additive appropriate laws for the thermodynamic properties U, S, F, etc.; these laws have the following form:

$$Y = \sum_{i=1}^{i=M} Y_i \tag{2.2}$$

where Y is any thermodynamic molar property, while Y_i is the contribution of the i-type motion to the value of the property Y.

Molecule Displacement and Motion

The additive law (2.2) for the composition of molar properties allows, therefore, a simplification of the thermodynamic function's statistical calculation: a separate calculation is possible for the contributions of each motion type of molecules within the *phasic*, abstract state (see Sect. 2.3) if the displacements are not influencing one another. This requirement, far from being accomplished in many real situations, is fulfilled by the fictional entity (model) designated in statistical physics as "ideal gas."

The possible independent motions are:

- Translation of molecules, namely the parallel molecule movement into space without any change in shape, dimensions, or orientation, as seen in Fig. 2.1a.
- Rotation of the molecule around its own center of gravity, as in Fig. 2.1b.
- Vibrations within molecule in Fig. 2.1c (where *A*, *B*, *C* are the nuclei, figured as material points), interpreted as periodic variations either of an internuclear distance (*AB* length) or of a valence angle (α).
- Variation of molecule's electronic state, representing the change of one of the electrons' quantum numbers.
- Variation of nuclear state of the molecule, designated as the change of one of the nuclear atomic quantum numbers.

The ideal gas' sum-over-states is, therefore, the product between five components, each one representing a molecule motion type within the *phase* space:

$$Z = Z_{tr} \cdot Z_{rot} \cdot Z_{vib} \cdot Z_{el} \cdot Z_{n} \tag{2.3}$$

If the motion takes place also in the *physical* space, it is simply called <u>*displacement*</u>: it is the case of transition, rotation, and the vibration from Fig. 2.1.

Rotational, vibrational, electronic, and nuclear components are sometimes embedded in a single *internal* movement of the molecule, which refers to changes *within* the molecule, $Z_{int} = Z_{rot} \cdot Z_{vib} \cdot Z_{el} \cdot Z_{n}$, so that the total sum-over-states can be expressed as the product $Z = Z_{tr} \cdot Z_{int}$ of the *external* sum-over-states (of translation) and internal one. Z_{tr} corresponds to a modified Boltzmann statistic (the molecules are indiscernible), but Z_{int} corresponds to a simple Boltzmann statistic, since the motion inside a molecule cannot be confused with the motion inside another molecule.

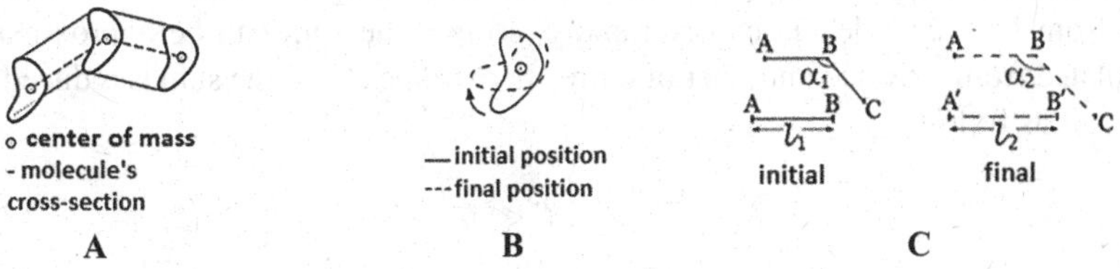

Fig. 2.1 Translational, rotational, and vibrational motions of molecules

Passing to the additive form of the composition law, the following is obtained:

$$\ln Z = \ln Z_{\mathrm{tr}} + \ln Z_{\mathrm{rot}} + \ln Z_{\mathrm{vib}} + \ln Z_{\mathrm{el}} + \ln Z_{\mathrm{n}} \tag{2.4a}$$

and similarly, for a certain thermodynamic property of the ideal gas, Y:

$$Y = Y_{\mathrm{tr}} + Y_{\mathrm{rot}} + Y_{\mathrm{vib}} + Y_{\mathrm{el}} + Y_{\mathrm{n}} \tag{2.4b}$$

Simplifications of Composition Laws

For certain cases, the laws of composition (2.3 and 2.4a and 2.4b) take simpler forms. Therefore, when a certain motion form i cannot take place, its thermodynamic probability and its corresponding partition function are equal to 1, so that the component can be neglected; for example, rotation and vibration are neglected for monoatomic molecules and therefore:

$$Z = Z_{\mathrm{tr}} \cdot Z_{\mathrm{el}} \cdot Z_{\mathrm{n}}$$

Additionally, there are situations where the partition function Z_i formula for a certain type of motion is simplified. Relation (1.43) for Z becomes, after replacing ε_j as a function of the characteristic temperature θ_j, defined by $\theta_j = \varepsilon_j \cdot k_B$: $Z = \sum_{j=0}^{j=K} g_j \cdot e^{-\theta_j/T}$ where $\theta_0 = 0$ ($\varepsilon_0 = 0$) so that:

$$Z = g_0 + g_1 \cdot e^{-\theta_1/T} + g_2 \cdot e^{-\theta_2/T} + \cdots \tag{2.5}$$

where θ_1, θ_2, ... are forming an increasing series of characteristic temperatures corresponding to the excited levels.

Sum-Over-States with Single Term and Integrals

If the temperature is much lower than the characteristic temperatures, the terms of sum from Eq. (2.5) decrease very rapidly, so that the sum can be approximately calculated from a reduced number of terms. In extreme cases, the sum is reduced to a constant as follows:

$$Z = g_0, T \ll \theta_1 \tag{2.6a}$$

because g_0 and g_1 have usually the same order of magnitude.

In the case of simplification (2.6a), the contribution of the motion with index "type" to the molar thermodynamic functions become, by considering Eqs. (1.58)–(1.61):

$$\begin{aligned} &U_{\mathrm{tip}} = 0; \\ &F_{\mathrm{tip}} = R \cdot T \cdot \ln g_0; S_{\mathrm{tip}} = R \cdot \ln g_0; (C_p)_{\mathrm{tip}} = 0 \end{aligned} \tag{2.6b}$$

Therefore, the motion types for which the excitation energy is high in relation to the product $k_B{\cdot}T$ (or, otherwise expressed: $\theta_1 \gg T$) bring a practically null contribution to the values of thermodynamic properties U and C_p (in whose statistical expressions only the derivatives of the property ln Z in relation to T are taken into account) and a constant contribution, respectively proportional to temperature, for properties S and F, that depend on the underived ln Z.

This situation occurs for the nuclear component of the sum-over-states, whose energy levels are very distanced from one another, the characteristic temperatures being of the order of millions of degrees. In many cases, such a simplification is also valid for the electronic sum-over-states Z_{el}: for most atoms or molecules, $\theta_{1,el}$ is of order of millions of degrees, so that, at usual temperatures, it can also be considered that $Z_{el} = 1$. For example, for $g_0 = g_1 = 1$, $T = 300$ K, and $\theta_{1,el} = 3000$ K, the sum Z_{el} becomes: $Z_{el} = 1 + e^{-10} + \cdots = 1 + 0.000046 + \cdots$, while the contribution of excited states to the electronic partition function is infinitesimal.

For the other limit case, $T \gg \theta$, a high amount of terms should be additional within the Eq. (2.5), but another simplification becomes possible, that is, shifting from the discontinuous form (2.5) to a continuous one, through replacement of the sum by an integral:

$$Z = \int e^{-\varepsilon/(k_B \cdot T)} dG \tag{2.7}$$

where G, the integral distribution function, represents the probability of finding the particle in any quantum state with energy lower than ε. G replaces the discrete weights set g_j from relation (1.43). This shifting always occurs at calculation of the translation energy, where differences from characteristic temperatures of energy levels are usually much lower than the actually achievable lowest temperature, namely 10^{-7} K.

For the case of rotation, the characteristic temperatures are of 1 K order and the addition can be replaced by integration within the gaseous state's field of existence, excepting the *quantum* gases (H_2, D_2, He) for which θ_{rot}—the rotational characteristic temperature—is significantly higher.

Perfect Gas and Ideal Gas

The *ideal* gas, as defined above (the different types of motion do not influence one another), is a notion that is different from the one of a *perfect* gas.

Unlike the thermodynamic definition—the perfect gas is defined as the body for which $F(V,T) = F^{\circ}(T) + R \cdot T \cdot \ln V$ or any other similar expression, so that $p \cdot V = R \cdot T$ together with $(\partial U/\partial V)_T = 0$—the perfect gas is defined in the case of statistical mechanics and molecular kinetic theory as the body for which a molecule state does not depend on the state of the other molecules from the system; this is equivalent to the lack of interactions between the internal motions of a molecule, *A*, and those of another molecule, *B*.

The ideal gas notion is narrower: for a gas to be ideal, the interactions between the four types of internal motions of the same molecule must also be missing. This condition is much more difficult to be accomplished: at $P \to 0$ the real gases are almost *perfect*, but their behavior even at zero pressure is far from being *ideal*.

Thus, the global sum-over-states decomposition over the motion type, as in Eq. (2.3), is only an approximate one; a more accurate calculation would also take into account the interactions between different motion types. For example, the total energy of a rotation–vibration level will no longer represent the sum $\varepsilon_{\text{rot}} + \varepsilon_{\text{vib}}$, but will differ from this sum by the value of a rotation–vibration interaction, $\varepsilon_{\text{rot \& vib}}$.

Therefore, the thermodynamic–statistical approach of gases as ideal gases is only a first approximation, which is, however, sufficiently precise under normal conditions. It is understood that, for condensed states, the thermodynamic–statistical approach becomes even more difficult because, in addition to the interactions within each molecule, it is also necessary to take into account the interactions between the internal motions of several molecules as well.

2.2 Nuclear Partition Function

Atom and Molecular Partition Functions

Since nuclear energy levels have high excitation energies (θ of the order of millions of degrees), the sum-over-states is reduced to a single term, except for the phenomena occurring in the case of nuclear explosions:

$$Z_n = g_{n,0} \tag{2.8}$$

where g is the multiplicity of the fundamental state (denoted so far with g_0). The property g is determined by the value of the nuclear quantum spin number from nucleus ground state, $s_{n,0}$:

$$g_{n,0} = 1 + 2 \cdot s_{n,0} \tag{2.9}$$

Table 2.1 Nuclear spin and multiplicity of the ground state

Nucleus	$^4He_2$, $^{12}C_6$, $^{16}O_8$	$^1H_1$, $^{19}F_9$, $^{31}P_{15}$	$^2D_1$, $^4N_7$	$^{23}Na_{11}$, $^{75}As_{33}$	$^{141}Ce_{58}$
s_n	0	1/2	1	3/2	5/2
g_n	1	2	3	4	6

The quantum number of *nuclear spin* $s_{n,0}$ is obtained by the vectorial addition of spin numbers of nucleus' protons and neutrons, numbers that are equal to 1/2. For nuclei having an even nucleons number, the result of the addition will be either an integer-positive or zero, and for the nuclei with an odd nucleons number it results $s_{n,0}$ as a half-integer number. The quantum number $s_{n,0}$ may therefore differ from one isotope to another of the same element.

By taking into account relation (2.9), the multiplicities are even for nuclei with half-integer spin and odd for the integer spin ones; in particular, at $s_{n,0} = 0$, the ground state is non-degenerate, with $g_{n,0} = 1$ and $\ln Z_n = 0$. In this case, the nuclear contribution to the thermodynamic functions is null. The spin numbers and multiplicities of several nuclei are presented in Table 2.1.

Nuclear Partition Function of Polyatomic Molecules

Atomicity μ represents an important characteristic of molecule from the statistical mechanics point of view, defined as molecule atom number. In the case of polyatomic molecules, each nucleus brings its own specific contribution to the nuclear sum-over-states; the sum-over-states Z_n of the polyatomic molecule therefore is:

$$Z_n = \prod_{i=1}^{i=\mu} (Z_n)_i \tag{2.10a}$$

where $(Z_n)_i$ is the nuclear sum-over-states of atom i. By considering the relations (2.8) and (2.9) and by denoting with $s_{n,i}$ the nuclear spin number of atom i, the nuclear sum-over-states of molecule becomes:

$$Z_n = \prod_{i=1}^{i=\mu} (2 \cdot s_{n,i} + 1)_i \tag{2.10b}$$

Nuclear Contribution to Thermodynamic Functions

When using Eqs. (1.58)–(1.61) and by observing that Z_n derivatives in relation to temperature are null, the following expressions for nuclear contribution to thermodynamic functions are obtained:

$$U_n = H_n = 0; (C_v)_n = (C_p)_n = 0 \qquad (2.11a)$$

$$S_n = R \cdot \sum_i \ln(2 \cdot s_{n,i} + 1) \qquad (2.11b)$$

$$F_n = G_n - R \cdot T \cdot \sum_i \ln(2 \cdot s_{n,i} + 1) \qquad (2.11c)$$

Although the thermodynamic functions U, H, C_v, and C_p are not influenced by the nuclear quantum state, S contains a constant nuclear contribution, and F and G contain nuclear contributions that are proportional to temperature.

The number and type of nuclei are conserved within chemical reactions, so that the total nuclear sum-over-states is identical for the members I and II of the chemical reaction. Consequently, the nuclear contribution is zero for all thermodynamic reaction functions—$\Delta^r V$, $\Delta^r C_P$, $\Delta^r H$, $\Delta^r S$, $\Delta^r G$, etc.—as well as for the equilibrium constant.

Practical Functions and Spectroscopic Functions

The fundamental properties of statistical thermodynamics—that is, the multiplicities and energy levels—are usually obtained by using a spectroscopic method. Yet determination is difficult in the case of g and ε nuclear contributions, due to the extremely low wavelength range (γ-rays) localization of nuclear energies' spectrum. For this reason, "corrected" sizes are often used in the sense that they do not include the nuclear contribution to thermodynamic functions.

The sum-over-states defined according to relation (2.3) as the product of the five sum-over-states on displacements within the *phase* space is called *absolute* or *spectroscopic* sum-over-states. A *practical* or *thermochemical* sum-over-states, which does not include Z_n, is also defined:

$$Z_{tch} = Z_{tr} \cdot Z_{rot} \cdot Z_{vib} \cdot Z_{el} \qquad (2.12a)$$

The thermodynamic properties can therefore have either spectroscopic or absolute values Y_{spc} (often denoted without indices) and practical values Y_{pr}, linked together by the following relation:

$$Y_{aspc} = Y_{tch} + Y_n \qquad (2.12b)$$

Practically, the values of thermodynamic functions are obtained from measurements of caloric capacities and latent heats, by applying the Third Thermodynamic Principle, due to the fact that Z_n takes values independent of temperature, not being zero nor to 0 K. For example, in standard conditions, the practical entropy for gaseous ethylene is $S^0_{298} = 219.5\text{J}/(\text{mol} \cdot \text{K})$ corresponding to the electronic, vibration, rotation, and translation contributions at 298 K; the nuclear sum-over-states can be calculated based on the data in Table 2.1 using Eq. (2.10a): $Z_n(C_2H_4) = [Z_{n,}C]^2 \cdot [Z_{n,}H]^4 = [gC]^2 \cdot [gH]^4 = 1^2 \cdot 2^4 = 16$. From here, using relation (2.11b): $S_n = R \cdot \ln Z_n = 8.314 \ln 16 = 23.1$ J/(mol·K).

The spectroscopic entropy is obtained by customizing relation (2.12b) for $Y = S$: $S_{spc} = S_{tch} + S_n$, from where: $S^0_{298} = 219.5 + 23.1 = 242.6\text{J}/(\text{mol} \cdot \text{K})$, the spectroscopic value therefore differing with more than 10% from the quadratic one.

2.3 Electronic Partition Functions

The electronic energy levels are less scattered than nuclear ones, so that seldomly the electronic contribution to thermodynamic quantities cannot be neglected, even at room temperature.

Information about electronic functions g and ε is obtained from the ultraviolet (UV), visible, or infrared (IR) optical, being usually easily accessible. In the case of molecules with many atoms, the electronic spectra interpretation may present some difficulties.

The Boltzmann factor is presented often in the form of $e^{-\theta/T}$ or $e^{-E/(R \cdot T)}$. Characterization of energy levels is realized in several equivalent ways, namely by wavelength λ, wavelength $\overline{\nu}$, frequency, one molecule energy ε, one mole energy E, or by the characteristic temperature θ. All these quantities are interdependent from one another:

$$E = N_A \cdot \varepsilon; \varepsilon = k_B \cdot \theta; \nu = \varepsilon/h; \overline{\nu} = \nu/c_0 = 1/\lambda \tag{2.13a}$$

For example, the characteristic temperature can be obtained from the experimental wavelength through $\theta = (h \cdot c_0/k_B)/\lambda$, which represents:

$$\theta/K = 0.014.39 \cdot \overline{\nu} \tag{2.13b}$$

The general relation (1.43) or Eq. (2.5) are used for electronic partition function calculation, keeping however a limited number of terms, due to the rapid decrease in Boltzmann's factor with the increase in ε, θ, E, $\overline{\nu}$, or ν. Together with temperature increase, it is necessary to include more terms of the sum; it is usual to keep the terms for which $\theta < 6 \cdot T$, which corresponds to Boltzmann factors surpassing 0.4%.

The multiplicity g_j of the j energy level is determined by the internal quantum number, J_j, corresponding to the considered level energy:

$$g_j = 2 \cdot J_j + 1 \tag{2.14}$$

The electronic sum-over-states will be:

$$Z_{\text{el}} = \sum\nolimits_{j=0}^{j=\infty} g_j \cdot \exp\left[E_j/(R \cdot T)\right] \tag{2.15a}$$

by possibly retaining, besides the fundamental term $g_0 = 2 \cdot J_0 + 1$, a number of excited terms depending on the molecule's nature and on temperature. By derivation, the dependence on electronic sum with temperature is obtained:

$$(\partial Z_{el}/\partial T)_V = \left[\sum\nolimits_{j=0}^{j=\infty} g_j \cdot E_j \cdot e^{-E_j/(R \cdot T)}\right]/(R \cdot T^2) \tag{2.15b}$$

Electronic Contribution to the Thermodynamic Functions

By introducing the electronic sum-over-states and its dependency on temperature from Eq. (2.15a and 2.15b) within the general relations (1.60) and (1.58), we obtain the following expressions for the electronic contributions to free energy and internal energy:

$$F_{\text{el}} = -R \cdot T \cdot \ln Z_{\text{el}} = -R \cdot T \cdot \ln\left(g_0 + g_1 \cdot e^{-\frac{E_1}{R \cdot T}} + g_2 \cdot e^{-\frac{E_2}{R \cdot T}} + \cdots\right) \tag{2.16a}$$

$$\begin{aligned} U_{\text{el}} &= \left(R \cdot T^2/Z_{\text{el}}\right) \cdot (\partial Z_{\text{el}}/\partial T)_V \\ &= \sum\nolimits_{j=0}^{j=\infty} g_j \cdot E_j \cdot e^{-E_j/(R \cdot T)} \Big/ \sum\nolimits_{j=0}^{j=\infty} g_j \cdot e^{-E_j/(R \cdot T)} \end{aligned} \tag{2.16b}$$

It can be observed from Eq. (2.16b) that, in the case of elevated temperatures, the excited terms contribute in a higher degree to the internal energy than to the free energy.

The expressions of electronic thermodynamic functions are simplified by introducing the sums Q_i with $i \in \{0, 1, 2\}$, defined by:

$$Q_i = \sum\nolimits_{j=0}^{j=K} g_j \cdot (E_j)^i \cdot \exp\left[-E_j/(R \cdot T)\right] \tag{2.17}$$

Therefore, Eqs. (2.16a and 2.16b) will become:

$$F_{\text{el}} = R \cdot T \cdot \ln Q_0; U_{\text{el}} = Q_1/Q_0 \tag{2.18a, b}$$

S_{el} and $(C_v)_{\text{el}}$ can be calculated from Eqs. (2.18a, b) using the phenomenological thermodynamic relations $F = U - T \cdot S$, respectively $(C_v)_{\text{el}} = (\partial U_{\text{el}}/\partial T)_V$:

$$S_{el} = R \cdot (\ln Q_0) + Q_1/(T \cdot Q_0); C_{v,el}$$
$$= \left[Q_0 \cdot Q_2 - (Q_1)^2\right] / \left[R \cdot (T \cdot Q_0)^2 \right. \qquad (2.18c, d)$$

If the first excited level is sufficiently elevated (ratio θ_i/T surpassing: 6 for F_{el}; 10 for S_{el} and U_{el}; 15 for $C_{v,el}$), then the electronic contributions can be calculated, using similar relations to the ones valid for nuclear contributions, namely it is considered that: $Z_{el} = g_o =$ const. Thus, for 1 mole of substance the following expression is obtained:

$$F_{el} = -R \cdot T \cdot \ln g_0; S_{el} = R \cdot \ln g_0; U_{el} = 0; C_{v,el} = 0 \qquad (2.18e)$$

Electronic Sum-Over-States for Monoatomic Molecules

The characteristics of j level (g_j and wavenumber $\overline{\nu_j}$, in cm^{-1}) depend on the quantum state of the atom and are presented in Table 2.2.

Spectral Term for Atoms

The type of energy level is indicated in the spectroscopic tables as a group of three properties, ${}^mL_J$, group that is called the spectral term for the *atom*, where "*L*" is the symbol of atomic *orbital* quantum number (symbols *S*, *P*, *D*, *F*, *G*, *H*, and *I* correspond to $L = 0, 1, 2, 3, 4, 5$, and 6); "*m*" is the *spectral term* multiplicity, an integer-positive number representing:

$$m = 1 + 2 \cdot \min(L, S) \qquad (2.19)$$

Table 2.2 The first energy levels from electronic spectra

j		F	Cl	Br	I	O	S	N	P	C	H
0	Type	${}^2P_{3/2}$				${}^3P_2$		${}^4P_{3/2}$	${}^4S_{3/2}$	${}^3P_0$	${}^2S_{1/2}$
	g	4				5		4		1	2
1	Type	${}^2P_{1/2}$				${}^3P_1$	${}^2P_1$			${}^3P_1$	${}^3S_{1/2}$
	g	2				3				3	2
	$\overline{\nu}$	407	881	3685	7680	158	398	28000	15700	148	82258
2	Type	${}^4P_{5/2}$				${}^3P_2$	${}^3P_0$			${}^1P_2$	
	g	6				5	1			5	
	$\overline{\nu}/1000$	102	72	63		0.228	0.57			10.2	

Table 2.3 Electronic spectrum of oxygen atom

J	0	1	2	3	4
Spectral term	$^3P_2$	$^3P_1$	$^3P_0$	$^1D_2$	$^1S_0$
g	5	3	1	5	1
θ	0	227.3	320.4	22.700	48.405

where min (L, S) is the lowest number among L and S, while "S" is the quantum atomic *spin* number, which is an integer number (0, 1, 2, ...) for atoms with even number of electrons and a half-integer one (1/2, 3/2, ...) for atoms with odd number of electrons; and the *internal* quantum number J is an integer or a half-integer number with S and takes all possible values between $|L - S|$ and $|L + S|$.

The *spectral term* multiplicity, m, is different from g multiplicity from *statistics* and represents the number of possible J values for an electronic state of the atom, that is, for a certain set (L, S).

The energy level's *statistical* multiplicity, g, is calculated based on the internal quantum number J using Eq. (2.14); for example, for the ground state of C: $J_0 = 0$ so $g_0 = 1$, while for the second excited level of Br: $J_2 = 5/2$, so $g_2 = 6$.

The spectral multiplicity m_0 of the ground level indicates the number of energy levels close to the ground level and which therefore must be taken into account when calculating the electronic sum-over-states. For example, the F atom $m_O = 2$ (fundamental spectral term $^2P_{3/2}$), indicating the existence of another excited level in the vicinity of the ground state; this level has the wavenumber $\overline{\nu}_1 = 407\ \text{cm}^{-1}$, from where using Eq. (2.13b) one obtains $\theta \cong 600$ K, meaning that the first excited level is important even at room temperature; the next excited level however can be usually neglected, as being more distant ($\overline{\nu}_2 \cong 102.000\ \text{cm}^{-1}$).

Similarly, for O atom: $m_0 = 3$ (fundamental spectral term $^3P_2$); therefore, the energies of the first three levels are closer to each other than the energies of the next levels, as seen in Table 2.3.

Consequently, when calculating the electronic sum-over-states of oxygen atom using Eq. (2.15a) only the first three levels are necessary to be taken into account:

$$Z_{\text{el}} = 5 + 3 \cdot \exp(-227.3/T) + \exp(-320.4/T).$$

Electronic Sum-Over-States of Polyatomic Molecules

In the process of molecule formation from atoms, atoms' electronic shells strongly interact; therefore, unlike Z_n, the Z_{el} of molecule cannot be calculated by multiplying the electronic sum-over-states of each atom. The molecule is characterized by its own electronic spectral terms, with symbols similar to the atom's electronic spectral terms, except that symbols S, P, D, ... for the values 0, 1, 2, ... of the L *atomic* orbital quantum number are replaced by the symbols Σ, Π, Δ, ... for the values 0, 1, 2, ... of the *molecular* orbital quantum number Λ.

Table 2.4 Electronic features of diatomic molecules

Molecule	Li_2	Na_2	K_2	H_2	Cl_2	I_2	O_2	NO
g_0	1	1	1	1	1	1	3	2
g_1	/	/	/	/	/	/	1	2
θ_1	20.192	21.005	16.711	129.612	26.084	16.966	11.330	173.9

Some electronic spectra features for diatomic molecules are presented in Table 2.4.

Unlike

atoms, for the majority of stable diatomic molecules the excitation energy is extremely high, and for Z_{el} calculation it is sufficient to take into account only the ground state $Z_{el} = g_0$. The only exception is the O_2 molecule for which the contribution of the first excited level starts to be important from 2000 K for F, from 1200 K for S or U, and from 800 K for $C_{v,el}$, according to the value $\theta_1 \approx 11.000$ K and to the rules accompanying Eq. (2.18e).

The excited levels are always important for diatomic molecules with a radical character, category in which are found, among others, all molecules with an odd number of electrons (NO, CaCl, CH), the molecules formed by one atom belonging to a group with an odd order number and one atom belonging to a group with even order numbers. For unstable molecules, with a free radical character, Z_{el} can bring an important contribution to thermodynamic properties, g_i and $\bar{\nu}$ being spectroscopically determined.

For the case of polyatomic molecules, the spectral terms are not usually listed in tables. For the majority of stable polyatomic molecules, only the electronic ground state is important; but $g_0 = 1$, so that $Z_{el} = 1$ and the electronic contribution to thermodynamic properties can be neglected.

2.4 Translational Motion

Physical Space and Phase Space

The state of each molecule of a system is illustrated in statistical mechanics by a point from a phase space μ differing from the effective Newton *physical* space from mechanics: molecule's *phase* space is an abstract space with $2f$ dimensions, f being the number of degrees of freedom of particle's motion.

Besides the physical space μ described and used in this chapter, another type of phase space, called the space γ, is also distinguished in statistical mechanics. This space has $2{\cdot}N{\cdot}f$ dimensions, N being the number of molecules. While a point from the phase space μ represents the characteristics of a molecule, a point from the phase space γ will represent the state of the whole system ("microstate").

To each degree of freedom from the μ phase space two conjugated dimensions correspond in this space. They are called generalized coordinate, q_i, and generalized force, F_i, of the respective motion type. A numeric index $i(i \in [1, 2, \ldots, f])$ differentiates the motion type.

For the case of transition motion, the generalized coordinate is the geometric position coordinate of the particle, and the generalized force is the transition motion momentum's coordinate along the respective direction. For example, in the case of a unidimensional motion along x-axis, the two conjugated dimensions are x and p_x—particle's momentum for its motion along Ox:

$$p_x = m \cdot v_x \tag{2.20}$$

where v_x is particle velocity along x-axis ($v_x = dx/dt$). Properties p_x and v_x have here a scalar feature and not a vectorial one.

Translational Distribution Function

A unique molecule translation state can be represented by a point in the μ phase space with six dimensions (because $f_{tr} = 3$): x, p_x, y, p_y, z, p_z. By using the phase space, the discontinuous distributions of microstates can be replaced by continuous distributions, which represent an advantage in the case of systems having many narrow energy levels. The sums are replaced by integrals, while the products by differential distribution functions, Γ, or integral functions, G.

The *differential* distribution function, Γ, of microstates in the phase space μ, a property also called *density distribution*, represents the number of microstates in the "volume" unit of this phase space.

The *integral* distribution function, G, represents the number of microstates from a certain part of this space. The number of microstates from the volume unit $d\tau$ is dG.

The volume element, $d\tau$, from molecule phase space represents the product of infinitesimal variations of all coordinates:

$$d\tau = (dq_1 \cdot dq_2 \cdot \ldots \cdot dq_f) \cdot (dF_1 \cdot dF_2 \cdot \ldots \cdot dF_f)$$

where $dq_1, dq_2, \ldots$ represent the generalized coordinates, and $dF_1, dF_2, \ldots$ represent the generalized forces. For the translational phase space case, the volume element is:

$$d\tau = dx \cdot dy \cdot dz \cdot dp_x \cdot dp_y \cdot dp_z \cdot \tag{2.21a}$$

where p_x, p_y, and p_z are particle momentum p projections on the three-coordinate axis. It can be observed that $d\tau$ differs from the usual volume element—the one from the physical space, $dv = dx \cdot dy \cdot dz$.

The integral distribution function G is the analogue of *statistical share* g. Therefore, the sum-over-states definition $z = \sum g_i \cdot e^{-\varepsilon_i/k_B \cdot T}$ for the case of discontinuous distributions, where g and ε are dependencies of the discontinuous argument i (level's number of order), becomes:

$$Z = \int K \int e^{-\varepsilon_i/k_B \cdot T} dG$$

where ε_i and G depend on the real arguments (continuous) $q_1 \ldots q_f$ and $F_1 \ldots F_f$. The differential distribution function Γ and the integral one G are correlated together by:

$$\Gamma = \frac{dG}{d\tau}; G = \int K \int \Gamma \cdot d\tau \tag{2.21b}$$

Translational Sum-Over-States

For the translational motion of a molecule, the sum-over-states will be:

$$Z_{tr} = \int_x \int_y \int_z \int_{p_x} \int_{p_y} \int_{p_z} e^{-\varepsilon_i/(k_B \cdot T)} \cdot dG \tag{2.21c}$$

where ε_i and G are the energy of i level, and its probability quantities depending on values x, y, z, p_x, p_y, and p_z of the molecule.

Expressed by Γ, the sum-over-states becomes: $Z = (0,5/f) \cdot \int \ldots \int \Gamma \cdot e^{-\varepsilon/(k_B \cdot T)} \cdot d\tau$, where Γ and ε depend on q_1, q_2, …, F_f. For the translation motion:

$$Z_{tr} = \int_x \int_y \int_z \int_{p_x} \int_{p_y} \int_{p_z} \Gamma \cdot e^{-\frac{\varepsilon}{k_B \cdot T}} \cdot d\tau \tag{2.21d}$$

The Γ function has the significance of probability for the particle to be found in a certain point of the phase space. Therefore, a law similar to the equiprobability law for microstates from Sect. 1.2 is valid for Γ-function, namely: a microstate can be found in any point from the phase space with an equal probability. Consequently, the distribution density Γ does not depend on the positioning within the phase space, but it is rather a constant quantity:

$$\Gamma = C \tag{2.22a}$$

The value of C is established by an analogy with the Heisenberg's uncertainty principle $\Delta q \cdot \Delta F = h$:

$$\int_{-\infty}^{\infty}\int_{-\infty}^{\infty} dq \cdot dF = h \tag{2.22b}$$

The sense of $-\infty$ and ∞ limits consists in the fact that Heisenberg's uncertainty principle is true for any possible values of q and F quantities. From Heisenberg's principle, $\Delta x{\cdot}\Delta p_x = \Delta y{\cdot}\Delta p_y = \Delta z{\cdot}\Delta p_z = h$, in the case of translation it results that:

$$\int_{-\infty}^{\infty}\int_{-\infty}^{\infty} dx \cdot dp_x = \int_{-\infty}^{\infty}\int_{-\infty}^{\infty} dy \cdot dp_y = \int_{-\infty}^{\infty}\int_{-\infty}^{\infty} dz \cdot dp_z = h$$

Equation (2.21b) becomes, when replacing quantity Γ by C from Eq. (2.22a) and by introducing $d\tau$ from Eq. (2.21a): $G = C{\cdot}\int\ldots\int dq_1{\cdot}\, dq_2{\cdot}\ldots dF_f$. After regrouping the factors $dq_1 \ldots dF_f$, the following is obtained:

$$G = C \cdot \left[\iint dq_1 \cdot dF_1\right] \cdot \left[\iint dq_2 \cdot dF_2\right] \cdot \left[\iint dq_3 \cdot dF_3\right] K \cdot \left[\iint dq_f \cdot dF_f\right] \tag{2.22c}$$

But, according to Eq. (2.22b), each of the above square brackets is equal to h.

Because in Eq. (2.22c) exists a number of f brackets of this type, and the total probability of particle phase space localization is $G = 1$, it results that $C = 1/h^f$ from where, by replacing C into (2.22a):

$$\Gamma = 1/h^f \tag{2.23}$$

Translational Partition Function

For the case of translational motion, the number of brackets is $f = 3$, therefore:

$$\Gamma = 1/h^3 \tag{a}$$

The energy corresponding to translation is the kinetic energy $\varepsilon = m{\cdot}w^2/2$, where m and w are the particle mass, respectively its velocity. By expressing w as a dependence on its components w_x, w_y, and w_z, the following is obtained: $w^2 = (w_x)^2 + (w_y)^2 + (w_z)^2$. But according to Eq. (2.20), it results that $w_x = p_x/m$ and similarly $w_x = p_x/m$, $w_x = p_x$/m, from where the following energy expression results:

$$\varepsilon = \left[(p_x)^2 + (p_y)^2 + (p_z)^2\right]/(2 \cdot m) \tag{b}$$

By replacing $d\tau$, Γ, and ε into Eqs. (2.21a), (a), and (b) from Eq. (2.21d):

$$Z_{\mathrm{tr}} = \left[\iiint dx \cdot dy \cdot dz\right] \cdot \left[\iiint e^{-\frac{\left(p_x^2+p_y^2+p_z^2\right)}{2 \cdot m \cdot k_B \cdot T}} dp_x \cdot dp_y \cdot dp_z\right] / h^3 \qquad \text{(c)}$$

where the first square bracket $\int\int\int dx \cdot dy \cdot dz = \int dv = v$ represents the part of system's volume corresponding to a single particle, $v = V/N_A$, where V is the molar volume and N_A the Avogadro's number.

The second square bracket is a product of three identical integrals: $\left[\int^{-p_x^2/2 \cdot m \cdot kT} \cdot dp_x\right] \cdot \left[\int^{-p_y^2/2 \cdot m \cdot k \cdot T} \cdot dp_y\right] \cdot \left[\int^{-p_z^2/2 \cdot m \cdot k \cdot T} \cdot dp_z\right]$, therefore:

$$Z_{\mathrm{tr}} = (V/N_A) \cdot (J/h)^3 \qquad \text{(d)}$$

where $J = \int e^{-p_x^2/2 \cdot m \cdot k \cdot T} \cdot dp_x$. The integration limits for J are $-\infty$ and ∞, because—unlike the geometric coordinate (the position)—the momentum coordinate axis is not subjected to any restriction: particle's position is situated inside the system's available volume, while the momentum projections—p_x, p_y, and p_z—can vary between $-\infty$ and ∞. Thus, the expression of the integral is: $J = \int_{p_x=-\infty}^{p_x=\infty} e^{-(p_x)^2/(2 \cdot m \cdot k_B \cdot T)} dp_x$.

By making the variable change $p_x = X \cdot \sqrt{2 \cdot m \cdot k_B \cdot T}$, the following is obtained, $dp_x = dx \cdot \sqrt{2 \cdot m \cdot k_B \cdot T}$, from where: $J = \sqrt{2 \cdot m \cdot k_B \cdot T} \cdot \int_{X=-\infty}^{X=-\infty} e^{-X^2} \cdot dx \cdot$. It is proved that the definite integral is equal to $\sqrt{\pi}$, so that:

$$J = \sqrt{2 \cdot \pi \cdot m \cdot k_B \cdot T} \qquad (2.24)$$

Thermodynamic Translational Functions

By introducing J from (2.24) into (d), and after replacing molecule mass m with the M/N_A ratio (where M is the molar mass) and the universal constant k_B with the ratio of other two constants, namely $k_B = R/N_A$ (where R is the perfect gases' constant, while N_A is Avogadro's number), it is obtained that:

$$Z_{\mathrm{tr}} = V \cdot (2 \cdot \pi \cdot M \cdot R \cdot T)^{3/2} / \left[h^3 \cdot (N_A)^4\right] \qquad (2.25)$$

From Eq. (2.25) it can be seen that Z_{tr} depends only on the macroscopic parameters V, T, and M. By replacing $m - M/N_A$ and by applying the logarithm, the following is obtained:

$$\ln Z_{tr} = K_v + \ln V + 1.5 \cdot \ln(T \cdot M); (\partial \ln Z_{tr}/\partial T)_V = 1.5/T \tag{2.26a}$$

where, by expressing R, h, and N_A constants through the S.I. fundamental units:

$$K_v \equiv 1.5 \cdot \ln\left[(2 \cdot \pi \cdot R/h^2\right] - 4 \cdot \ln N_A = 16.105 \tag{2.26b}$$

By replacing ($\ln Z_{tr}$) and its derivative into Eqs. (1.58) and (1.60), it follows that:

$$F_{tr} = -R \cdot T \cdot [K_v + \ln V + 1.5 \cdot \ln(M \cdot T)]; Utr = 1.5 \cdot R \cdot T \tag{2.27a, b}$$

from where, by using relation $S = (U - F)/T$, the translational entropy is obtained:

$$\begin{aligned} S_{tr} &= R \cdot \left[\ln Z_{tr} + T \cdot (\partial \ln Z_{tr}/\partial T)_V\right] \\ &= R \cdot (K_v + \ln V + 1.5 \cdot \ln T + 1.5 \cdot \ln M + 1.5) \end{aligned}$$

in which <M> = kg/mol, <V> = m^3/mol, while $R = 8.314$ J/(mol·K). But the ideal gas is a perfect gas, so that in the last equation $V = R \cdot T/P$ can be replaced, passing therefore from the $S_{tr}(V,T)$ dependency to the $S_{tr}(P,T)$ dependency. With dimensions J/(mol·K), Pa, K, and kg/mol for S, P, T, and M, it results that:

$$S_{tr} = (R/2) \cdot \ln\left(T^5 \cdot M^3/P^2\right) + 163.99 \tag{2.28a}$$

The entropy in standard conditions is obtained by replacing R from above, $P = 101{,}325$ Pa (=1 atm) and $T = 298.15$ K into Eq. (2.28a):

$$S^0_{298,tr} = 6.236 \ln M + 186.58 \tag{2.28b}$$

Quantity $C_{v,tr}$ can be obtained directly from Eq. (2.27b):

$$(C_v)_{tr} = (\partial U_{tr}/\partial T)_V = 1.5 \cdot R \tag{2.28c}$$

Unlike nuclear and electronic contributions to the sum-over-states, Z_{tr} and the translation contributions to thermodynamic properties depend not only on temperature but also on gas volume or pressure.

The ideal gas translational properties can be calculated without prior knowledge of the molecule's structure, the molar mass being the only characteristic differentiating the substances, and it is macroscopically determinable (for example, from substance density ρ; through $M = \rho \cdot R \cdot T/P$).

2.5 Thermodynamics of Monoatomic Ideal Gas

For the case of monoatomic gases, rotation and vibration are not possible (due to atom's spherical symmetry, the different rotation positions do not differ from one another), so only the nuclear, electronic, and translational motions contribute to the thermodynamic functions.

The statistical calculation of the thermodynamic functions for monoatomic gas case is realized by summing the previously calculated contributions. The nuclear component is usually neglected, obtaining thus the *practical thermodynamic* functions (thermochemical) differing from the *absolute* (*spectroscopical*) ones in the case of F and S, but not in the case of U and C_V. When denoting by Q_0, Q_1, and Q_2 the expressions related to Z_{el} and defined by Eq. (2.17):

$Q_0 = \sum_j g_j \cdot e^{-E_j/R \cdot T}$, $Q_1 = \sum_j E_j \cdot g_j \cdot e^{-E_j/R \cdot T}$, $Q_2 = \sum_j (E_j)^2 \cdot g_j \cdot e^{-E_j/R \cdot T}$

it results, for the values—undetermined—of the free energy F (or of the internal energy U), thermochemical, F_{tch} (index «tch» is usually neglected), and respectively spectroscopically, F_{spc}:

$$F_{tch} = F = F_0 - R \cdot T \cdot (\ln Q_0 + K_V + \ln V + 1.5 \cdot \ln T + 1.5 \cdot \ln M) \quad (2.29a)$$

$$F_{spc} = F_{tch} - R \cdot T \cdot \ln(2 \cdot s_n + 1) \quad (2.29b)$$

the last expression taking into account the nuclear contribution from Eq. (2.14). For the internal thermochemical energy (the one used usually in the thermodynamic calculations), the following expression is true:

$$U (= U_{tch}) = U_0 + Q_1/Q_0 + 1.5 \cdot R \cdot T \quad (2.29c)$$

where both Q_1 and Q_0 depend only on temperature. It can be therefore observed that the internal energy of the ideal monoatomic gas does not depend on the volume at T = const., rediscovering therefore *Joule*'s law, empirically established for the perfect gas.

For the case of entropy, the following expression is found:

$$S = \frac{Q_1}{T \cdot Q_0} + R \cdot \left(K_V + \ln V + \frac{3}{2} \cdot \ln T + \frac{3}{2} \cdot \ln M + \ln Q_0\right) \quad (2.29d)$$

S_{abs} is given by the same expression, but with $R \cdot \ln(1 + 2 \cdot s')$ as an additional term in its right member. Because this term does not depend on T, and the *practical* entropy S corresponds to the thermochemically determined value (to which it is considered that entropy is zero at 0 K, according to the Third Principle of Thermodynamics), it results that the absolute value S_{spc} is obtained by considering a non-zero entropy, equal to $R \cdot \ln(1 + 2 \cdot s')$ of any substance at 0 K.

For the C_V case it is obtained that:

$$C_v = \left[Q_0 \cdot Q_2 - (Q_1)^2\right] / \left[R \cdot T^2 \cdot (Q_0)^2\right] + 1.5 \cdot R \tag{2.29e}$$

At much smaller temperatures than the characteristic temperature θ_1 of the first electronic level, (Q_0, Q_1, Q_2) become equal respectively to (g_0, 0, 0) and the practical thermodynamic functions will include only the translational terms.

2.6 Rotational and Vibrational Motions

Besides the nuclear, electronic, and translational components of the molecule energy, in the case of polyatomic gases two additional components occur, due to rotational and vibrational motions. The polyatomic molecule is represented for both types of motion as a set of material points, located within the atom nuclei. The set of these points is characterized by atomic masses $m_i = A_i/N_A$ (where A_i is the atomic mass of atom I and $i \in [1, 2, \ldots]$) assumed to be located within the nuclei, by the internuclear distances, by the angles between the straight segments joining the nuclei, and by the force constants of the possible oscillations.

The simplest model of treating the diatomic molecule is the one considering it as both a *rigid rotor* (dumbbell composed of masses m_1 and m_2 situated at the fixed distance D, rotating around the axis perpendicular to the internuclear axis and which passes through the rotor's center of gravity) and a *harmonic oscillator* (a material body characterized by the variation of D distance around an equilibrium value, and where force is proportional with elongation at each moment). While molecule's nuclear and electronic energies are *potential* energies and the translational one is only a *kinetic* energy, the vibrational energy has both a potential and a kinetic component. The rotational energy is however only a kinetic one.

Rigid Rotor Geometry

For a body with rotational motion around a Δ axis, *the kinetic energy* ε_c is given by the following relation:

$$\varepsilon_c = I \cdot \omega^2 / 2 \tag{2.30a}$$

where I is the moment of inertia relative to the Δ axis, and ω is the rotational angular velocity, $\omega = w/r$, w being the linear velocity of a point situated at distance r from the axis).

The moment of inertia around a Δ axis for a system of material points is given by the following relation:

$$I = \sum_i m_i \cdot (r_i)^2 \quad (2.30b)$$

where m_i and r_i represent the mass and respectively the distance between i point and Δ axis.

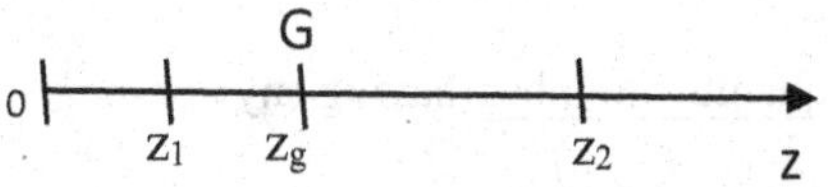

For the case of diatomic molecule, I can be expressed as a function of the internuclear distance D; it is assumed that nuclei 1 and 2 are located on the Oz axis at the coordinate points z_1 and z_2. The center of gravity coordinate, z_g, is obtained from the equality $m_1 \cdot (z_g - z_1) = m_2 \cdot (z_2 - z_g)$, from where it results to be: $z_g = (m_1 \cdot z_1 + m_2 \cdot z_2)/(m_1 + m_2)$. Distances r_1 and r_2 are the distances between points 1 and 2 and the diatomic molecule center of gravity G, since the rotation axis Δ is perpendicular to the Oz axis and passes through the G point. Therefore: $r_1 = z_g - z_1$, $r_2 = z_g - z_2$ or by replacing z_g from the last relation: $r_1 = \frac{m_1 \cdot z_1 + m_2 \cdot z_2}{m_1 + m_2} - z_1 = \frac{m_2 \cdot (z_2 - z_1)}{m_1 + m_2}$; $r_2 = \frac{m_1 \cdot (z_2 - z_1)}{m_1 + m_2}$.

But $(z_2 - z_1) = D$, thus: $r_1 = \frac{m_2 \cdot D}{m_1 + m_2}$; $r_2 = \frac{m_1 \cdot D}{m_1 + m_2}$. For a Δ axis passing through the center of gravity of two material points, the expression (2.30b) becomes $I = m_1 \cdot r_1^2 + m_2 \cdot r_2^2$ or $I = \frac{m_1 \cdot m_2^2 \cdot D^2}{(m_1 + m_2)^2} + \frac{m_2 \cdot m_1^2 \cdot D^2}{(m_1 + m_2)^2}$, from where:

$$I = \mu \cdot D^2 \quad (2.31a)$$

where μ is diatomic molecule's *reduced mass*, defined by:

$$\mu \equiv m_1 \cdot m_2 / (m_1 + m_2).$$

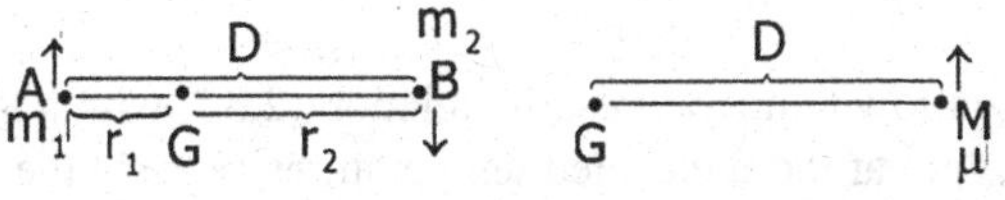

The *reduced* mass μ can be calculated from the atomic masses A_1 and A_2; because $m_i = A_i/N_A$ (where N_A is the Avogadro's number). Therefore:

$$\mu = A_1 \cdot A_2 / [N_A \cdot (A_1 + A_2)] \quad (2.31b)$$

Sum-Over-States of Molecule Rotation

In quantum mechanics it is shown that rotational characteristic quantities (ω and total energy ε) are quantified. By replacing into Schrödinger's equation, the expression $\varepsilon_0 = I \cdot \omega^2/2$—relation (2.30a)—the total rotational energy permitted values are obtained, as given by the following expression:

$$\varepsilon_r(r) = r \cdot (r + 1) \cdot (h/\pi)^2/(8 \cdot \mathrm{I}) \tag{2.32}$$

where r is a rotational quantum number, having any non-negative integer value: 0, 1, 2, ..., ∞.

But a rotational energy value corresponds to several wave functions, so that the rotational energy levels are usually degenerated; the number of wave functions having the energy ε_r is:

$$g_r = 2 \cdot r + 1$$

so only the ground state is simple ($r = 0 \Rightarrow g_0 = 1$), the other levels being degenerate (multiple).

By replacing ε_r and g_r into the expression (1.43) of the partition function, it results that:

$$Z_{\mathrm{rot}} = \sum_{r=0}^{r=\infty} (1 + 2 \cdot r) \cdot e^{-\gamma \cdot r \cdot (r+1)} \tag{2.33}$$

where r replaces the j index from relation (1.43) and γ is grouping the properties independent of r; with ε_r from (2.32) the following is obtained:

$$\gamma = (h/\pi)^2/(8 \cdot I \cdot k_B \cdot T) \tag{2.34a}$$

Effect of Temperature on Rotational Sum-Over-States

For expressing z_{rot} at low temperatures, the relation (2.33) with a low finite number of terms is used, while at more elevated temperatures (γ <<1) the sum from (2.33) can be replaced by the integral $Z_{\mathrm{rot}} = \int_{r=0}^{r=\infty} (2 \cdot r + 1) \cdot e^{-\gamma \cdot r \cdot (r+1)} \cdot dr$.

By making the variable change $\gamma \cdot r \cdot (r + 1) = x$, and by applying the differentiation, it is obtained that $\gamma \cdot (2 \cdot r + 1) \cdot dr = dx$ and the integral becomes: $Z_{\mathrm{rot}} = \int_{x=0}^{x=\infty} \gamma^{-1} \cdot e^{-x} \cdot dx = 1/\gamma$ or

$$Z_{\mathrm{rot}} = 8 \cdot I \cdot k_B \cdot T \cdot (\pi/h)^2 \tag{2.34b}$$

Often, instead of the γ non-dimensional parameter, its proportional parameter is used, that is, the characteristic rotational temperature θ_{rot}, defined as the $\gamma{\cdot}T$ product, where T is the temperature, so that:

$$\theta_{\text{rot}} = (h/\pi)^2/(8 \cdot I \cdot k_B) \tag{2.34c}$$

Relations (2.34a and 2.34b) are valid only for heteronuclear diatomic molecules, that is, composed of different atoms or isotopes (HCl, CO, ^{16}O, ^{18}O, etc.).

Specific restrictions for the rotational motion occur in the case of homonuclear molecules (Cl_2, O_2, N_2, etc.). Indeed, the *wave function parity* from Schrödinger's rotational equation must be of opposite sign to the parity of electronic and nuclear wave functions product: if g_n and g_{el} (multiplicities of the nuclear and electronic ground levels) are both even and odd as in the case of homonuclear molecules, then in the sum from Eq. (2.33) only the r odd value terms will exist:

$$Z_{\text{rot}} = \sum\nolimits_{r=1,3,5\ldots} (2 \cdot r + 1) \cdot e^{-r\cdot(r+1)\cdot\gamma} \tag{2.35a}$$

and if one of g_n and g_{el} properties is even while the other odd, as in the heteronuclear molecules, the sum will contain only terms with r even values:

$$Z_{\text{rot}} = \sum\nolimits_{r=2,4,6\ldots} (2 \cdot r + 1) \cdot e^{-r\cdot(r+1)\cdot\gamma} \tag{2.35b}$$

At elevated temperatures, both cases can be expressed through the unique relation:

$$Z_{\text{rot}} = 2 \cdot k_B \cdot T \cdot I \cdot (2 \cdot \pi/h)^2/\sigma \tag{2.36}$$

The σ parameter, molecule's *symmetry number*, represents the number of equivalent positions obtained by overlapping the molecule over itself; $\sigma = 1$ for heteronuclear molecule and $\sigma = 2$ for homonuclear one.

Symmetry number occurrence reflects the fact that the number of microstates is two times smaller in the case of homonuclear molecules, due to the identity of the two molecule positions obtained by their rotation with 180 degrees.

The Harmonic Oscillator as a Vibrator

Nuclei vibrations around equilibrium positions are reflected into the internuclear distance periodic variation. Similarly, as to the rigid rotor case, the two nuclei vibrations can be reduced to a single material point vibration having its mass equal to the μ reduced mass.

The simplest oscillator model is the harmonic oscillator, which has the force proportional to the difference between instantaneous length (L) and equilibrium length (L_e): $f = -k_f \cdot (L - L_e)$.

By denoting with $x \equiv L - L_e$ the spring elongation from the equilibrium position, and with k_f, a constant invariant in time, the spring *strength*, force can be expressed by:

$$f = -k_f \cdot x \tag{2.37a}$$

By equaling $f = \mu \cdot d^2x/dt^2$, Eq. (2.37) becomes: $\mu \cdot d^2x/d\,t^2 = k_f x$. The solution x (t) of this homogeneous second-order differential equation, namely:

$$x(t) = C_1 \cdot \sin\left[2 \cdot \pi \cdot (\mu/k_f) \cdot 0.5 \cdot (t - C_2)\right]$$

where C_1 and C_2 are integration constants, represents a periodic time variation of the elongation, whose frequency ν is given by the expression:

$$\nu = (k_f/\mu)^{0.5}/(2 \cdot \pi) \tag{2.37b}$$

Vibrational Sum-Over-States

The potential energy of the oscillator becomes, using f expression from relation (2.37a): $\varepsilon_p = -\int_0^x f \cdot dx - k_f \cdot \int_0^x x \cdot dx = k_f \cdot x^2/2$ By replacing k_f expression from relation (2.37b), it results that: $\varepsilon_p = 2 \cdot \mu \cdot (\pi \cdot \nu \cdot x)^2$. Finally, by replacing ε_p within Schrödinger's equation for the case of one dimension (motion along the Ox axis), the following is obtained:

$$d^2\psi/dx^2 + \left(\varepsilon - 2 \cdot \pi^2 \cdot \nu^2 \cdot \mu \cdot x^2\right) \cdot 8 \cdot \pi^2 \cdot m/h^2 = 0$$

where ε is the total energy (kinetic + potential). By solving Schrödinger's equation, it is found that all admissible ε eigenvalues depend on a quantum vibrational non-negative integer number v, ($v = 0, 1, 2, \ldots$),

$$\varepsilon_v = h \cdot \nu \cdot (v + 1/2) \tag{2.38a}$$

and that no level is degenerated: $g_v = 1$ for all v.

The total energy ε dependency on internuclear distance D is given in Fig. 2.2, as well as the positionings of vibrational energy levels of the harmonic oscillator.

From relation (2.38a), it can be seen that the vibrational energy ε_V is non-zero even for the case of vibrational ground state:

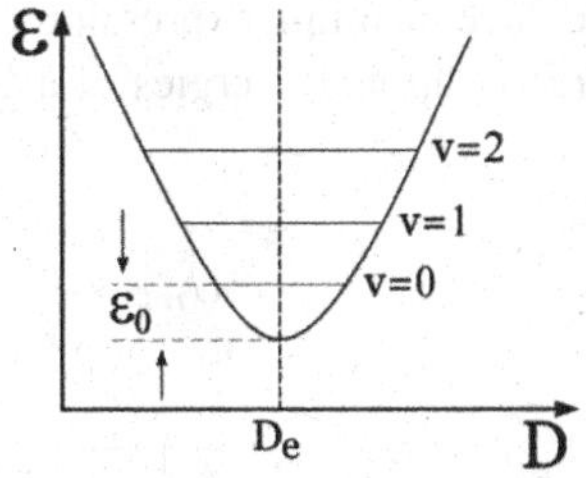

Fig. 2.2 Vibrational energy levels

$$\varepsilon_0 = h \cdot \nu/2 \tag{2.38b}$$

By combining the relations (2.38a and (2.38b), level vibrational energy becomes:

$$\varepsilon_v = \varepsilon_o + h \cdot \nu_v \tag{2.38c}$$

By introducing ε_v from (2.38c) in the sum-over-states from expression (1.43), and by defining a characteristic vibrational temperature θ_{vib} (a material constant), through:

$$\theta_{\text{vib}} \equiv h \cdot \nu/k_B \tag{2.38d}$$

the level vibrational energy and the vibrational sum-over-states become:

$$Z_{\text{vib}} = e^{-\varepsilon_0/(k_B \cdot T)} \cdot \sum_{v=0}^{v=\infty} e^{-v \cdot \theta_{\text{vib}}/T} \tag{2.39a}$$

The sum from relation (2.39a) is a decreasing infinite geometric progression, having the first term $a_o = 1$ and the ratio $q = \exp(-a_0/T)$, so that the following is obtained:

$$Z_{\text{vib}} = [\exp(-\varepsilon_0/(k_B \cdot T))]/[1 - \exp(-\theta_{\text{vib}}/T)] \tag{2.39b}$$

Usually the factor containing ε_o is excluded from Z_{vib} and included into free energy's component F_0, or into the zero entropy. Then it results that:

$$Z_{\text{vib}} = 1/[1 - \exp(-\theta_{\text{vib}}/T)] \tag{2.39c}$$

2.7 Thermodynamics of Diatomic Ideal Gas

Contribution of Rotation to Thermodynamic Functions

At temperatures not much higher than θ_{rot}, the thermodynamic contributions F_{rot}, S_{rot}, U_{rot}, and $C_{v,\text{rot}}$ are calculated using relations (2.34a–c)–(2.36), followed by their replacement into expressions (1.58)–(1.61). For example, for the case of internal and free rotational energies of a *heteronuclear* molecule:

$$U_{rot} = E' \cdot \frac{\sum_{r=0}^{r=\infty}(2r+1)\cdot r\cdot(r+1)\cdot e^{-\frac{E'_{r(r+1)}}{R\cdot T}}}{\sum_{r=0}^{r=\infty}(2r+1)\cdot e^{-\frac{E'_{r(r+1)}}{R\cdot T}}} \quad (2.40a)$$

$$F_{\rm rot} = -R\cdot T\cdot \ln\left\{\sum_{r=0}^{r=\infty}(2\cdot r+1)\cdot \exp\left[-E'\cdot r\cdot(r+1)/(R\cdot T)\right]\right\} \quad (2.40b)$$

where, according to relation (2.32):

$$E' = N_A \cdot (h/\pi)^2/(8\cdot I) \quad (2.40c)$$

At $T >> \theta_{\rm rot}$, when relation (2.36) is valid for $Z_{\rm rot}$, by noting $A \equiv \ln(8\cdot\pi^2\cdot k_B/h^2)$ it results that:

$$F_{\rm rot} = -R\cdot T\cdot[A + \ln(T\cdot I/\sigma)]; \qquad S_{rot} = -R\cdot[A+\ln(T\cdot I/\sigma)] \quad (2.41a, b)$$

$$U_{\rm rot} = R\cdot T; \qquad C_{v,\rm rot} = R \quad (2.41c, d)$$

At $T >> \theta_{\rm rot}$, it is therefore observed that both $U_{\rm rot}$ and $C_{v,\rm rot}$ are independent of the molecular characteristics, while $F_{\rm rot}$ and $S_{\rm rot}$ depend on the ratio of the two molecular properties—*moment of inertia* (I) and symmetry number (σ). For the $S_{\rm rot}$ numerical calculations, relation (2.41b) becomes:

$$S_{\rm rot} = 8.314\cdot \ln\left(10^{46}\cdot T\cdot I/\sigma\right) - 11.58 \quad (2.41e)$$

where the moment of inertia is obtained through combination of relations (2.31a and 2.31b) into relation $\mu = D^2\cdot A_1\cdot A_2/[N_A\cdot(A_1+A_2)]$, where the *internuclear distance* D is obtained from internuclear distance tables or, approximately, by summing the two atoms' radius.

Experimental Determination of the Rotational Contribution

The internuclear distance into a given molecule is experimentally obtained using electron diffraction methods, but the commonly used method is the spectroscopic one. By spectroscopic way, property I is directly determined by using either the pure rotational spectrum (located in the far infrared—FIR), or the rotational–vibrational spectrum (near infrared—NIR), or the Raman spectrum (combined light dispersion spectrum) from the visible field. For example, for determination of I from the pure rotational spectrum (existing only for heteronuclear molecules), conservation of energy law is used, from which it is concluded that spectral line frequency depends on the energies of initial levels (ε' and ε''), namely through: $\pm h \cdot \nu = \varepsilon'' - \varepsilon'$. The (+) sign is used for the absorption spectrum, while (-) sign for emission spectrum.

By denoting with r′ and r″ the initial and final rotational quantum numbers, the absorption spectrum using relation (2.32) becomes:

$\varepsilon = h \cdot \nu = [r'' \cdot (r'' + 1) - r' \cdot (r' + 1)] \cdot h^2/(8 \cdot \pi^2 \cdot I)$ from where, using the relation $\overline{\nu} = \nu / C_0$ it results that:

$$\overline{\nu} = [r'' \cdot (r'' + 1) - r' \cdot (r' + 1)] \cdot h/\left(8 \cdot \pi^2 \cdot I \cdot C_0\right) \tag{2.41f}$$

Spectral Features of Rotation

In Table 2.5 are presented *rotational* spectrum features of several diatomic molecules. The value of θ_{rot} is highly inferior to room temperature (excepting the cases of H_2 and D_2), and therefore the simplified formulas (2.36, 2.41) are used for both Z_{rot} and thermodynamic contributions.

Not any rotational transition is possible, but only the transitions where $\Delta r = \pm 1$. The pure rotational spectrum exists only at absorption, therefore $r'' > r'$, that is, $\Delta r = 1$ or $r' = r'' - 1$; by replacing r' into Expression (2.41f), the following is obtained:

$$\overline{\nu} = B \cdot r'', r'' = 1, 2, \ldots; \qquad B = h/\left(4 \cdot \pi^2 \cdot I \cdot c\right). \tag{2.42}$$

The rotational spectrum is composed therefore of lines situated at an equal distance from one other: $\overline{\nu} = B$ (transition of r from 0 to 1), $\overline{\nu} = 2 \cdot B(1 \to 2)$, $\overline{\nu} = 3 \cdot B(2 \to 3)$, etc. The characteristic rotational constant, *B*, is therefore the distance between two neighboring lines from the rotational spectrum. From B can

Table 2.5 Rotation of diatomic molecules

Molecule	σ	10^{26} μ, kg	10^{10} *D*, m	10^{46} *I*, kg·m^2	θ_{rot}, *K*
H_2	2	0.0837	0.741	0.463	174
D_2	2	0.1674	0.741	0.926	87
N_2	2	1.162	1.098	13.84	5.8
O_2	2	1.328	1.207	19.23	4.4
F_2	2	1.577	1.418	27.8	2.9
Cl_2	2	2.94	1.989	113.9	0.7
I_2	2	10.54	2.667	741.5	0.11
OH	1	0.1571	0.971	1.50	27.0
HF	1	0.1589	0.971	1.35	30.0
CO	1	1.139	1.125	14.43	2.8
NO	1	1.240	1.151	16.51	2.4

subsequently be obtained I_0 through the relation $I_0 = h/(4 \cdot \pi^2 \cdot B)$, and then from I_0 the value of D property can be calculated, as following: $D = (I_0/\mu)^{1/2}$.

Vibrational Einstein Functions

The vibrational thermodynamic functions depend only on the Θ_{vib}/T ratio: by using relation (2.39c) $z_{vib} = 1/\left(1 - e^{-\theta_{vib}/T}\right)$ and by calculating the thermodynamic contributions from Z with relations (1.58)–(1.61), the following expressions are obtained:

$$U_{vib} = R \cdot \theta_{vib}/\left(e^{\theta_{vib}/T} - 1\right); \quad F_{vib} = R \cdot T \cdot \ln\left(1 - e^{-\theta_{vib}/T}\right) \quad (2.43a, b)$$

$$S_{vib} = -R \cdot \left[(\theta_{vib}/T)/\left(e^{\theta_{vib}/T} - 1\right) + \ln\left(1 - e^{-\theta_{vib}/T}\right)\right] \quad (2.43c)$$

$$C_{v,vib} = R \cdot (\theta_{vib}/T)^2 e^{\theta_{vib}/T}/\left(e^{\theta_{vib}/T} - 1\right)^2 \quad (2.43d)$$

To facilitate calculation, three non-dimensional functions of argument $x = \theta_{vib}/T$ have been introduced, referred to as Einstein functions: $\xi_U(x)$, for the internal energy; $\xi_S(x)$, the one for entropy; and $\xi_C(x)$, for the caloric capacity:

$$\xi_U = \frac{x}{e^x - 1}; \quad \xi_S = \frac{x}{e^x - 1} - \ln(1 - e^{-x}); \quad \xi_C = x^2 \cdot e^x/(e^x - 1)^2 \quad (2.44a, b, c)$$

The expressions of thermodynamic functions become therefore:

$$F_{vib} = R \cdot T \cdot [\xi_U(\theta_{vib}/T) - \xi_S(\theta_{vib}/T)]; U_{vib} = R \cdot T \cdot \xi_U(\theta_{vib}/T) \quad (2.44d, e)$$

$$S_{vib} = R \cdot \xi_S(\theta_{vib}/T); \quad C_{v,vib} = R \cdot \xi_C(\theta_{vib}/T) \quad (2.44f, g)$$

At lower temperatures ($T << \theta_{vib}$), the argument of Einstein functions x is higher than 1, so that $e^x - 1 \cong x$. The Einstein functions become:

$$\xi_U(x) = \xi_S(x) = x/\exp(x); \quad \xi_C(x) = x^2/\exp(x) \quad (2.45a, b, c)$$

Therefore the values of F, U, S, and C_V vibrational contributions are at very low temperatures, but they have an exponential increase with temperature.

At elevated temperatures ($T >> \theta_{vib}$, thus $x << 1$), Einstein functions become:

$$\xi_U = 1; \quad \xi_S = 1 - \ln x; \xi_C = 1 \quad (2.46a, b, c)$$

while the vibrational thermodynamic contributions will be:

$$F_{vib} = R \cdot T \cdot \ln(\theta_{vib}/T); \quad U_{vib} = R \cdot T \qquad (2.47a, b)$$

$$S_{vib} = R \cdot [1 - \ln(\theta_{vib}/T)]; \quad (C_v)_{vib} = R \qquad (2.47c, d)$$

Characteristic Vibrational Temperatures

In Table 2.6 the values of characteristic *vibrational* temperatures are given, as obtained from the rotational–vibrational spectra fundamental frequency or from the *Raman* spectrum band structure, through relation (2.13b) applied for vibration, $\theta_{vib} = (h \cdot c_0/k_B) \cdot \overline{\nu}_{vib} = 0.01439 \cdot \overline{\nu}_{vib}$.

Through values comparison from Tables 2.5 and 2.6, it is shown that $\theta_{vib} \gg \theta_{rot}$. Usually, at room temperature, vibrations are non-excited, vibrational thermodynamic function values representing at room temperature a small fraction from the limiting values $U = R{\cdot}T$ or $S = C_V = R$ reached at elevated temperatures. It is also observed that θ_{vib} greatly differs between molecules: differences occur between isotope molecules (H_2, HD, D_2, OH, OD), as well as between the neutral and ionized molecules (H_2 and H_2^+or O_2 and O_2^+).

The values of θ_{vib} increase with covalent bond multiplicity, they decrease with molecular mass increase, and they increase with bond strength increase.

Total Thermodynamic Functions of Diatomic Molecule

These functions are obtained by summing up the five types of contributions, among which the nuclear one, S_n (which is neglectable at temperatures below 10^5 K), the electronic one, S_{el}, and the transitional one, S_{tr}, given by relation (2.28a), also occur in the case of the monoatomic molecule. The rotational contribution S_{rot} and the vibrational one S_{vib}, obtained from relations (2.41e) and (2.47c), are specific to diatomic molecules. Therefore, diatomic gas thermochemical entropy, in the case where all properties are expressed using S.I. fundamental units, has the following form:

$$S = R \cdot \left\{ \xi_S(\theta_{vib}/T) + \ln\left[T^{\frac{7}{2}} \cdot M^{\frac{3}{2}} \cdot 10^{54} \cdot I/(\sigma \cdot P) \right] \right\} + S_{el} - 0.757 \qquad (2.48)$$

with Einstein function $\xi_S(x)$ obtained from relation (2.44b).

Table 2.6 Characteristic vibrational temperatures

Molecule	H_2	**HD**	D_2	H_2^+	Na_2	**CaF**	Hg_2	**BO**	
θ_{vib}/K	6320	5486	5173	4269	228.8	843.9	51.7	2616	
Molecule	**CO**	N_2	**NO**	O_2	$(O_2)^+$	**OH**	**OD**	**HCl**	Cl_2
θ_{vib}/K	3120	3391	2737	2271	2696	5373	3911	3293	812

Relation (2.48) refers to molecule's *practical* entropy, used in the thermodynamic data tables. For the case of "absolute" entropy (spectroscopical), an additional "nuclear" term is entered in the right member of relation (2.11b), $S_n = R \cdot \ln [(2 \cdot S_{n,1} + 1) \cdot (2 \cdot S_{n,2} + 1)]$.

Diatomic gas heat capacity C_p is obtained from C_v contributions summation for the five types of motion, by adding R for the difference $(C_p - C_v)$. The result is:

$$C_p = C_{v,\text{el}} + R \cdot [3.5 + \xi_C(\theta_{\text{vib}}/T)] \quad (2.49\text{a})$$

where $\xi_C(\theta_{\text{vib}}/T)$ is defined by relation (2.44c). It can be seen that C_p does not depend on P (or V) at $T = \text{const.}$; this conclusion is valid only for the ideal gas case. On the other hand, C_p depends on T even when $C_{v,\text{el}}$ is neglectable, because ξ_C is a dependent of temperature (varying between 0 and 1).

Temperature Dependence on Heat Capacity

The quantity ξ_c is negligible at low temperatures ($T \ll \theta_{\text{vib}}$); the following is obtained:

$$C_p = 3.5 \cdot R \cong 29.1\ \text{J}/(\text{mol} \cdot \text{K}) \cong 7\ \text{cal}/(\text{mol} \cdot \text{K}) \quad (2.49\text{b})$$

At room temperature, many stable-enough covalent molecules verify relation (2.49b), which predict a heat capacity practically independent of temperature (BC level of Fig. 2.3) and higher by R than the one of monatomic molecules, to which this level is situated at the value $C_p \cong 16.6$ J/(mol·K) = 5 cal/(mol·K).

For somewhat higher temperatures, where the condition $T \ll \theta_{\text{vib}}$ is no longer fulfilled, C_p becomes dependent on T, namely C_p increases with temperature, because the quantity ξ_c decreases when the (θ_{vib}/T) ratio increases. This conclusion is confirmed by the calorimetric determinations, which show for the diatomic gases an increase in C_p with temperature.

At temperatures sufficiently elevated as compared to θ_{vib}, $\xi_c \to 1$, so that C_p becomes again independent of T and namely:

$$C_p = 4.5 \cdot \text{R} \cong 37\ \text{J}/(\text{mol} \cdot \text{K}) = 9\ \text{cal}/(\text{mol} \cdot \text{K}) \quad (2.49\text{c})$$

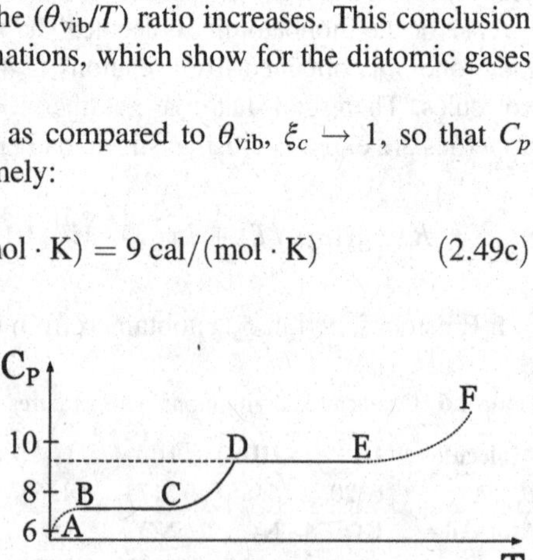

Fig. 2.3 Temperature T dependence on isobaric heat capacity C_P expressed in cal/(mol·K), for ideal gas diatomic molecules

Statistical thermodynamics therefore conducts for diatomic gas cases to the conclusion that $C_p(T)$ dependency presents two steps, at 7 and 9 cal/(mol·K) respectively, according to Fig. 2.3. The AB part corresponds to very low temperatures, where $T \ll \theta_{rot}$, while the EF part to the very elevated temperatures, where $T \gg \theta_{el}$. The same variation way can be qualitatively obtained from the theorem of equipartition of energies on degrees of freedom (see Sect. 2.9.).

2.8 Thermodynamics of the Polyatomic Ideal Gas

The contributions of nuclear, electronic, and translational partition functions to the ideal gas thermodynamic properties having polyatomic molecules ($\mu > 2$) are obtained from the already discussed relations in the case of monoatomic gas. The rotational and vibrational are calculated based on the model from Sect. 2.6—rigid rotator and simultaneously harmonic oscillator—but several differences to the diatomic molecule occur: molecule's rotational motion around *three* axes is considered, while at vibration the molecule is equivalent to a group of harmonic oscillators instead of a single one.

Thermodynamics of Rotation for Polyatomic Molecules

Using a similar reasoning to the one used in the case of diatomic gas, the following expression is obtained for the sum-over-states, Z_{rot}, of a molecule with nonlinear structure:

$$Z_{rot} = \left(2 \cdot \pi \cdot k_B \cdot T/h^2\right)^{3/2} \cdot (I_1 \cdot I_2 \cdot I_3)^{1/2} 8 \cdot \pi^2/\sigma \tag{2.50a}$$

where I_1, I_2, I_3 are molecule's moments of inertia related to three perpendicular axes, intersecting in the molecule's center of gravity.

In mechanics it is shown that $I_1 \cdot I_2 \cdot I_3$ product does not depend on the molecule's positioning so that, when calculating the moments of inertia from atomic masses and the interatomic distances, it is practically usual to orient the molecule in a way that a large number of molecules will be situated on the three axes.

Into relation (2.50a), σ represents molecule symmetry number. Therefore, for CH_4 molecule, having a tetrahedral structure, 12 identical molecule positionings exist, thus $\sigma = 12$; the BF_3 molecule, having the F atoms arranged at the tops of an equilateral triangle and the B atom in the triangle plane, has the value of $\sigma = 6$; the H_2O molecule has the value of $\sigma = 2$; and the totally non-symmetrical molecules have the value of $\sigma = 1$.

If all molecule atoms are collinear (with nuclei on the same line) as by CO_2, HCN, or C_2H_2, molecule rotation around the internuclear axis no longer contributes to Z, because these molecules have a cylindrical symmetry (like diatomic molecules). In this case, relation (2.50a) is replaced by relation (2.36) for diatomic molecules: $z_{rot} = 2 \cdot k_B \cdot T \cdot I_m \cdot (2 \cdot \pi / h)^2 / \sigma$, where σ takes either value 1 (for asymmetric linear molecules, like HCN) or value 2 (for symmetrical linear molecules, like CO_2 or C_2H_2). The moment of inertia is calculated using relation (2.30b).

It is observed that for any polyatomic molecule the following relation is true:

$$Z_{rot} = \left(8 \cdot \pi^3 \cdot k_B \cdot T \cdot I_m / h^2\right)^{1+\alpha/2} / (\sigma \cdot \pi) \quad (2.50b)$$

Into relation (2.50b), α is a parameter of *nonlinearity*, equal to 0 for linear molecules (or diatomic ones) and with 1 for nonlinear molecules, and I_m is molecule mean moment of inertia.

Moment of Inertia for a Polyatomic Molecule

- For the nonlinear molecule case, I_m is the geometric mean of the non-zero I_1, I_2, and I_3 moments of inertia related to the three perpendicular axes drawn through the center of gravity, namely:

$$I_m = (I_1 \cdot I_2 \cdot I_3)^{1/3} \quad (2.50c)$$

- For the linear molecule case, two of the three moments (related to the same perpendicular axes drawn through the center of gravity) are zero, and the remaining moment is the mean moment, $I_1 = I_2 = 0$, $I_m = I_3$.

The contribution of rotation to the thermodynamic functions of a polyatomic molecule is obtained using relations (1.58)–(1.61). Thus, for S_{rot} and $C_{v,rot}$ the following expressions are obtained, respectively:

$$S_{rot} = R \cdot \left\{(1 + \alpha/2) \cdot \left[1 + \ln\left(8 \cdot \pi^3 \cdot k_B \cdot e \cdot T \cdot I_m / h^2\right)\right] - \ln(\pi \cdot \sigma)\right\} \quad (2.51a)$$

$$C_{v,rot} = R \cdot (1 + \alpha/2) \quad (2.51b)$$

For the numeric calculation using S.I. fundamental units, the following expressions can be used:

- (2.41e) $S_{rot} = 8.31446 \cdot \ln(10^{46} \cdot T \cdot I / \sigma) - 11.584$ for linear molecules case.
- For nonlinear molecules case:

$$S_{rot} = 8.31446 \cdot \ln\left(\left[\left(10^{46} \cdot T \cdot I\right)^{1,5} / \sigma\right) - 12.616 \quad (2.51c)$$

Vibrational Sum-Over-States for Polyatomic Molecules

The number of vibrational motions is greater than 1 in a molecule with more than two atoms. This number is calculated by taking into account molecule's total number of atoms, μ; as the position of each atom is described by three coordinates, the total number of coordinates will be $N_{total} = 3\ \mu$. Three of the coordinates are describing the center of gravity positioning (corresponding to the translational motion, $N_{transl} = 3$) and other three are describing molecule orientation (considered as a rigid assembly) in relation to the center of gravity; these last three coordinates correspond to the rotational motion. If the molecule is linear, the rotational state is characterized using only two coordinates, so that generally the number of rotational motions is $N_{rot} = 2 + \alpha$, where α is the nonlinearity parameter.

Therefore, the remaining number of degrees of freedom available for vibration will be:

$$N_{vib} = 3 \cdot \mu - 5 - \alpha \tag{2.52}$$

The N_{vib} vibrations are independent, the *total* vibrational sum-over-states being:

$$Z_{vib} = \prod_{i=1}^{i=N_{vib}} Z_{vib,i} \tag{2.53a}$$

where $Z_{vib,i}$ is the sum-over-states of the vibration number i, given by relation (2.39c):

$$Z_{vib} = \prod_{i=1}^{i=N_{vib}} \left(1 - e^{-\theta_i/T}\right) \tag{2.53b}$$

where θ_i is the characteristic vibrational temperature of vibration i, linked through relation (2.38d), $\theta_i = h{\cdot}\nu_i/k_B$, by the fundamental frequency ν_i of this vibration. It can be seen that the vibrational sum-over-states increases as temperature increases.

The vibrational frequencies are obtained from the rotational–vibrational spectra located in the infrared range. For example, for the nonlinear molecule HONO, by introducing into relation (2.52), $\mu = 4$ and $\alpha = 1$, it results that: $N_{vib} = 3{\cdot}4\text{–}5 - 1 = 6$. The six wavenumbers $\overline{\nu}_i$ (defined by the ratio between vibrational frequency and speed of light in vacuum) determined from the infrared spectrum are, in cm^{-1}: 540, 593, 791, 1265, 1699, and 3588. Therefore:

$$Z_{vib} = \left[\left(1 - e^{-1.4388 \cdot 540/T}\right) \cdot \left(1 - e^{-1.4388 \cdot 593/T}\right) \cdot \ldots \cdot \left(1 - e^{-1.4388 \cdot 3588/T}\right)\right]$$

In the case of molecules having symmetry elements, some vibrations are *degenerated*. For example, the nonlinear molecule SiF_3 has, besides the *simple* vibrations with $\overline{\nu}_i$ of 430 and 830, two more *double* vibrations, with $\overline{\nu}_i$ of 390 and 980 cm^{-1}. The total number of vibrations remains six, $N_{vib} = 12\text{–}5 - 1 = 6$, according to relation (2.52). The sum-over-states expression contains terms of type $[1 - \exp(-\theta_i / T)]^2$ for each of the double vibrations.

Vibrational Thermodynamics for Polyatomic Molecules

The contribution of vibration to thermodynamic functions is obtained using relations (1.58)–(1.61). The vibrations being independent, the total contribution of vibration Y_{vib} to the Y thermodynamic property will be the sum of individual vibrational contributions:

$$Y_{\mathrm{vib}} = \sum_{i=1}^{i=N_{\mathrm{vib}}} Y_{\mathrm{vib},i} \tag{2.53c}$$

where $Y_{\mathrm{vib},i}$ is obtained from relations (2.43). For example, for the case of CO_2, where the vibrational wavenumbers $\overline{\nu}_i$ (cm^{-1}) are 673 (double), 1355, and 2396, the following expression is obtained:

$$U_{\mathrm{vib},i} = R \cdot \left[\frac{2 \cdot 673}{e^{\frac{1.4374 \cdot 673}{T}} - 1} + \frac{1355}{e^{\frac{1.4374 \cdot 1355}{T}} - 1} + \frac{2396}{e^{\frac{1.4374 \cdot 2396}{T}} - 1} \right]$$

by using relation (2.53c) where $U_{\mathrm{vib},\,i} = R \cdot \theta_i / [\exp(-\theta_i / T) - 1]$.

At sufficiently low temperatures, vibrations do not bring a significant contribution to thermodynamic functions. At moderate temperatures, only low-frequency vibrations will be taken into account, while at elevated temperatures all vibrations contribute to thermodynamic functions. And if the temperature is much higher than all the characteristic temperatures, each vibrational motion makes a contribution that is consistent with relations (2.47), for example:

$$U_{\mathrm{vib}} = (3 \cdot \mu - 5 - \alpha) \cdot R \cdot T; \qquad C_{v,\mathrm{vib}} = (3 \cdot \mu - 5 - \alpha) \cdot R \tag{2.54a, b}$$

2.9 Equipartition of Energy Over the Degrees of Freedom

Statistical thermodynamic reasonings allowed also the establishment of approximate calculation rules for thermal energies and heat capacities, rules based on the *equipartition of energy per degrees of freedom* theorem, according to which when calculating the energy from the molecular motions, each type of motion is considered only by the corresponding degrees of freedom number.

The statement of the theorem is: "To every fully excited degree of freedom corresponds a contribution of R·T/2 to the internal energy of one mole, respectively a contribution of R/2 to the molar heat capacity."

Therefore, the total internal energy will be:

$$U = 0.5 \cdot R \cdot T \cdot \sum_{i=1}^{i=K} f_i \cdot \beta_i \tag{2.55a}$$

where i is the index corresponding to motion type; K represents the number of motion types; β_i, the excitation parameter ranging from zero (in the case of low temperatures where the i motion is not excited) up to 1 (for elevated temperatures where the i motion is completely excited); f_i, the degrees of freedom number corresponding to all motions of i type. For C_v:

$$C_V = 0.5 \cdot R \cdot \sum_{i=1}^{i=K} f_i \cdot \beta_i \tag{2.55b}$$

Usually, the translational, rotational, and vibrational motions are taken into account. The equipartition of energy theorem can be compared to the results obtained for $C_{v,tr}$, $C_{v,rot}$, $C_{v,vib}$ at elevated temperatures: $C_{v,tr} = 1.5 \cdot R$, $C_{v,rot} = R \cdot (1 + \alpha/2)$, $C_{v,vib} = R \cdot (3 \cdot \mu - 5 - \alpha)$ and with the number of motions from each type, namely: $N_{tr} = 3$, $N_{rot} = 2 + \alpha$, $N_{vib} = 3 \cdot \mu - 5 - \alpha$.

Degrees of Freedom for Energy Equipartition

It can be seen that relation (2.55b) is verified if an assumption is made, namely that each independent translational or rotational motion corresponds to *one* degree of freedom, while each independent vibrational motion corresponds to *two* degrees of freedom. The difference is explained by the fact that translational and rotational energies have only a kinetic nature, while in the case of vibrational motion, potential energy also intervenes. The relation between the number of degrees of freedom f_i and the number of Ni motions will be given by the following expression:

$$f_i = N_i \cdot k_i \tag{2.55c}$$

where k_i represents motion's number of degrees of freedom, namely: $k = 1$ (for any translation or rotation), but $k = 2$ (for every vibration).

By replacing f_i from (2.55c) into (2.55a and 2.55b) it finally results that:

$$U = (R \cdot T/2) \cdot \sum_{i=1}^{i=K} N_i \cdot k_i \cdot \beta_i; \quad C_V = (R/2) \cdot \sum_{i=1}^{i=K} N_i \cdot k_i \cdot \beta_i \tag{2.56a, b}$$

If only the three discussed types of motions intervene (as in the ideal gas case, at sufficiently low temperatures for the electronic motion to bring no contribution), it is found that:

$$C_V = (N_{tr} \cdot k_{tr} \cdot \beta_{tr} + N_{rot} \cdot k_{rot} \cdot \beta_{rot} + N_{vib} \cdot k_{vib} \cdot \beta_{vib}) \cdot R/2 \text{ or}$$

$$C_V = [3 \cdot 1 \cdot \beta_{tr} + (2 + \alpha) \cdot 1 \cdot \beta_{rot} + (3 \cdot \alpha \cdot \mu - 5 - \alpha)2 \cdot \beta_{vib}) \cdot R/2 \qquad (2.56c)$$

Since β_{rot} is much lower than room temperature, the rotation can be considered completely excited at ordinary temperatures; translation is completely excited at any temperature. By replacing $\beta_{tr} = \beta_{rot} = 1$ into relation (2.56c) and taking into account that $C_P = C_v + R$, the following expression is obtained:

$$C_p = R \cdot [(7 + \alpha)/2 + (3 \cdot \mu - 5 - \alpha) \cdot \beta_{vib}] \qquad (2.56d)$$

which is valid at room temperature. For molecules sufficiently small and stable, the characteristic vibrational temperatures are much higher than room temperature, so that:

$$Cp = R \cdot (7 + \alpha)/2 \qquad (2.56e)$$

Through temperature raising, β_{vib} increases also; when all vibrations are completely excited, $\beta_{vib} = 1$ and relation (2.56d) become:

$$Cp = R \cdot (6 \cdot \mu - 3 + \alpha)/2 \qquad (2.56f)$$

The C_p value given by relation (2.56f) is reached in rare cases: at the temperature required for all vibrations' excitement, the molecule breaks down into atoms.

2.10 Five Worked Examples

Ex. 2.1
The electronic levels of gaseous Hg are known (Sansonetti 2005, page 1838):

Level no. j	0	1	2	3
Symbol	$^1S^0$	$^3P^0$		
Quantum numbers				
n. c. main, L	1	2		
n. c. of spin, s	0	2	1	0
n. c. internal, J	0	0	1	2
Multiplicity, g_j	1	1	3	5
Wavenumber, $\overline{\nu}$, cm^{-1}	0	37645	39412	44043

For functions F, U, S, and C_v of Hg (molar mass being $M = 0.20061$ kg/mol) gas, at $T = 10000$ K and $P = 1$ atm (or 101325 Pa), the following are required to be calculated:

1. The electronic contribution.
2. The translational contribution.
3. The total value.
4. Estimate for the given conditions the share of electronic contribution from the total value of the thermodynamic functions U, H, F, G, S, Cv, and Cp of Mercury.

Solution

1. The characteristic temperatures of the three excited levels are calculated using relation (2.13b) $\theta_i = 1.4388 \cdot (\overline{\nu}_i/\text{cm}^{-1})$, from where:

$\theta_1 = 1.4388 \cdot \overline{\nu}_i = 1.4388 \cdot 37645 = 54163\,\text{K}$ and similarly
$\theta_2 = 1.4388 \cdot 39412 = 56706$ K, $\theta_3 = 1.4388 \cdot 44043 = 63369$ K.

The molar energies of the levels are obtained from relations (2.13a) $E = N_A \cdot \varepsilon$ and $\varepsilon = k_B \cdot \theta$, from where $E_j = N_A \cdot k_B \cdot \theta_j$; but $N_A \cdot k_B = R = 8.31446$ J/(mol·K).

From the θ_j values it results: $E_1 = 8.31446 \cdot \theta_1 = 8.31446 \cdot 54163 = 450336$ J/mol; similarly $E_2 = 8.31446 \cdot 56706 = 471480$ J/mol and the same $E_3 = 526880$ J/mol.

Relation (2.17), $Q_i = \sum_{j=0}^{j=3} (E_j)^i \cdot g_j \cdot \exp(-\theta_j/T)$, is transformed by replacing $E_j/R = \theta_j$ into exponents.

The three known excited levels are considered, thus: $k = 3$.

By replacing with $T = 10000$ and with values: $g_0 = 1$, $g_1 = 1$, $g_2 = 3$, $g_3 = 5$, tabulated in the statement, the following sums are obtained:

$$Q_0 = 1 + \exp(-\theta_1/1000) + 3 \cdot \exp(-\theta_2/1000) + 5 \cdot \exp(-\theta_3/1000)$$
$$= 1 + \exp(-5.42) + 3 \cdot \exp(-5.67) + 5 \cdot \exp(-6.34) = 1.025$$

$$Q_1 = E_1 \cdot \exp(-\theta_1/1000) + 3 \cdot E_2 \cdot \exp(-\theta_2/1000) + 5 \cdot E_3 \cdot \exp(-\theta_3/1000)$$
$$= 12070\,\text{J/mol}$$

$$Q_2 = (E_1)^2 \cdot \exp(-\theta_1/1000) + 3 \cdot (E_2)^2 \cdot \exp(-\theta_2/1000) + 5 \cdot (E_3)^2$$
$$\cdot \exp(-\theta_3/1000) = 5.66 \cdot 10^8 (\text{J/mol})^2$$

From sums Q_0, Q_1, Q_2, when using relations (2.18), the following contributions are found:

$$F_{\text{el}} = -R \cdot T \cdot \ln Q_0 = -8.31 \cdot 10000 \cdot \ln 1.0247 = -2020\,\text{J/mol}$$

$$U_{\text{el}} = Q_1/Q_0 = 12070/1.025 = 11780\ \text{J/mol}$$

$$C_{V,\text{el}} = (Q_2 \cdot Q_0 - Q_1^2)/\left[R \cdot (T \cdot Q_0)^2\right]$$

$$= (5.66 \cdot 10^8 \cdot 1.0247 - 12070^2)/\left[8.31 \cdot (10000 \cdot 1.0247)^2\right] = 0.497\text{J}/(\text{mol} \cdot \text{K})$$

The contribution into S is calculated by relation $F = U - S \cdot T$, from where:

$$S_{\text{el}} = (U_{\text{el}} - F_{\text{el}})/T = (11780 + 2020)/10000 = 1380\ \text{J}/(\text{mol}\cdot\text{K}).$$

2. With relations (2.27b), (2.28c), and (2.28a), the following values are obtained:

$$U_{\text{tr}} = 1.5 \cdot R \cdot T = 1.5 \cdot 8.31 \cdot 10^4 = 124700\ \text{J/mol};$$
$$C_{v,\text{tr}} = 1.5 \cdot \text{R} = 1.5 \cdot 8.31 = 12.47\ \text{J}/(\text{mol}\cdot\text{K});$$
$$S_{\text{tr}} = (R/2) \cdot \ln\left(T^5 \cdot M^3/P^2\right) + 164 = \ldots$$
$$(8.31/2) \cdot \ln\left(10^{5.4} \cdot 0.207^3/101300^2\right) + 164 = 240.9\ \text{J}/(\text{mol}\cdot\text{K})$$

F contribution is:

$$F_{\text{tr}} = U_{\text{tr}} - T \cdot S_{\text{tr}} = 124700 - 10^4 \cdot 2409 = -2283000\ \text{J/mol}.$$

3. For F and U, by summation of contributions, the following values are obtained:

$$F = F_{\text{el}} + F_{\text{tr}} = -2.02 - 2283 = -2285\ \text{kJ/mol}$$
$$U = U_{\text{el}} + U_{\text{tr}} = 11.8 + 124.7 = 136.5\ \text{kJ/mol}$$

Both energies represent in fact the differences between the current values of the free energy or the internal energy in the given conditions and its values at 0 K, $F_0 = U_0$.

Similarly, for C_v and S:

$$C_v = C_{v,\text{el}} + C_{v,\text{tr}} = 0.4971 + 12.4717 = 12.9688\ \text{J}/(\text{mol}\cdot\text{K})$$
$$S = S_{\text{el}} + S_{\text{tr}} = 1.380 + 240.808 = 242.418\ \text{J}/(\text{mol}\cdot\text{K})$$

4. Entities H, G, and C_p are thermodynamically obtained, through: $H = U + P{\cdot}V$ or (since the gas is perfect) $H = U + R{\cdot}T$, from where:

$H = 136493 + 8.31446{\cdot}10^4 = 219638$ J/mol. Similarly, it is found that:

$$G = F + P \cdot V = F + R \cdot T = -2285388 = 8.31446 \cdot 10^4 = -2202243\ \text{J/mol}$$
$$C_p = C_v + R = 13.0 + 8.31446 \cdot 10^4 = 13157.6\ \text{J}/(\text{mol}\cdot\text{K}).$$

All three supplements (for H, G, and C_p) are added to the translation part of the thermodynamic function, and not to the electronic part.

Therefore, the share of electronic contribution to the total thermodynamic property Y can be defined as the ratio π_Y obtained by dividing Y_{el} to Y. Thus, the following values are obtained:

$$\pi_U = U_{el}/U = 11776/136493 = 8.6\%$$
$$\pi_H = H_{el}/H = 126493/219638 = 62.1\%$$
$$\pi_F = 2025/2283363 = 0.09\%;\ \pi_G = 2285388/2202243 = 103.8\%;$$
$$\pi_S = 1.380/242.418 = 0.56\%,\ \pi_{Cv} = 0.4971/12.9688 = 3.8\%\text{and}$$
$$\pi_{CP} = 13.0/13157.6 = 0.1\%$$

In the given case, the electronic contribution is negligible for C_p, S, and F (less than 1%), but it becomes more and more important for C_v or U (almost 4% and 9%, respectively). It is crucial to enthalpy (over 60%) and especially for the free enthalpies, where the share of electronic contribution exceeds 100%, as the translation contribution is of opposite sign.

Ex. 2.2

For the gaseous phase reaction $N_2 + O_2 \rightleftharpoons 2\ NO$, the following two wavelengths are known for each of the reaction participants: $\lambda'(n)$ for the limit to the elevated frequency of the vibrational spectrum and $\lambda''(n)$ for the fundamental vibrational frequency.

For N_2, O_2, respectively NO, the following values (expressed in m) have been spectroscopically determined: λ': $1.27005{\cdot}10^{-7}$; $2.42298{\cdot}10^{-7}$; $1.90509{\cdot}10^{-7}$ and λ'': $4.23830{\cdot}10^{-6}$; $6.33000{\cdot}10^{-6}$; $5.25186{\cdot}10^{-6}$.

It is required $\Delta^r F^{\circ}{}_0$—the standard value at 1 atm and 0 K of the free energy for this reaction.

Solution

In order to convert the wavelengths into corresponding molar energies, E' and E'', relations (2.13a), $E = N_A{\cdot}\varepsilon$, $\varepsilon = h{\cdot}\nu$ and $\nu = c_0/\lambda$ are combined, leading to $E = f_t/\lambda$, with the multiplier $f_t = N_A{\cdot}h{\cdot}c_0$, that is:

$$\begin{aligned} f_t &= \left(6.02214 \cdot 10^{23}\right) \cdot \left(6.62607 \cdot 10^{-34}\right) \cdot \left(2.99792 \cdot 10^{8}\right) \\ &= 0.1196266\ \mathrm{J/(mol \cdot K)} \end{aligned}$$

The following values are therefore obtained: $E'(N_2) = 0.1196266/1.27005{\cdot}10^{-7} = 941{,}905$ J/mol, $E'(O_2) = 0.1196266/2.42298{\cdot}10^{-7} = 493{,}717$ J/mol, and $E'(NO) = 0.1196266/1.90509{\cdot}10^{-7} = 627{,}931$ J/mol.

By dividing the transformation factor to the λ'' values, it similarly results for E'' expressed in J/mole: 28,225 (N_2), 18,898 (O_2), and 22,778 (NO).

But the λ' limits to UV of the vibration spectrum correspond for the diatomic molecule case to the shifting from the minimum energy state (corresponding to potential curve minimum in Fig. 2.2.) to the isolated atoms state. The *spectroscopic* dissociation energy, D_{sp}, is the difference $E' - E''$.

The equilibrium state of the diatomic molecule is, however, at 0 K, the energy level 0 from the same figure. Therefore, for equilibrium calculation the important element is not the spectroscopic value (absolute) D_{spc}, but the thermochemical dissociation energy, which corresponds to molecule transformation found on the vibrational level 0 into two free atoms. According to relations ((2.38b and 2.38c), the difference between the ground state ($U = 0$) and the potential curve minimum is, for one molecule, the energy of zero, $\varepsilon_0 = h{\cdot}\nu''/\ 2$, where it is denoted by ν'', the fundamental vibration frequency. The molar energy of the equilibrium level is thus $E_0 = E''/2$, from where $D_{ch} = E'–E''/2$.

By combining through Hess law from § 4.3. from Daneş et al. 2013 [1] the dissociation processes into atoms for each of the four molecules participating to the reaction from the statement, it results that at 0 K, where the translational and rotational sum-over-states are zero—expressions (2.27b) and (2.41c)—the following equality is true: $\Delta^r U^o{}_0 = -\Sigma_i\ \nu_i{\cdot}D_{ch,I}$, where: ν_i (−1, −1, 2 for N_2, O_2, NO) is the stoichiometric coefficient of the i reaction participant, and $\Delta^r U_0{}^o$ is the standard internal energy of reaction at 1 atm and 0 K; after replacement of D_{ch} by $(E'–E''/2)$, it results that: $\Delta^r U^o{}_0 = \Sigma_i\ \nu_i{\cdot}(E_i''/2 - E_i') = 0.5{\cdot}(-28{,}225–18{,}898 + 2 \cdot 22{,}778) - (-941{,}905–493{,}717 + 2 \cdot 627{,}931) = 178{,}977$ J/mol.

At 0 K, the $T{\cdot}S$ difference between the internal and free energies is canceled, from where $\Delta^r F^o{}_0 = \Delta^r U^o{}_0 \Rightarrow$ $\mathbf{\Delta^r F^o{}_0 = 179}$ **kJ/mol.**

Ex. 2.3

For N_2, O_2, and NO, the following are known: the interatomic distance expressed in $m{\cdot}10^{13}$ of 1097, 1207, and 1151 respectively, the electronic ground state multiplicity g_0 of 1, 3, and 2 respectively, as well as the number of excited states that intervene in the sum-over-states: 0, 1, 1, respectively. The excited state multiplicity is 1 (O_2) and 2 (NO), and its characteristic temperature θ_1 expressed in K is 11,330 (O_2) and 173.9 (NO). The atomic mass (kg/mol) are: 0.014 (N) and 0.016 (O). The standard free energy at 0 K is (see the previous example) $\Delta^r F^o{}_0 = 179$ kJ/mol. It is required to determine the equilibrium practical constant K_{sp} of the reaction: $N_2 + O_2 \rightleftharpoons 2\ NO$ at 3000 K.

Solution

Firstly, the variation of the free energy of reaction is calculated for each type of motion, as a function of temperature.

(a) The nuclear component does not intercede into the chemical equilibrium (Sect. 2.1.).
(b) The electronic component of reaction-free energy will be:
$\Delta^r F_{el} = -\Sigma_i\ \nu_i{\cdot}F_{el,i}$; by replacing $F_{el,i} = -R{\cdot}T{\cdot}\ln Z_{el,i}$ (relation 2.16a) and the stoichiometric coefficients ν_i with −1, −1, and 2 for N_2, O_2, and NO, the following expression is obtained:

$$\Delta^r F_{el} = R \cdot T \cdot \ln\left\{[Z_{el} \cdot (N_2)] \cdot [Z_{el}(O_2)]/[Z_{el}(NO)]^2\right\}.$$

For $Z_{el,i}$ relation (2.15a) is used, limiting it at the lowest levels:
$Z_{el,i} = \sum_{j=0}^{j=J} g_{i,j} \exp\left(-E_{i,j}/R \cdot T\right)$, or $Z_{el,i} = \sum_{j=0}^{j} g_{i,j} \exp\left(-\theta_{i,j}/T\right)$ if the characteristic temperatures are introduced; index i is referring to the gas nature and j to the number of order of the level; J is the *maximum* number of order of the considered levels. For each gas, the following are calculated:

- N_2: $J = 0$, $g_0 = 1$ from where $Z_{el}(N_2) = 1$
- O_2: $J = 1$, $g_0 = 3$, $g_1 = 1$, $\theta_1 = 11{,}330$ K; at $T = 3000$ K it results that: Z_{el} $(O_2) = 3 + e^{-11{,}330/3000} = 3.02$
- NO: $J = 1$, $g_0 = 2$, $g_1 = 2$, $\theta_1 = 173.9$ K from where:
- Z_{el} (NO) $= 2 + 2 \cdot e^{-173.9/3000} = 3.89$

From Z_{el} values: $\Delta^r F_{el} = 8.31 \cdot 3000 \cdot \ln\,(1 \cdot 3.02/3.89^2) = -40{,}140$ J/mol.

(c) The translational component is obtained from relation (2.27a):

$F_{tr} = -R \cdot T \cdot [K_v + \ln V + 1.5 \cdot \ln\,(M \cdot T)]$; by multiplying with ν_i, summing for the three reaction participants, and observing that $\Sigma_i\, \nu_i = 0$, it results that:
$\Delta^r F_{tr} = -1.5 \cdot R \cdot T \cdot \Sigma_i\,(\nu_i \cdot \ln M_i)$. Dar $M(O_2) = 2 \cdot A_O = 0.032$; $M(N_2) = 2 \cdot A_N = 0.028$ and $M(NO) = A_O + A_N = 0.030$, from where:

$$\Delta^r F_{tr} = -1.5 \cdot 8.31 \cdot 3000 \cdot \ln\left(0.032 \cdot 0.028/0.030^2\right) = -153\ \mathrm{J/mol}$$

(d) The rotational component is calculated using relation (2.41a):

$F_{rot,\,i} = -R \cdot T \cdot [A + \ln\,(I \cdot T/\sigma)]$ or, after the addition using ν_i weight:

$$\Delta^r F_{rot} = R \cdot T \cdot \left(\sum_i \nu_i \cdot \ln\,\sigma_i - \sum \nu_i \cdot \ln\,I_i\right) \qquad \text{(a)}$$

I_i is obtained from relation (2.31): $I = \mu \cdot d^2$, $\mu = A_1 \cdot A_2/N_A(A_1 + A_2)$, from where $I_i = A_{1i} \cdot A_{2i} \cdot d_i^2/N_A(A_{1i} + A_{2i})$; the second in the relation (a) becomes:

$$\sum_i \nu_i \cdot \ln I_i = 2 \cdot \ln\left[d^2(NO)/d(O_2) \cdot d(N_2)\right]$$
$$+2 \cdot \ln\left[A_N \cdot A_O/N_A(A_N + A_O)\right] - \ln\left[A_N^2/2N_A \cdot A_N\right] - \ln\left[A_O^2/2N_A \cdot A_O\right]$$

or, after simplifying:

$$\sum_i \nu_i \cdot \ln I_i = 2 \cdot \ln\left[d^2(NO)/d(O_2) \cdot d(N_2)\right] + \ln\left[4A_N \cdot A_O/(A_N + A_O)^2\right]$$
$$= 2 \cdot \ln\left(1{,}151^2/1{,}097 \cdot 1{,}207\right) + \ln\left[4 \cdot 14 \cdot 16/(14 + 16)^2\right] = -0{,}0033$$

For the first sum from (a) the symmetry numbers $\sigma = 2$ are used for N_2 and O_2 (homonuclear) and 1 for NO (heteronuclear). By replacing into (a):

$$\Delta^r F_{rot} = 8.31 \cdot 3000 \cdot (2 \cdot \ln 1 - \ln 2 - \ln 2 + 0.0033) = -34500\ \mathrm{J/mol}$$

For calculating the vibrational sum, the characteristic vibrational temperatures are first calculated, from the λ'' values of the previous example, these corresponding to: $1/\overline{\nu}$ or with relations (2.13a and 2.13b), to: $\theta/K = 1.4388 \cdot (\overline{\nu}/\mathrm{cm}^{-1})$.

Therefore, $\theta/K = 1.439/\ \lambda''$, with λ'' expressed in cm. From the values $\lambda''\cdot 10^4$ (cm): 4.238 (N_2), 6.33 (O_2), and 5.252 (NO), the θ_{vib} values of 1.439/ 0.0004238 = 3395 (N_2) are obtained and similarly: 2.273 (O_2), 2.740 (NO).

From relation (2.43b), $F_{\mathrm{vib},\,i} = R \cdot T \cdot \ \ln\,[1 - \ \exp\,(-\theta_{\mathrm{vib},\,i}/T)]$, it is found that:

$$\Delta^r F_{\mathrm{vib}} = R \cdot T \sum \nu_i \cdot \ \ln\,[1 - \exp\,(-\theta_{\mathrm{vib},i}/T)] =$$

$$\ldots 8.31 \cdot 3000 \cdot \ \ln\left\{[1 - \exp\,(-2740/3000)]^2 \cdot [1 - \exp\,(-3390/3000)]/..\right\}$$

$$\ldots [1 - \exp\,(-2280/3000)] = -96.4\ \mathrm{J/mol}$$

By summing the contributions from points *a*, *b*, *c*, *d*, *e* with the value $\Delta^r F^{\circ}$ at 0 K, $\Delta^r F^{\circ}{}_0 = 178{,}977$ J/mol from the problem statement, the following is obtained for the standard reaction free energy at the given temperature:

$$\Delta^r F^{\circ}{}_{3000} = \Delta^r F^{\circ}{}_0 + \Delta^r F_n + \Delta^r F_{\mathrm{el}} + \Delta^r F_{\mathrm{tr}} + \Delta^r F_{\mathrm{rot}} + \Delta^r F_{\mathrm{vib}} \text{ or}$$

$$\Delta^r F^{\circ}{}_{3000} = 178977 + 0 - 40138 - 153 - 34496 - 96 = 104.094\ \mathrm{J/mol}.$$

From $G = F + P{\cdot}V$ it results $\Delta^r G^{\circ}{}_{3000} = \Delta^r F^{\circ}{}_{3000} + R{\cdot}T{\cdot}\Sigma_i\,\nu_i$ for the perfect gases. But $\cdot\Sigma_i\,\nu_i = 0$, therefore: $\Delta^r G^{\circ}{}_{3000} = \Delta^r F^{\circ}{}_{3000} = 104.094$ J/mole.

For reactions occurring into perfect gas, the practical chemical equilibrium constant *K* is equal (§11.4 din [02]) to the thermodynamic equilibrium constant K_a:

$$K = K_a = \exp\,[-\Delta^r G^{\circ}/(R \cdot T)] = e^{-104100/(8.31\cdot 3000)} \Rightarrow \mathbf{K = 0.0154}$$

The concordance of spectroscopic *K* value 0.0154 with the experimental one, of 0.012 ÷ 0.013, is satisfactory, considering the very elevated temperature. As the most important contribution was brought by the zero energy, followed by the electronic and rotational sum-over-states, the departure is probably due to the neglected superior excited electronic levels.

Ex. 2.4

The following characteristics of the vibrational levels from IR and Raman spectra of the gaseous methyl chloride CH_3Cl are known:

Number of order, *i*	1	2	3	4	5	6
ν_i, cm^{-1}	732.1	1015.0	1355.5	1454.6	2923.5	3041.8
g_i	1	2	1	2	1	2

The following are required:

1. To demonstrate that methyl chloride molecule is nonlinear.
2. To calculate the molar heat capacity, C_P, at 600 K.

Solution

1. The number of molecule atoms is $\mu = 5$, and the number of vibrational motions is $N_{\text{vib}} = \sum_{i-1}^{6} g_i = 1 + 2 + 1 + 2 + 1 + 2 = 9$.

From relation (2.52), $\alpha = 3 \cdot \mu - 5 - N_{\text{vib}} = 3 \cdot 5 - 5 - 9 = 1$, therefore the molecule is nonlinear (which also results directly from the structural data).

2. The contribution for *vibration* to the heat capacity: from the table the ratios $\theta_{\text{vib},i}/T$ are calculated and noted by x_i; but $\theta_{\text{vib,i}} = 1.439 \cdot (\overline{\nu}_i/\text{cm}^{-1})$ from relation (2.13b), so that $x_i = 1.439 \cdot \overline{\nu}_i/600 = 0.00240 \cdot \overline{\nu}_i$.

For $i = 1, \ldots, 6$, the following values x_i are obtained: 1.756; 2.434; 3.25; 3.49; 7.01; and 7.29. When using relation (2.44c), $E_c = e^x \cdot [x/ (e^x - 1)]^2$, the Einstein functions for the heat capacity are calculated. It results that: $E_{c,1} = e^{1.756} \cdot [1.756/ (e^{1.756} - 1)]^2 = 0.778$. For $E_{C,2}, E_{C,3}, \ldots, E_{C,6}$, the following values are found: 0.624; 0.443; 0.396; 0.044; and 0.036. When summing relation (2.44g) for all the vibrations, it results that: $C_{\text{vib}} = R \cdot \sum_{i=1}^{6} E_{C,i} = 8.31 \cdot (0.778 + 2 \cdot 0.624 + 0.443 + 2 \cdot 0.396 + 0.044 + 2 \cdot 0.036) = 28.1\ \text{J}/(\text{mol} \cdot \text{K})$.

Translation: From relation (2.28c), the translational contribution is $C_{\text{tr}} = 3 \cdot R/2$.

Rotation: From $\alpha = 1$, we find using relation (2.51c): $C_{\text{rot}} = R \cdot (1 + \alpha/2) = 1.5 \cdot R$.

By neglecting the nuclear and electronic contributions and by introducing the Robert Mayer difference $(C_P - C_v) = R$, it results through addition that: $Cp = C_{\text{vib}} + C_{\text{tr}} + C_{\text{rot}} + (C_P - C_v) = 28.1 + R \cdot (1.5 + 1.5 + 1) = 28.08 + 8.31 \cdot 4 \Rightarrow$ $\mathbf{Cp = 61.3\ J/\ (mol \cdot K)}$.

The difference from the thermochemical experimental value being of 3–4%.

Ex. 2.5

The gaseous natrium chloride dimer Na_2Cl_2 has a rhombic structure of its molecule, with the symmetry number $\sigma = 4$, the ions Cl^- and Na^+ being situated at distances $d_{\text{Cl}} = 2 \cdot 10^{-10}$ m and respectively $d_{\text{Na}} = 1.42 \cdot 10^{-10}$ m from the rhomb's center. The characteristic vibrational temperatures are: 201, 201, 292, 305, 372, and 384 K. The atomic masses are: $A_{\text{Na}} = 0.023$ kg/mol and $A_{\text{Cl}} = 0.0355$ kg/mol.

It is required to calculate, at $P = 1$ bar and $T = 1000$ K: (1) the mean moment of inertia, I_m; (2) rotational entropy; (3) vibrational entropy; and (4) total entropy.

Solution

1. A practical axis system in which moments of inertia can be considered is the following: the Ox axis, through the Cl atoms; the Oy axis, through the Na atoms; and the Oz axis, perpendicular to molecule's plan and intersecting the other two axes in the point O, the rhombus' center.

According to relation (2.30b), $I_x = 2 \cdot m_{\text{Na}} \cdot (d_{\text{Na}})^2$, $I_y = 2 \cdot m_{\text{Cl}} \cdot (d_{\text{Cl}})^2$, $I_z = 2 \cdot m_{\text{Na}} \cdot (d_{\text{Na}})^2 + 2 \cdot m_{\text{Cl}} \cdot (d_{\text{Cl}})^2$.

From definition (2.50b), $I_m = (I_x \cdot I_y \cdot I_z)^{1/3}$ where, according to relation (2.30b): $I_x = 2 \cdot m_{Na} \cdot d_{Na}^2$; $I_y = 2 \cdot m_{Cl} \cdot d_{Cl}^2$; $I_z = 2 \cdot m_{Na} \cdot d_{Na}^2 + 2 \cdot m_{Cl} \cdot d_{Cl}^2$. From the Expression (2.50c): $I_m = (I_x \cdot I_y \cdot I_z)^{1/3}$ or, expressing m_{Cl} and m_{Na} through the use of atomic masses, $m_i = A_i / \cdot N_A$:

$$I_m = 2 \cdot \left[A_{Na} \cdot d_{Na}^2 \cdot A_{Cl} \cdot d_{Cl}^2 \cdot \left(A_{Na} \cdot d_{Na}^2 + A_{Cl} \cdot d_{Cl}^2\right)\right]^{1/3} / N_A =$$
$$= 2 \cdot \left[23 \cdot 1.42^2 \cdot 35.5 \cdot 2^2 \cdot \left(23 \cdot 1.42^2 + 35.5 \cdot 2^2\right)\right]^{1/3} \cdot 10^{-23} / 6.02 \cdot 10^{23}$$
$$\Rightarrow \mathbf{I_m = 35.7 \cdot 10^{-46} kg \cdot m^2}$$

2. From relation (2.51c): $S_{rot} = 8.31 \cdot \ln([(10^{46} \cdot T \cdot I)^{1.5} / \sigma]) - 12.6$ that is

$$S_{rot} = 8.31 \cdot \ln\left[35600^{1.5}/4\right] - 12.6 \Rightarrow \mathbf{S_{rot} = 107\ J/(mol \cdot K)}$$

3. From x_i ratios, $i \in [1, 6]$ defined by $x_i = \theta_{vib,\,i}/T$, using relation (2.44b), $E_{S,\,i} = [x_i/(\exp x_i - 1)] - \ln[1 - \exp(-x_i)]$, the vibrational Einstein functions for entropy can be calculated, for example: $x_1 = \theta_{vib,1}/T = 201/1000 = 0.201$, then $E_{S,1} = \frac{x_1}{e^{x_1} - 1} - \ln(1 - e^{-x_1}) = 2.61$. Similarly, it is obtained that: $E_{S,\,2} = E_{S,\,1} = 2.61$; $E_{S,3} = 0.292/(e^{0.292} - 1) - \ln(1 - e^{-0.292}) = 2.23$; $E_{S,4} = 2.19$; $E_{S,5} = 1.99$; and $E_{S,6} = 1.96$.

Using relation (2.44f) and after summation, it is found that: Svib $= R \cdot \Sigma_{i = 1\,..0.6} E_{S,i} = 8.31 \cdot (2 \cdot 2.61 + 2.23 + 2.19 + 1.99 + 1.96) \Rightarrow$ **Svib = 128 J/(mol·K)**

4. $M = 2 \cdot (A_{Na} + A_{Cl}) = 0.002 \cdot (23 + 35.46) = 0.117$ kg/mol.

The translational entropy is calculated from relation (2.28a): $S_{tr} = (R/2) \cdot \ln(T^5 \cdot M^3 / P^2) + 164 = (8.31/2) \cdot \ln(10^{3 \cdot 5} \cdot 0.1172^3 / 10^{5 \cdot 2}) + 164 = 185$ J/(mol·K).

By neglecting S_n and S_{el} contributions, the total entropy is obtained: $S = S_{rot} + S_{vib} + S_{tr} = 107 + 128 + 185 \Rightarrow$ **S = 420 J/(mol·K).**

Chapter 3
Distribution of Molecular Properties in Gases

Ideal Gas Kinetic Theory The kinetic theory of *matter* aims to establish relations between the macroscopic variables (pressure, volume, temperature, thermodynamic quantities such as the caloric ones) that characterize matter from different aggregation states and the quantities characterizing the molecules motion, such as the kinetic energy or the amount of motion (momentum).

Under this theory two complementary study methods are used:

(a). The *statistical mechanical* method, introduced by Gibbs, treats the state of aggregation by considering a large number of particles from the very beginning and applies probabilities calculation to a statistical mechanics type of treatment. The method is much more abstract, but the results are much safer. Statistical mechanics methods have been systematically applied in Chaps. 1 and 2 for steady states study.

(b). The *kinetic-molecular* theory, initiated by Boltzmann, uses molecular models—it is plausible but not absolutely accurate—regarding molecules arrangement and interactions; every particle is taken into account and an elementary probability calculation is applied in the end. Its advantage is that it can be also applied to processes from non-equilibrium systems, such as transport or phase transfer processes. This chapter—the third—is devoted to the kinetic-molecular theory basics.

From particle motion and interaction point of view, the states of aggregation are distinguished by the ratio between the kinetic energy and the average potential energy of particles ($E_{c,m}$ and $E_{p,m}$). While *kinetic* energy is due to molecules thermal agitation and depends on T, *potential* energy is due to forces of attraction—electrostatic, chemical, Van der Waals, etc.

- gases have $E_{c,m} \gg E_{p,m}$ because molecules' strong agitation overcomes their cohesion force
- *condensed* states—that is, liquids and solids—have, on the contrary, the potential energy comparable to the kinetic one, or even higher

F. E. Daneş et al., *Molecular Physical Chemistry for Engineering Applications*,
https://doi.org/10.1007/978-3-030-63896-2_3

Assumptions of Kinetic-Molecular Theory of Gases The kinetic-molecular theory of gases is based on a set of simplifying assumptions, namely:

(a) the gas comprises within its structure a very large number of molecules, placed at long distances one from another (relative to molecules dimensions)
(b) molecules are in constant motion; none of the motion directions is preferential, the motion being chaotic, disordered
(c) gas molecules collide one with the other or with vessel walls and consequently, the motion direction and molecular velocities are constantly changing
(d) intermolecular and wall collisions are considered to be perfectly elastic, the molecule being a perfectly elastic rigid sphere
(e) the interaction forces between molecules are exerted only in the moment of collision and do not modify the potential energy of the molecule, but only the kinetic one

Based on these assumptions and by applying either the kinetic method or the statistical one, the kinetic-molecular theory offers a structural interpretation of gases pressure, evaluates gas molecules motion velocity and mean free path, even studies non-equilibrium phenomena such as mass and heat transport, flow process, friction process, or physical transformations velocities and chemical reactions.

3.1 Elements of the General Theory of Distribution

A Y variable *distribution* is the dispersal pattern of values taken by this variable within a given situation.

Velocities, momentums, and translational kinetic energies, as well as velocities and momentums projections on coordinate axes, vary from one molecule to another. Moreover, for an individual molecule this variation is unpredictable over time, changing at each collision with other molecules or vessel walls. However, if the gas is in equilibrium and contains a sufficiently high number of molecules, then statistical mechanics allows the distribution of these units determination, which is invariant in time and space.

Distribution for the Reduced Size Sample

For the *reduced* size sample case (i.e., constituted of a relatively low number of elements from a measurable real variable, Y), the distribution is described as following:

- either by its global value (*integral* distribution) Γ_{red} defined as the total number of elements found between limits Y' and Y'' of an Y range:

$$\Gamma_{red} = N_{Y\in(Y'...Y'')}$$

- either by *density* distribution (also called *differential* distribution) G_{red}, which is the ratio between this number of elements and the length $\Delta Y = Y'' - Y'$ of the distribution interval:

$$G_{red} = N_{Y\in(Y'...Y'')}/\Delta Y.$$

For example, for a set of 14 Y values {3, 5, 5, 7, 8, 10, 10, 10, 12, 12, 12, 12, 12, 23}, if both interval endpoints are $Y' = 9$ and $Y'' = 11$, then $\Gamma_{red} = N_{Y\in(Y'...Y'')} = \underline{3}$, because three particles have the values $Y \in (9 \ldots 11)$, $\Delta Y = 11–9 = \underline{2}$, and the mean density distribution will therefore be: $G_{red} = \underline{3/2} = 1.5$.

Distribution Functions

Velocities, momentums, and translational kinetic energies, as well as velocities and momentums projections on coordinate axes, vary from one molecule to another. Moreover, for an individual molecule this variation is unpredictable over time, changing at each collision with other molecules or vessel walls. However, if the gas is in equilibrium and contains a sufficiently high number of molecules, then statistical mechanics allows the distribution of these units determination, which is invariant in time and space.

Differential Distribution Function

For a large number of elements set, instead of density distribution for a certain distribution Γ_{red} given above, which depends on interval's two endpoints, a similar function can be used, denoted by Γ_Y. It is a simpler function because it depends on a single point: the *differential* distribution function of Y property at Y' point, defined as the ratio of elements number $d\,N_{Y\in(Y'...Y'+dY)}$ found between Y' and $Y' + dY$, at dY length of this interval:

$$\Gamma_{Y=Y'} \equiv dN_{Y\in(Y'...Y'+dY)}/dY \qquad (3.1a)$$

Therefore, Γ_Y has a "density" meaning on Y axis of points representing those particles with a certain given value of the Y feature.

Integral Distribution Function

For similar assemblies comprising an important number of elements, the *integral* distribution function G_Y can be defined at any given point, which is the number of assembly elements having Y property's value inferior to the limit-value Y′:

$$G_{Y=Y} = N(Y < Y') \tag{3.1b}$$

Link between Differential and Integral Distribution Functions

Differential and integral distribution functions are correlated through:

$$G_Y = \int_{Y=-\infty}^{Y=Y\prime} \Gamma_y \cdot dY \tag{3.1c}$$

$$\Gamma_Y = lim_{\,Y\to Y'}(dG_Y/dY) \tag{3.1d}$$

Normalization of Distributions

The *normalized* distribution functions are typically used—differential *normalized* distribution function φ and integral *normalized* distribution function F—and are obtained from the preceding distribution functions (unnormalized) that have been denoted by Γ and G, through their division to the total number of objects (molecules), N:

$$\Phi = \Gamma/N \tag{3.2a}$$

$$F = G/N \tag{3.2b}$$

The *differential* normalized distribution function from Fig. 3.1a represents the probability for particle Y value to be placed inside the interval between Y and Y + *d* Y, reported to *d* Y interval length:

$$\Phi_Y(Y') \equiv [Prob(Y' < Y < Y' + dY)]/dY \tag{3.3a}$$

The surface area between the curve $\Phi_Y = f(Y)$ and the abscissa axis is equal to 1.

The *integral* normalized distribution function represents the probability that particle Y value is lower than a given limit:

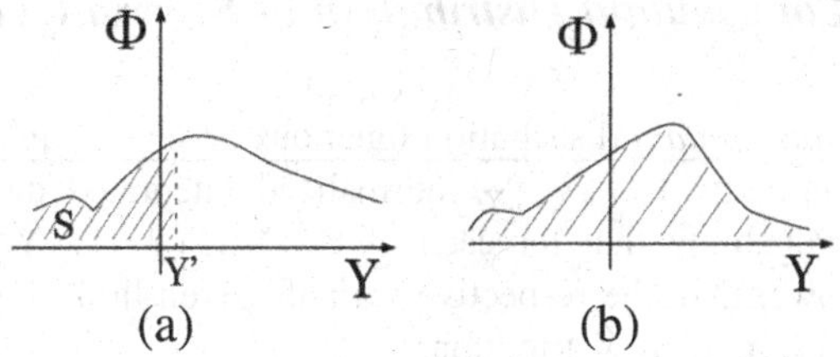

Fig. 3.1 Normalized density distribution function (**a**) Translation to the integral function.(**b**). Normalized density of distribution plot

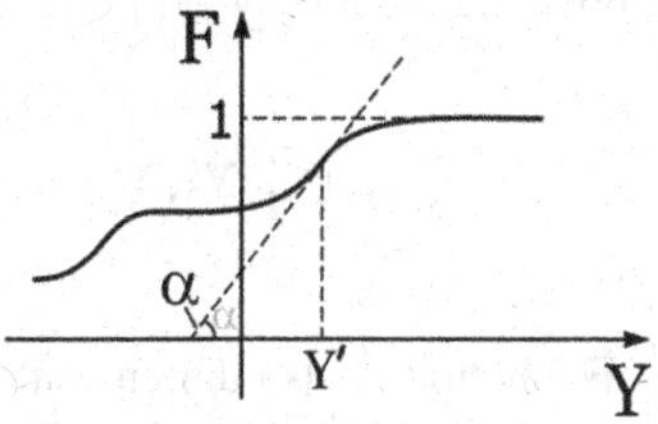

Fig. 3.2 Integral normalized distribution function plot

$$F(Y = Y') = Prob(Y < Y') \tag{3.3b}$$

From the probability definition it follows that F (Y) is undecreasing, not necessarily continuous and takes values in the [0 . . . 1] interval. Also, for any distribution type $F_{Y=-\infty} = 0$ and for $F_{Y=\infty} = 1$.

By integrating and differentiating it is possible to pass from differential to integral normalized distribution function and vice versa:

$$F_Y = \int_{Y=Y'}^{Y=Y''} \Phi_Y \cdot dY \tag{3.3c}$$

$$\Phi_Y = dF_Y/dY \tag{3.3d}$$

The integral distribution function can be graphically obtained from the differential one Φ(Y) as in Fig. 3.1b, by measuring the area bounded to the right by $Y = Y'$ ordinate and comprised between curve and abscissa axis: $F_{Y=Y'} = S$.

The normalized integral distribution function F_Y of Y property is represented by F (Y) decreasing curve from Fig. 3.2, defined by any value Y and having the limits 0 and 1 at $Y \to -\infty$ respectively $Y \to \infty$.

From this plot differential function Φ_Y can be obtained, for a given value Y′ of Y, as slope of F(Y) representation in the point with $Y = Y'$: $\Phi_Y(Y') = tg\ \alpha$.

The Φ and Γ differential distribution functions have Y^{-1} as dimensions, while the integral distribution functions, F and G, are dimensionless.

Concomitant Distribution of Several Quantities

Concomitant distribution functions of several quantities $Y_1, Y_2, \ldots, Y_n$ are defined similarly: $F_{Y1,Y2,\ldots,Yn}$ normalized integrated distribution function represents the probability that for each of the $Y_1, Y_2, \ldots, Y_n$ features the particle has values lower than the respective variable given limit; for example, for the case of 3 units (Y_1, Y_2, Y_3) distribution:

$F_{Y_1Y_2Y_3}(Y'_1, Y'_2, Y'_3) = \mathit{Prob.}(Y_1 < Y'_1 \cap Y_2 < Y'_2 \cap Y_3 < Y'_3)$, and function Φ is obtained from F by applying the derivative n times—one time for each variable:

$$\Phi_{Y_1Y_2Y_3}(Y'_1, Y'_2, Y'_3) = \left[\frac{d^3F_{Y_1Y_2Y_3}(Y'_1, Y'_2, Y'_3)}{dY_1 \cdot dY_2 \cdot dY_3}\right]_{Y_1=Y'_1, Y_2=Y'_2, Y_3=Y'_3}$$

F function is also dimensionless in this case, and Φ function has the dimensions $\langle\Phi\rangle = Y_1^{-1} \cdot Y_2^{-1} \cdot \ldots \cdot Y_n^{-1}$. The reverse translation—from Φ to F—is made by applying the integral n times, one time for each variable:

$$F_{Y_1Y_2Y_3}(Y'_1, Y'_2, Y'_3) = \int_{Y_1=-\infty}^{Y_1=Y'_1} \int_{Y_2=-\infty}^{Y_2=Y'_2} \int_{Y_3=-\infty}^{Y_3=Y'_3} \Phi_{Y_1Y_2Y_3} \cdot dY_1 \cdot dY_2 \cdot dY_3 \tag{3.4a}$$

For the case of distribution functions, both changing the arguments (distributed variables) and reducing their number are possible.

Distributed variables number reduction is realized by integration for all possible values of disappearing variables.

For example, shifting from three variables Y_1, Y_2, Y_3 distribution function to two variables Y_1, Y_2 one is made through relation:

$$\Phi_{Y_1Y_2}(Y_1, Y_2) = \int_{Y_3=-\infty}^{Y_3=\infty} \Phi_{Y_1Y_2Y_3}(Y_1, Y_2, Y_3) \cdot dY_3 \tag{3.4b}$$

Change of arguments: The differential normalized functions $\Phi_Y(Y)$ and those of certain interdependent variables **Y** and **X,** respectively, can be converted one into another by:

$$\Phi_X = \Phi_Y \cdot (dY/dX) \tag{3.4c}$$

The translation from concomitant distribution of variables $Y_1, Y_2, \ldots, Y_n$ to variables $X_1, X_2, \ldots, X_n$ distribution (in equal number) is made using the following relation:

$$\Phi_{X_1X_2K\,X_n} = \Phi_{Y_1Y_2K\,Y_n} \cdot \frac{D(Y_1, Y_2, \ldots, Y_n)}{D(X_1, X_2, \ldots, X_n)} \quad (3.4d)$$

where determinant $D(Y_1, Y_2, \ldots, Y_n)/D(X_1, X_2, \ldots, X_n)$ also denoted by $J(X_1, X_2, \ldots, X_n \parallel Y_1, Y_2, \ldots, Y_n)$ represents $Y_1, Y_2, \ldots, Y_n$ group of variables Jacobian toward $X_1, X_2, \ldots, X_n$ group of variables and is defined by:

$$\frac{D(Y_{1,2,\ldots,n})}{D(X_{1,2,\ldots,n})} = \begin{vmatrix} \frac{\partial Y_1}{\partial X_1} & \frac{\partial Y_2}{\partial X_1} & \cdots & \frac{\partial Y_n}{\partial X_1} \\ \frac{\partial Y_1}{\partial X_2} & \frac{\partial Y_2}{\partial X_2} & \cdots & \frac{\partial Y_n}{\partial X_2} \\ \cdots & \cdots & & \cdots \\ \frac{\partial Y_1}{\partial X_n} & \frac{\partial Y_2}{\partial X_n} & \cdots & \frac{\partial Y_n}{\partial X_n} \end{vmatrix}.$$

The above exposed *general* statistical theory of distribution has been applied by Maxwell to *molecular features* distribution study (velocity and its projections, energy) in the ideal case ⇒ at equilibrium, as it will be outlined in 3.2 and 3.4.

Boltzmann already obtained the same results through kinetic molecular theory, which is more intuitive.

3.2 Molecular Velocities Distributions

Concomitant Distribution of Position and Momentum Coordinates

The three position coordinates (x, y, z) and the corresponding momentums (p_x, p_y, p_z) represent the six basic characteristics of gas particles translational motion. Microstates differential distribution function has been introduced in 2.4, for three possible directions in space:

$$\Gamma' = h^{-3}$$

The N particles number with a given ε energy is connected to Z sum over states through relation (1.45), where it was admitted that translational energy levels are equivalent from occupancy rate point of view: $N = e^{-\varepsilon_j/(k_B \cdot T)}/Z$.

By normalizing distribution through relation (3.2a) it results, for particles normalized differential distribution function over the six features that: $\Phi_{x,y,z,p_x,p_y,p_z} = e^{-\varepsilon/(k_B \cdot T)}/(Z \cdot h^3)$. But: $\varepsilon = [(p_x)^2 + (p_y)^2 + (p_z)^2]/(2 \cdot m)$ and

$Z = (2 \cdot \pi \cdot m \cdot k_B \cdot T/h^2)^{3/2} \cdot V/N_A$ according to relation (2.25) and, therefore, by denoting with V the molar volume it results that:

$$\Phi_{x,y,z,p_x,p_y,p_z} = \frac{N_A}{V \cdot (2 \cdot \pi \cdot m \cdot k_B \cdot T)^{3/2}} \cdot e^{-\left[(p_x)^2+(p_y)^2+(p_z)^2\right]/(2 \cdot m \cdot k_B \cdot T)}.$$

Concomitant Distribution of the Three Velocities Projections

By reducing the spatial coordinates of relation (3.4b) through the method introduced in 3.1, a transition can be made from the distribution function over the six features to the distribution function over the three momentum projections, as can be seen below:

$$\Phi_{p_x,p_y,p_z} = \iiint_{x,y,z} \Phi_{x,y,z,p_x,p_y,p_z} \cdot dx \cdot dy \cdot dz.$$

Since Φ_{x,y,z,p_x,p_y,p_z} function does not depend on x, y, z spatial coordinates, momentum variables can be pulled out from the integral leading to:

$$\Phi_{p_x,p_y,p_z} = \frac{N_A \cdot e^{-\left[(p_x)^2+(p_y)^2+(p_z)^2\right]/(2 \cdot m \cdot k_B \cdot T)}}{V \cdot (2 \cdot \pi \cdot m \cdot k_B \cdot T)^{3/2}} \cdot \int_x \int_y \int_z \cdot dx \cdot dy \cdot dz.$$

In the above expression, the triple integral represents the single molecule corresponding volume,
$v_{1\ molec} = V/N_A$, so that by simplifying it results that:

$$\Phi_{p_x,p_y,p_z} = e^{-\left[(p_x)^2+(p_y)^2+(p_z)^2\right]/(2 \cdot m \cdot k_B \cdot T)} \cdot /(2 \cdot \pi \cdot m \cdot k_B \cdot T)^{3/2}.$$

By replacing momentum projections as variables through velocities projections: $p_x = m \cdot v_x$, $p_y = m \cdot v_y$, $p_z = m \cdot v_z$, it results the Jacobian expression $\frac{D}{D}\left(\frac{p_x, p_y, p_z}{v_x, v_y, v_z}\right)$, namely:

$$\frac{D}{D}\left(\frac{p_x, p_y, p_z}{v_x, v_y, v_z}\right) = \begin{vmatrix} \frac{\partial p_x}{\partial v_x} & \frac{\partial p_y}{\partial v_x} & \frac{\partial p_z}{\partial v_x} \\ \frac{\partial p_x}{\partial v_y} & \frac{\partial p_y}{\partial v_y} & \frac{\partial p_z}{\partial v_y} \\ \frac{\partial p_x}{\partial v_z} & \frac{\partial p_y}{\partial v_z} & \frac{\partial p_z}{\partial v_z} \end{vmatrix} = \begin{vmatrix} m & 0 & 0 \\ 0 & m & 0 \\ 0 & 0 & m \end{vmatrix} = m^3$$

from where by replacing momentum projections from exponential expression, the following is obtained:

$$\Phi_{v_x,v_y,v_z} = e^{-\left[(v_x)^2+(v_y)^2+(v_z)^2\right]\cdot m/(2\cdot k_B\cdot T)}/(2\cdot\pi\cdot k_B\cdot T)^{3/2} \tag{3.5}$$

Therefore, particle distribution after the three components is *uniform*, namely it does not depend on x, y, z spatial coordinates.

Velocity Projection Distribution after a Given Direction

Particles velocity distribution function after v_x projection on Ox axis is obtained from relation (3.5) by variables v_y and v_z reduction:

$$\Phi_{v_x} = \int_{v_y=-\infty}^{v_y=\infty}\int_{v_z=-\infty}^{v_z=\infty}\Phi_{v_x,v_y,v_z}\cdot dv_y\cdot dv_z$$

or

$$\Phi_{v_x} = \int_{-\infty}^{\infty}\int_{-\infty}^{\infty}\left(\frac{m}{2\cdot\pi\cdot k_B\cdot T}\right)^{3/2}\cdot e^{-\left[(v_x)^2+(v_y)^2+(v_z)^2\right]\cdot m/(2\cdot k_B\cdot T)}\cdot dv_y\cdot dv_z =$$

$$(2\cdot\pi\cdot k_B\cdot T/m)^{3/2}\cdot e^{-m\cdot(v_x)^2/(2\cdot k_B\cdot T)}\cdot\int_{-\infty}^{+\infty}e^{\frac{-m\cdot(v_y)^2}{2\cdot k_B\cdot T}}\cdot dv_y\cdot\int_{-\infty}^{+\infty}e^{\frac{-m\cdot(v_z)^2}{2\cdot k_B\cdot T}}\cdot dv_z$$

Since the two definite integrals are equal, it results:

$$\Phi_{v_x} = J^2\cdot(2\cdot\pi\cdot k_B\cdot T/m)^{-3/2}\cdot e^{-m\cdot(v_x)^2/(2\cdot k_B\cdot T)} \tag{3.2a2}$$

where $J = \int_{-\infty}^{\infty}e^{-m\cdot\lambda^2/(2\cdot k_B\cdot T)}\cdot d\lambda$, and λ denote either v_y either v_z. The integrand *exp* $[m\cdot\lambda^2/(2\cdot k_B\cdot T)]$ is a symmetric function having λ as integration variable, so that J is the double of integral having zero inferior limit $\int_0^{\infty}e^{-m\cdot\lambda^2/(2\cdot k_B\cdot T)}\cdot d\lambda$ and obtained through expression (A.2b) from the mathematical annex, $\int_0^{\infty}x^{2\cdot k}\cdot e^{-\alpha\cdot x^2}\cdot dx = \frac{(2\cdot k)!}{k!}\cdot(4\cdot\alpha)^{-(k+1/2)}\cdot\sqrt{\pi}$, with $k = 0$ and $\alpha = m/(2\cdot k_B\cdot T)$. It results that $J = \sqrt{2\cdot\pi\cdot k_B\cdot T/m}$. By replacing J into relation (3.2a2), the following expression is obtained:

$$\Phi_{v_x} = (2\cdot\pi\cdot k_B\cdot T/m)^{-1/2}\cdot e^{-m\cdot(v_x)^2/(2\cdot k_B\cdot T)} \tag{3.6}$$

Density distribution is therefore a symmetric dependency function of velocity projection over the considered axis, and it depends only on the projection over the

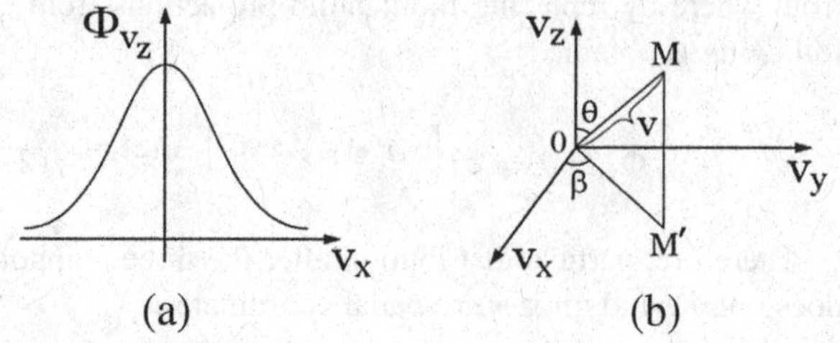

Fig. 3.3 Velocity projection over an axis. (**a**) Distribution of velocity projection. (**b**) Velocity and its components

respective direction. As observed from expression (3.6), v_x projection distribution is Gaussian. Similarly, it results:

$$\Phi_{V_y} = (2 \cdot \pi \cdot k_B \cdot T/m)^{-1/2} \cdot e^{-m \cdot (v_y)^2/(2 \cdot k_B \cdot T)}$$

or

$$\Phi_{V_z} = (2 \cdot \pi \cdot k_B \cdot T/m)^{-1/2} \cdot e^{-m \cdot (v_z)^2/(2 \cdot k_B \cdot T)}.$$

The graphical representation of *Gaussian* or *normal* distribution is through the "error curve" from Fig. 3.3a. The average is always zero, being here equal to the most probable velocity v_x projection value, denoted by $(v_x)_W$:

$$(v_x)_W = 0 \tag{3.7}$$

Dispersion (mean squared deviation) is $(k_B \cdot T/m)^{1/2}$.

Particles Velocity Distribution

Particles velocity distribution obtaining is based on the three projections distribution— relation (3.5)—with previous variable change from v_x, v_y, v_z to v, β, θ, respectively, which correspond to spherical coordinates, as in Fig. 3.3b.

The transition relations to the new arguments are as following:

$$v_x = v \cdot \sin \boldsymbol{\theta} \cdot \cos \boldsymbol{\beta};\ v_y = v \cdot \sin \boldsymbol{\theta} \cdot \sin \boldsymbol{\beta};\ v_z = v \cdot \cos \boldsymbol{\theta} \tag{3.2b2}$$

and the corresponding Jacobian will be, by replacing the partial derivatives obtained from (3.2b2):

$$\frac{D}{D}\left(\frac{v_x, v_y, v_z}{v, \theta, \beta}\right) = \begin{vmatrix} \frac{\partial v_x}{\partial v} & \frac{\partial v_y}{\partial v} & \frac{\partial v_z}{\partial v} \\ \frac{\partial v_x}{\partial \theta} & \frac{\partial v_y}{\partial \theta} & \frac{\partial v_z}{\partial \theta} \\ \frac{\partial v_x}{\partial \beta} & \frac{\partial v_y}{\partial \beta} & \frac{\partial v_z}{\partial \beta} \end{vmatrix}$$

$$= \begin{vmatrix} \sin\theta \cdot \cos\beta & \sin\theta \cdot \sin\beta & \cos\theta \\ v \cdot \cos\theta \cdot \cos\beta & v \cdot \cos\theta \cdot \sin\beta & -v \cdot \sin\theta \\ -v \cdot \sin\theta \cdot \sin\beta & v \cdot \sin\theta \cdot \cos\beta & 0 \end{vmatrix} = v^2 \cdot \sin\theta;$$

Therefore, function distribution corresponding to the new variables becomes:

$$\Phi_{v,\theta,\beta} = \Phi_{v_x,v_y,v_z} \cdot \frac{D}{D}\left(\frac{v_x, v_y, v_z}{v, \theta, \beta}\right).$$

When using ϕ_{v_x,v_y,v_z} from relation (3.5) and considering that $(v_x)^2 + (v_y)^2 + (v_z)^2 = v^2$, the following expression is obtained:

$$\Phi_{v,\theta,\beta} = (2 \cdot \pi \cdot k_B \cdot T/m)^{-3/2} \cdot e^{-m \cdot v^2/(2 \cdot k_B \cdot T)} \cdot v^2 \cdot \sin\theta.$$

The Φ_v distribution function can be calculated from $\Phi_{v,\theta,\beta}$ for the absolute velocity value case, by reducing θ and β variables:

$$\Phi_v = \int_{\theta=0}^{\theta=\pi} \int_{\beta=0}^{\beta=2\cdot\pi} \Phi_{\theta,\beta,v} \cdot d\theta \cdot d\beta$$

or

$$\Phi_V = V^2 \cdot (2 \cdot \pi \cdot k_B \cdot T/m)^{-3/2} \cdot \exp\left[-m \cdot v_y^2/(2 \cdot k_B \cdot T)\right] \cdot I_\theta \cdot I_\beta,$$

where

$I_\theta = \int_{\theta=0}^{\theta=\rho} \sin\theta \cdot d\theta = -\cos\pi + \cos 0 = 2$, and
$I_\beta = \int_{\beta=0}^{\beta=2\cdot\pi} d\beta = 2 \cdot \pi$,

therefore:

$$\Phi_V = (2/\pi)^{1/2} \cdot (k_B \cdot T/m)^{-3/2} \cdot v^2 \cdot e^{-m \cdot v^2/(2 \cdot k_B \cdot T)} \tag{3.8}$$

This distribution function depends on v but not on the angles θ and β. Therefore, velocity distribution is not only *uniform* (independent on x, y, z particle positioning) but also *isotropic*, namely it does not depend on velocity direction.

Velocity Distribution Function Form

Velocity distribution function form can be studied by shifting from variable v to a proportional X variable, through the following relation:

$$v = X \cdot (2 \cdot k_B \cdot T/m)^{1/2} \quad (3.2c2)$$

By replacing v from (3.2c2) into equality (3.8) it results that:

$$\Phi_v = 2 \cdot \sqrt{\frac{2 \cdot m}{\pi \cdot k_B \cdot T}} \cdot X^2 \cdot e^{-X^2} \quad (3.2d2)$$

Function $\Phi_v(v)$ only makes sense for positive v values and becomes zero at both $v = 0$ and $v = \infty$. Its derivative over X is as following:

$$d\Phi_V/dX = 4 \cdot \sqrt{2 \cdot m/(\pi \cdot k_B \cdot T)} \cdot X \cdot (1 - X^2) \cdot exp(-X^2),$$

expression that becomes zero at $X = 0$, $X = \pm 1$ and $X = \infty$. Since Φ_v is always positive, the density distribution maximum—its most probable value—will be found at $X = 1$, which corresponds in relation (3.2c2) to the following most probable value v, denoted by v_w:

$$v_w = (2 \cdot k_B \cdot T/m)^{1/2} \quad (3.9)$$

Thus, velocity distribution φ_v differs from velocity projection distribution over an axis, Φ_{V_X}: the most probable value of velocity, v_W, is different from 0, and the most probable value v_X is zero.

The Φ_v distribution form, illustrated in Fig. 3.4, is not Gaussian but asymmetric: density distribution decrease is more readily to maximum's left side than to its right one.

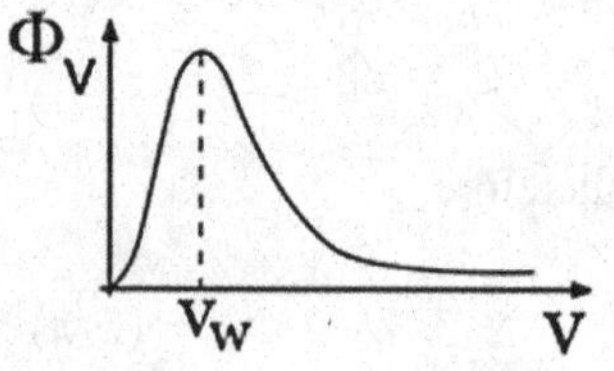

Fig. 3.4 Velocity distribution

3.3 Features of Velocity and Its Projections

Any distribution Φ_Y of a single property (Y) can be globally described by several characteristic values, like *mode* (the most probable value of the distributive property) and different means, among which *arithmetic mean* and *quadratic mean* are the most used for distributions study.

The Most Probable Value

The most probable value Y_W is defined as the value corresponding to differential distribution function maximum:

$$Y_W = Y'|\Phi_Y(Y') = \max$$

For velocities projections and velocities distributions respectively, it has been seen from relations (3.8) and (3.9) that: $(v_X)_W = 0$ and $v_W = (2 \cdot k_B \cdot T/m)^{1/2}$

Mean Values

A mean value can characterize not only the distribution functions or functions, but also any finite or infinite set of real numbers. Unless otherwise specified, the word "*mean*" designates the *arithmetic* mean value.

Generalized k-mean value, $<x>_k$, of a set {X} of N real numbers (x_1, x_2, ..., x_{N-1}, x_N) is:

$$<x>_k \equiv \left[N^{-1} \cdot \sum_{n=1}^{n=N} (x_n)^k\right]^{1/k} \tag{3.10a}$$

where k is a real number. Its definition does not allow the most probable value x expression, but includes minimum and maximum values for the {X} set:

$$x_{min} = <x>_{-\infty}; x_{max} = <x>_{\infty}.$$

It can be shown that *geometric* mean value x_{geom} corresponds to zero generalized k-mean value:

$$x_{geom} \equiv \left[\prod_{n=1}^{n=N} (x_n)\right]^{1/N} = \lim_{k \to 0} <x>_k.$$

while k can be a non-integer number, the finite nonpositive k-means values—like the *harmonic* mean value $x_{harm} \equiv <x>_{-1}$ and the geometric one—do not make sense; moreover, the fractionate order means values cannot even be calculated, unless all elements of {X} set are positive.

Jensen's theorem allows the set of means values ranking for the case of a given set. It states that generalized mean value increases in the same direction as mean's value k order:

$$d < x>_k/dk > 0 \tag{3.10b}$$

As an example, the harmonic mean value is lower than the geometric one, and this latter is inferior to the *arithmetic* mean value, $<x>_1$, denoted by $\bar{x}$.

Arithmetic Mean Value

The *arithmetic* mean value $\overline{Y}$ is defined for discrete distributions by the following relation:

$$\overline{Y} = N^{-1} \cdot \sum_{i=1}^{i=N} Y_i \tag{3.11a}$$

Since molecules number N from the usually considered systems is very elevated, the discrete distribution is replaced by continuous distributions—according to relations (3.6) for ϕ_{v_x} and (3.8) for Φ_v—where summation becomes an integration, $\overline{Y} = \int_{Y=Y'}^{Y=Y''} Y \cdot \Phi_Y \cdot dY / \int_{Y=Y'}^{Y=Y''} \Phi_Y \cdot dY$, where Y′ and Y″ are minimum and maximum possible distributed variable Y values. However, when φ_v function is normalized, denominator integral is equal to 1, according to normalized distribution functions quantities. It results that:

$$\overline{Y} = \int_{Y=Y'}^{Y=Y''} Y \cdot \Phi_Y \cdot dY \tag{3.11b}$$

The arithmetic mean value V_X projection

$$\overline{V_X} = \int_{V_X=-\infty}^{V_X=\infty} V_X \cdot \Phi_{V_X} \cdot dv_X$$

will therefore be:

$$\overline{V_X} = \int_{V_X=-\infty}^{V_X=\infty} (2 \cdot \pi \cdot k_B \cdot T/m)^{-1/2} \cdot e^{-m \cdot (v_x)^2/(2 \cdot k_B \cdot T)} \cdot v_X \cdot dv_X \tag{3.11c}$$

Indefinite integration $[k_B \cdot T/(2 \cdot \pi \cdot m)] \cdot exp[-m \cdot (v_x)^2/(2 \cdot k_B \cdot T)]$ result takes the same value—zero—both at upper integration limit ($v_X \to \infty$) and lower one ($v_X \to -\infty$), so that definite integral will be zero. Therefore:

$$\overline{v}_X = 0 \tag{3.12a}$$

By replacing Φ_v from (3.8) into the mean value's velocity:

$$\overline{v} = (2/\pi)^{1/2} \cdot (k_B \cdot T/m)^{-3/2} \cdot I \tag{3.3f2}$$

where:

$$I = \int_{v=0}^{v=\infty} v^3 \cdot e^{-m \cdot v^2/(2 \cdot k_B \cdot T)} \cdot dv \tag{3.3g2}$$

The integral from (3.3g2) is calculated using (A.2c) formula from the mathematical annex:

$$\int_0^{\infty} x^{2 \cdot k+1} \cdot e^{-\alpha \cdot x^2} \cdot dx = k!/\left(2 \cdot \alpha^{k+1}\right)$$

with $k = 1$ and $\alpha = m/(2 \cdot k_B \cdot T)$.

It results $I = 1/(2 \cdot \alpha^2) = 2 \cdot (k_B \cdot T/m)^2$. By replacing I into relation (f), the following is found:

$$\overline{v} = \sqrt{8 \cdot k_B \cdot T/(\pi \cdot m)} \tag{3.12b}$$

where molar mass M can be introduced through $m = M/N_A$ and $k_B = R/N_A$.

Quadratic Mean Value of Velocity Projections

The following relation defines Y_p *quadratic mean value* for the discrete distributions case:

$$Y_p = \left[\sum\nolimits_{i=1}^{i=N} (Y_i)^2/N\right]^{1/2} \tag{3.13a}$$

The squared Y_p value represents square of Y property arithmetic mean value: $(Y_p)^2 = \overline{Y^2} = \sum (Y_i)^2/N$.

If Y_i values are not equal to each other, then $Y_p > \overline{Y}$, according to Jenkins' theorem: from relation (3.10b), $d < x >_k / d\,k > 0$, it is concluded that $<x>_2 > <x>_1$, also Y_p and $\overline{Y}$ are defined as $<x>_2>$ and $<x>_1$, respectively.

For continuous distributions case, the quadratic mean value will not be given by (3.13a), but:

$$Y_p = \left(\int_{Y=Y'}^{Y=Y''} Y^2 \cdot \Phi_Y \cdot dY \right)^{1/2} \tag{3.13b}$$

For velocity projection case $\Phi_{v_x} = (2 \cdot \pi \cdot k_B \cdot T/m)^{-1/2} \cdot e^{-m \cdot (v_x)^2/(2 \cdot k_B \cdot T)}$ from relation (3.6), the following quadratic mean value is resulting from relation (3.13b):

$$(v_x)_p = [2 \cdot m/(\pi \cdot k_B \cdot T)]^{1/4} \cdot \sqrt{L} \tag{3.3h2}$$

where L is the integral $\Psi_2 = \int_{v_x=0}^{v_x=\infty} (v_x)^2 \cdot e^{-m \cdot (v_x)^2/(k_B \cdot T)} \cdot d v_x$, obtained through formula (A.2b) from the mathematical annex.

With $\Psi_{2 \cdot k(\alpha)} = \int_0^\infty x^{2 \cdot k} \cdot e^{-\alpha \cdot X^2} \cdot dx = \frac{(2 \cdot k)!}{k!} \cdot (4 \cdot \alpha)^{-(k+1/2)} \cdot \sqrt{\pi}$, where it is considered that $x = v_x$, $k = 1$ and $\alpha = m/(2 \cdot k_B \cdot T)$. It results that $L = 0.25 \cdot (\pi/\alpha^3)^{1/2}$ or $L = (\pi/2)^{1/2} \cdot (k_B \cdot T/m)^{3/2}$. By replacing L into relation (h) it follows that:

$$(v_X)_P = (k_B \cdot T/m)^{1/2} \tag{3.14a}$$

Velocity Quadratic Mean Value

Relation (3.13b) is applied, with (0 and ∞) integration limits, corresponding to molecular velocity v possible value interval; it therefore results for velocity v_p quadratic mean value:

$$v_p = \left(\int_{v=0}^{v=\infty} v^2 \cdot \Phi_v \cdot dv \right)^{1/2},$$

with density normalized by distribution from relation (3.8),

$$\Phi_v = (2/\pi)^{1/2} \cdot (k_B \cdot T/m)^{-3/2} \cdot v^2 \cdot e^{-m \cdot v^2/(2 \cdot k_B \cdot T)}$$

which leads to: $v_p = (2/\pi)^{1/4} \cdot (k_B \cdot T/m)^{-3/4} \cdot \sqrt{I}$, where I is the integral:

$$I = \Psi_4 = \int_{v=0}^{v=\infty} v^4 \cdot e^{-m \cdot v^2/(2 \cdot k_B \cdot T)} \cdot dv$$

obtained through (A.2b) formula from the mathematical annex, with:

$$\Psi_{2\cdot k}(\alpha) = \int_0^{\infty} x^{2\cdot k} \cdot e^{-\alpha \cdot x^2} \cdot dx = \frac{(2 \cdot k)!}{k!} \cdot (4 \cdot \alpha)^{-(k+1/2)} \cdot \sqrt{\pi}$$

where it is considered that:$x = v$, $k = 2$ and $\alpha = m/(2 \cdot k_B \cdot T)$, resulting: $I = (3/8) \cdot (\pi/\alpha^5)^{1/2}$ or $I = 3 \cdot (\pi/2)^{1/2} \cdot (k_B \cdot T/m)^{5/2}$.

By replacing I into the above relation, it results that:

$$v_p = \sqrt{3 \cdot k_B \cdot T/m} \tag{3.14b}$$

Table 3.1 centralizes the characteristic velocities—and velocities projections—according to relations (3.7, 3.9, 3.12a, 3.12b, 3.14a, 3.14b).

All velocities are proportional to $\sqrt{k_B \cdot T/m}$, having therefore the following form (where f is a numeric factor):

$$v_{caract} = \sqrt{f \cdot k_B \cdot T/m} \tag{3.15a}$$

For any thermomechanical conditions—that is, any (P, T) values—and any gas (any M) the ratio of different characteristic velocities is constant: $v'/v'' = (f'/f'')^{1/2}$. For example, for total velocity case, the numerical factors from this table are: $f_\alpha = 2$; $f_v = 8/\pi$; $f_v = 3$; therefore, $\overline{v}/\alpha = \sqrt{8/\pi}/\sqrt{2} = 1.128$ namely $\overline{v}$ is of ~13% higher than α, and the quadratic mean value's velocity is 9% higher than velocities arithmetic mean values and 23% than the most probable velocity.

It is also observed from this table that: $(v_p/v_{x,p}) = \sqrt{3}$.

Factors Influencing Velocities Distribution

For a given composition and temperature gas, the characteristic velocities do not depend on gas pressure, as seen in Table 3.1.

The characteristic velocities increase in the same way as T, which modifies also v_x or v quantities distributions: when T increases, the maximum is moved towards higher velocities (since $\alpha : \sqrt{T}$) and a curve flattening occurs (Φ_v frequency is proportional with $T^{-1/2}$), as in Fig. 3.5.

Table 3.1 Characteristic velocities

Velocity type	Projection on an axis	Absolute value
Notation	v_x	v
The most probable value	0	$(2)^{1/2} \cdot (k_B \cdot T/m)^{1/2}$
Mean value (arithmetic)	0	$(8/\pi)^{1/2} \cdot (k_B \cdot T/m)^{1/2}$
Quadratic mean value	$(k_B \cdot T/m)^{1/2}$	$(3)^{1/2} \cdot (k_B \cdot T/m)^{1/2}$

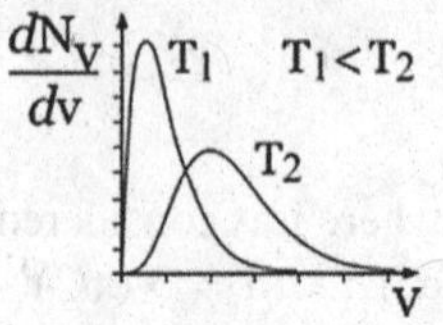

Fig. 3.5 Temperature effect over velocities distribution

For numerical calculations purposes, k_B/m ratio from (3.15a) can be replaced by R/M, since $R = N_A \cdot k_B$ and $M = N_A \cdot m$:

$$v_{caract} = \sqrt{f \cdot R \cdot T/m} \tag{3.15b}$$

The characteristic velocities are of the order of hundreds of m/s. For example, for O_2 (M = 0.032 kg/mol) at room temperature (T ≅ 300 K), the mean velocity is

$$\bar{v} = [8 \cdot 8.314 \cdot 300/(0.032 \cdot \pi)]^{1/2} = 446\ \text{m/s}.$$

From relation (3.15b) can be observed that characteristic velocities decrease with gas molecular weight increase: $v_{caract} \sim M^{-1/2}$.

3.4 Molecular Energies Distribution

Translational Energy Distribution

A transition can be made from particles velocity distribution to translational (kinetic) energy distribution, using the relation between energy and velocity:

$$\varepsilon = m \cdot v^2/2 \tag{3.16a}$$

$$\varepsilon_x = m \cdot (v_x)^2/2 \tag{3.16b}$$

where ε represents particle total kinetic energy, and ε_x is the kinetic energy corresponding only to particle motion along Ox axis. Therefore, this ε corresponds to a three degrees of freedom translational motion while ε_x is a single degree of freedom motion.

A transition can be made from velocities distribution functions to energies distribution functions using the v to ε change of variables: $\Phi_\varepsilon = \Phi_v \cdot \frac{D(v)}{D(\varepsilon)}$. In order to change only one variable, the Jacobian is reduced to the derivative: $D(v)/D(\varepsilon) = d(v)/d(\varepsilon)$, thus: $\Phi_v = \Phi_\varepsilon \cdot (d\varepsilon/dv) - 1$.

But from (3.16a) it results that: $d\varepsilon/dv = m\cdot v$ and according to relation (3.8):

$$\Phi_V = (2/\pi)^{1/2} \cdot (k_B \cdot T/m)^{-3/2} \cdot v^2 \cdot e^{-m\cdot v^2/(2\cdot k_B\cdot T)},$$

so that:

$$\Phi_\varepsilon = (2 \cdot m/\pi)^{1/2} \cdot (k_B \cdot T)^{-3/2} \cdot v \cdot e^{-m\cdot v^2/(2\cdot k_B\cdot T)}.$$

By replacing $v = \sqrt{2 \cdot \varepsilon/m}$ according to relation (3.18a), the following expression is obtained:

$$\Phi_\varepsilon = 2 \cdot (k_B \cdot T)^{-3/2} \cdot (\varepsilon/\pi)^{1/2} \cdot e^{-\varepsilon/(k_B\cdot T)} \tag{3.17a}$$

Relation (3.17a) represents particles energy distribution for their three-dimensional space motion (a space with three degrees of freedom).

Similarly, for ε_x energies distribution (over a single direction), relation (3.16b), and particles distribution (3.6) over velocities projection on Ox axis are used: $\Phi_{\varepsilon_x} = \Phi_{v_x} \cdot (d\varepsilon_x/dv_x)^{-1}$. By deriving equality (3.16b) it results that: $d\varepsilon_x/dv_x = m\cdot v_x$ and relation (3.6) can be put under the following form:

$$\Phi_{V_x} = (2 \cdot \pi \cdot k_B \cdot T/m)^{-1/2} \cdot e^{-m\cdot (v_x)^2/(2\cdot k_B\cdot T)},$$

from where:

$$\Phi_{\varepsilon_x} = (2 \cdot \pi \cdot m \cdot k_B \cdot T)^{-1/2} \cdot (1/v_x) \cdot e^{-m\cdot (v_x)^2/(2\cdot k_B\cdot T)}.$$

But $v_x = \sqrt{2 \cdot \varepsilon_x/m}$ and thus:

$$\Phi_{\varepsilon_x} = 0.5 \cdot (\pi \cdot k_B \cdot T \cdot \varepsilon_x)^{-1/2} \cdot e^{-\varepsilon_x/(k_B\cdot T)} \tag{3.17b}$$

Degrees of Freedom for Molecules Energy Distribution

Both energy distributions from relations (3.17a) and (3.17b) contain a Boltzmann factor, $e^{-\varepsilon/(k_B\cdot T)}$ and $e^{-\varepsilon_x/(k_B\cdot T)}$. However, they differ by the pre-exponential factor value: both numerical coefficient and power to which the factor ($k_B\cdot T$) is raised or the power to which ε is raised are different. It can be, however, shown that, for a certain number, f, of degrees of freedom, the energies distribution function is:

$$\Phi_{f,\varepsilon} = \left(\frac{2 \cdot \varepsilon}{k_B \cdot T}\right)^{f/2} \cdot \left(\frac{2}{\pi}\right)^{(mod_2 f)/2} \cdot e^{-\varepsilon/(k_B\cdot T)}/[2 \cdot \varepsilon \cdot (f-2)!!] \tag{3.18a}$$

where *mod*$_2$f is the rest of f integer division by 2, while (f – 2)! represents the product of odd integer numbers that are lower or equal to (f – 2). From relation (3.18a), relation (3.17a) is obtained using f = 3, and relation (3.17b) with f = 1.

The simpler form of energies distribution function is in the case of 2 degrees of freedom motions:

$$\Phi_{2,\varepsilon} = e^{-\varepsilon/k_B \cdot T}/(k_B \cdot T) \tag{3.18b}$$

Due to its simplicity, relation (3.18b) is used for calculating the number of N* molecules having the energy higher than a given ε* value. These "hot" molecules are the molecules capable of reacting chemically.

Making the transition from differential distribution function to the integral one, the following expression is obtained: $F_{2,\varepsilon*} = \int_{\varepsilon=0}^{\varepsilon=\varepsilon*} \Phi_{2,\varepsilon} \cdot d\varepsilon = 1 - e^{-\varepsilon/k_B \cdot T}$. Since $F_{2,\varepsilon*}$ represents the probability that a molecule has $\varepsilon < \varepsilon*$, the probability that the molecule has $\varepsilon > \varepsilon*$ is obtained by subtracting $F_{2,\varepsilon*}$ from 1:

$$Prob(\varepsilon > \varepsilon*) = 1 - Prob(\varepsilon < \varepsilon*) = 1 - F_{2,\varepsilon*} = exp\left[-\varepsilon * /(k_B \cdot T)\right]$$

or, by passing to molar activation energy $E* = N_A \cdot \varepsilon$: $Prob(E > E*) = e^{-E*/(R \cdot T)}$.

If the total number of molecules is N, the number of activated molecules, N*, will therefore be $N* = N \cdot Prob(E > E*)$ or:

$$N* = N^{-E*/(R \cdot T)} \tag{3.19}$$

Mean Energies

Translational energy arithmetic mean, $\overline{\varepsilon}_f$, is calculated using relation (3.11b) which becomes, when Y = ε: $\overline{\varepsilon}_f = \int_{\varepsilon=0}^{\varepsilon=\infty} \varepsilon \cdot \Phi_{f,\varepsilon} \cdot d\varepsilon$, where $\Phi_{f,\varepsilon}$ is energies distribution function for f degrees of freedom.

For f = 3 and from relation (3.17a) it is obtained that: $\overline{\varepsilon}_3 = \int\limits_0^\infty \varepsilon \cdot \frac{2}{\sqrt{\pi}} \cdot \frac{1}{(k_B \cdot T)^{3/2}} \cdot \sqrt{\varepsilon} \cdot e^{-\varepsilon/k_B \cdot T} \cdot d\varepsilon$, from where the mean is: $\overline{\varepsilon}_3 = 1.5 \cdot k_B \cdot T$.

Similarly, for $\overline{\varepsilon}_1$ mean results the following expression, with φ_{ε_x} from relation (3.17b): $\overline{\varepsilon}_1 = 0.5 \cdot k_B \cdot T$.

Generally, for f degrees of freedom it is found that:

$$\overline{\varepsilon}_f = (f/2) \cdot k_B \cdot T \tag{3.20}$$

from relation (3.18a), and for one mole of gas it is found that: $\overline{E}_f = N_A \cdot \overline{\varepsilon}_f = (f/2) \cdot R \cdot T$ namely to each degree of freedom corresponds a molar energy (R·T/2), according to equipartition of energy on degrees of freedom theorem from 2.9.

3.5 Wall Collision of Gaseous Molecules

Molecular Number Density

The *molecular number density* μ (differs from *density* ρ itself, which is the mass of the unit of volume) is the ratio between number of molecules N and volume v where they are contained:

$$\mu \cong N/v \qquad (3.21a)$$

by replacing N with the product $n \cdot N_A$ of property n (number of moles of substance) and Avogadro number N_A, while volume v is expressed by the ideal gas law:

$v = n \cdot R \cdot T/P$, and therefore the following expression is obtained:

$$\mu = N_A \cdot P/(R \cdot T) \qquad (3.21b)$$

Molecular number density depends thus on the gas state thermo-mechanical macroscopic quantities—pressure and temperature.

For a multi-component gas, relation (3.21b) offers the total molecular number density—that is, the sum of molecular number densities μ_i of all gas species, calculable by:

$$\mu_i = N_A \cdot P \cdot X_i/(R \cdot T) \qquad (3.21c)$$

where X_i is the molar fraction of gaseous mixture i component.

Wall Collisions Frequency

Wall surface s will be collided within a second by all cross section layer molecules, section equal to surface s and of thickness equal to $\bar{u}$, where $\bar{u}$ is the arithmetic mean of velocity projections on the Ox axis for molecules with positive velocity projection: these molecules are approaching the wall, while molecules with $v_x < 0$ are getting away from the wall and cannot therefore be considered for collisions calculation. This layer's volume will thus be $s \cdot \bar{u}$, while the number of molecules collided to the wall in the time unit is: $N_1 = \mu \cdot s \cdot \bar{u}$, where μ is the molecular number density.

The collision density Z_s, defined as the number of gas molecules collided on wall unit of area within the time unit, is computed starting from the distribution of the velocities projected on the Ox axis, which is perpendicular to the wall. The collision density is obtained by dividing the collision number by the area, $Z_s = N_1/s$, and is thus given by:

$$Z_S = \mu \cdot \bar{u} \tag{3.5a2}$$

Calculation of $\bar{u}$ is to be made according to arithmetic mean expression (3.11b), where $Y' = 0$ and $Y'' = \infty$: $\bar{u} = \int_{x=0}^{x=\infty} v_x \cdot \Phi_{v_x} \cdot d\,v_x$. If the inferior limit would be $Y' = -\infty$ as in relation (3.11c), value $\bar{v}_x = 0$ would be obtained as a result of averaging process.

By replacing into relation (3.5a2) the $\Phi_{v,x}$ from relation (3.6), the following results:

$$\bar{u} = \int_{v_x=0}^{v_x=\infty} \sqrt{m/(2 \cdot \pi \cdot k_B \cdot T)} \cdot e^{-m \cdot (v_x)^2/(2 \cdot k_B \cdot T)} \cdot v_x \cdot d\,v_x$$

where, with $k = 0$ and $\alpha = m/(2 \cdot k_B \cdot T)$, relation (A.2c) is used, $\Psi_{2 \cdot k+1}(\alpha) \equiv \int_0^\infty x^{2 \cdot k+1} \cdot e^{-\alpha \cdot x^2} \cdot dx = k!/(2 \cdot \alpha^{k+1})$, which leads to:

$$\bar{u} = \sqrt{k_B \cdot T/(2 \cdot \pi \cdot m)} \tag{3.22a}$$

Thus, $\bar{u}$ is 4 times lower than the arithmetic mean velocity $\bar{v}$ from relation (3.12b). By replacing into relation (a) μ and $\bar{u}$ expressions from (3.21a) and (3.22a), respectively, it results that:

$$Z_s = \sqrt{k_B \cdot T/(2 \cdot \pi \cdot m)} \cdot P \cdot N_A/(R \cdot T)$$

or, with $k_B/m = R/M$:

$$Z_s = P \cdot N_A/\sqrt{2 \cdot \pi \cdot M \cdot R \cdot T} \tag{3.22b}$$

Molecule-Wall Collisions in Physics and Chemistry

Vapor Condensation Rate at Liquid Free Surface The number of molecules in vapor phase that condensate in the unit of time is given by the following expression:

$$dN/dt = Z_s \cdot s \cdot \alpha \tag{3.22c}$$

The *accommodation* coefficient α is subunitary and not all molecules reaching the surface condensate, some of them reflecting back into the gaseous phase. If vapors and liquid are in equilibrium, liquid vaporization rate will be equal to vapors condensation rate, and expression (3.22c) is also valid for vaporization rate case.

Corrosion Rate Under Gas Conditions Z_s represents the superior limit of metal corrosion rate by an aggressive gas, like O_2 or F_2. The corrosion rate is expressed by dividing the corroded metal mass to the time unit and the unit of area.

If every gas molecules collision with metal surface removes one metal atom, the corrosion rate will be $v_{cor} = Z_s \cdot m_{at}$, where m_{at} is the metal atom mass.

The Z_s property is used also for calculating gases maximum reaction rate on a solid catalyst surface, for *effusion* phenomenon (gas leakage through holes) studying purposes, etc.

3.6 Intermolecular Collisions within Gases

Simplifying hypotheses are used for calculating the number of collisions between gas molecules. They are the same hypotheses that underline the kinetic molecular theory of gases, namely: the molecule is a rigid and perfectly elastic sphere, and the repulsion forces between molecules occur only during collision.

Two molecules' (denoted by a and b) collision takes place when their centers are δ_{ab} distance one from another, δ_{ab} being the colliding molecules diameters (δ_a and δ_b) mean:

$$\delta_{ab} = (\delta_a + \delta_b)/2 \tag{3.23a}$$

The molecular diameter δ_{ab} from relation (3.21a) represents the *collision diameter*, obtained through the *macroscopic* methods detailed in 3.7. These methods apply approximation models which start from kinetic or thermodynamic measurements. Molecule real dimensions (which actually is not spheric and not even has well defined spatial limits) are determined through *microscopic* procedures of structural physics like X-ray diffraction or electron diffraction.

In order to simplify the calculation procedure, Maxwell considered that molecules position are fixed, excepting one molecule, which moves at a $\overline{v_r}$ velocity, that is, the mean relative molecular velocity of one molecule to the other. Collisions frequency between the mobile molecule a and the immobile molecules b is denoted by $z_{a,b}$, and given by:

$$z_{a,b} = \mu_b \cdot v_{cc} \tag{3.23b}$$

which is a product between μ_b (collided particles molecular density) on the one hand, and v_{cc} volume, on the other hand. Collided particles' molecular number density is defined through relation (3.21a) and represents the number of collided molecules from the unit of volume. Volume v_{cc} is equal to:

$$V_{cc} = \pi \cdot \delta^2 \cdot \overline{v_r} \tag{3.23c}$$

for the so-called cylinder of collision—the right cylinder of radius δ and length $\overline{v_r}$, and having the mobile molecule center moving on its axis. The centers of all collided fixed molecules are found within.

Identical Type Molecules Collisions

When two molecules of same type collide—furtherly referred to by index 1—their kinetic diameters have the same value, denoted by symbol δ without indices. By introducing $\delta_a = \delta_b = \delta$ into relation (3.23a), a *maximum collision distance* δ equal to molecules' kinetic diameter is obtained for such collisions.

Therefore, the collision cylinder diameter in this case is 2·δ (therefore a radius equal to δ) and length is $\overline{v_r}$. Intermolecular collisions number in the unit of time, denoted here $z_{1,1}$, will be obtained from relation (3.23a) by customizing μ_b and δ with μ_1, respectively δ_1:

$$z_{1,1} = \pi \cdot (\delta_1)^2 \cdot \mu_1 \cdot \overline{v_r} \tag{3.24}$$

The previous reasoning considered the same motion direction for all molecules. Actually, molecules' motion directions are random, with the possibility of comprising the angles between them in the 0°–180° range.

Relative velocity v_r module mean, $\overline{v_r}$, is equal to velocity projection module mean $\overline{v_1}$ through the expression (proven in Example 3.3)

$$\overline{v_r} = \overline{v_1} \cdot \sqrt{2} \tag{3.25a}$$

After introducing average relative velocity expression (3.25a) into relation (3.22b), collisions frequency $z_{1,1}$ of type 1 molecules is obtained:

$$z_{1,1} = \sqrt{2} \cdot \pi \cdot (\delta_1)^2 \cdot \overline{v_1} \cdot \mu_1 \tag{3.25b}$$

Different Type Molecules Collisions

For chemical kinetics calculation purposes, collisions frequency of different type molecules $z_{1,2}$ is also of interest.

For the two molecule species gaseous mixture case, marked by indices 1 and 2, molecular diameters δ_a and δ_b from relation (3.23a) become respectively δ_1 and δ_2, so that collision diameter becomes:

$$\delta = (\delta_1 + \delta_2)/2 \tag{3.26a}$$

By introducing δ into (3.23c) formula, the collision cylinder volume becomes:

$$v_{cc} = (\pi/4) \cdot (\delta_1 + \delta_2)^2 \cdot \overline{v_r} \tag{3.26b}$$

It can be proven that mean relative velocity $\overline{v_r}$ is linked to $\overline{v_1}$ and $\overline{v_2}$ mean velocities of the two molecule types, through relation:

$$\overline{v_r} = \sqrt{(\overline{v_1})^2 + (\overline{v_2})^2} \tag{3.27a}$$

Relation (3.27a) customization for identical type molecule collisions, denoted by 1, leads to expression (3.25a).

Elimination of $\overline{v_r}$ between relations (3.26b, 3.27a) leads to:

$$v_{cc} = (\pi/4) \cdot (\delta_1 + \delta_2)^2 \cdot \sqrt{(\overline{v_1})^2 + (\overline{v_2})^2} \tag{3.27b}$$

and collisions frequency $z_{1,2}$ of species 1 molecule (considered mobile) and species 2 molecules (immobile) will be, according to relation (3.23b): $z_{a,b} = \mu_b \cdot v_{cc}$, where v_{cc} is given by formula (3.26b) and index b of molecular number density μ_b for the immobile type of molecules is replaced by index 2 of their species, thus:

$$z_{1,2} = (\pi/4) \cdot (\delta_1 + \delta_2)^2 \cdot \mu_2 \cdot \sqrt{(\overline{v_1})^2 + (\overline{v_2})^2} \tag{3.27c}$$

The expression "*density of intermolecular collisions*" designates the total number of these collisions from the unit of volume and time.

Density of Intermolecular Collisions in Pure Gases

Intermolecular collisions density $Z_{1,1}$ of identical type molecules (with index 1) will be equal to the product between collisions frequency $z_{1,1}$ of one molecule and *half* of molecules density μ_1 of species 1 molecule (for not considering *twice* each collision):

$$Z_{1,1} = z_{1,1} \cdot \mu_1/2 \tag{3.28a}$$

where $z_{1,1}$ is given by relation (3.25b), where the arithmetic mean of the molecular velocity becomes $\overline{v}_1 = \sqrt{8 \cdot k_B \cdot T/(\pi \cdot m_1)}$, according to expression (3.12b) customized for type "1″ of molecules. It results that:

$$Z_{1,1} = 2 \cdot (\delta_1 \cdot \mu_1)^2 \cdot \sqrt{\pi \cdot k_B \cdot T/m_1} \tag{3.28b}$$

Density of Intermolecular Collisions in Multicomposant Gases

Collisions density between molecule species with indexes 1 and $j \neq 1$ is:

$$\forall j \neq 1 : Z_{1,j} = z_{1,j} \cdot \mu_j \tag{3.28c}$$

where, unlike relation (3.28a), molecules belong to different species, ½ factor no longer appears and therefore a collision is taken into account only once.

By expressing $z_{1,j}$ from relation (3.28c) through expression (3.27c) and taking into account relation (3.12b) for the mean velocity $\overline{v}$, the following relation is obtained:

$$Z_{1,j} = \mu_1 \cdot \mu_j \cdot (\delta_1 + \delta_j)^2 \cdot \sqrt{\pi \cdot k_B \cdot T \cdot (1/m_1 + 1/m_j)/2} \tag{3.28d}$$

For the general case, the gas contains C molecular species, having indices $i \in [1, 2, \ldots, C]$. Collision density Z_i of all i type molecules is:

$$Z_i = \sum\nolimits_{j=1}^{j=C} Z_{i,j} \tag{3.28e}$$

with $Z_{i,j}$—collisions density between type i molecule and type j one, and therefore z_i—the frequency of all collisions suffered by a molecule of i species, will be:

$$z_i = \sum\nolimits_{j=1}^{j=C} z_{i,j} \tag{3.29a}$$

where $z_{i,j}$ represent collisions frequency of an i molecule with a j molecule. When replacing index 2 by j relation (3.27c) becomes:

$$z_{i,j} = (\pi/4) \cdot (\delta_i + \delta_j)^2 \cdot \mu_j \cdot \sqrt{(\overline{v}_1)^2 + (\overline{v}_2)^2}, \forall j$$

By expressing the mean velocities through type (3.12b) relations, relation (3.29a) becomes:

$$z_i = \sqrt{\pi \cdot k_B \cdot T/2} \cdot \sum_{j=1}^{j=C} \left[\mu_j \cdot (\delta_i + \delta_j)^2 \cdot \sqrt{(1/m_i + 1/m_j)} \right] \qquad (3.29b)$$

where μ_j, the molecular density of j species, is given by relation (3.21c).

Macroscopic Factors Effect on Collisions

For a typical gas, with $M = 0.03$ kg/mol and $\delta = 3.5 \cdot 10^{-10}$ m, found in temperature and pressure standard conditions (1 atm and 300 K), collisions frequency $z_{1,1}$ is 10^9–10^{10} s^{-1} (the mean time range between two successive collisions is 10^{-9} s or 10^{-10} s) and collision density $Z_{1,1}$ is 10^{34}–10^{35} $m^{-3} \cdot s^{-1}$.

Indeed, relations (3.12b) and (3.22b), respectively, are leading here to $\bar{v} \cong 460$ m/s and $\mu \cong 2.4 \cdot 10^{25}$ molecules/m^3, and from relations (3.24) and (3.27a) the following values are obtained: $z_{1,1} \cong 6.1 \cdot 10^9$ s^{-1} and $Z_{1,1} \cong 7.5 \cdot 10^{34}$ m^{-3} s^{-1}.

Temperature and pressure influence over collision density is found by expressing the collision density from the relation (3.27b) depending on P and T:

$$Z_{1,1} \sim P^2/T^{3/2} \qquad (3.30)$$

This means that collision number increases for *isobaric* temperature decrease case and for *isothermal* pressure increase.

If volume is considered as a physical quantity, when replacing within expression (3.30) the T/P ratio with V/R one, it results that:

$$Z_{1,1} \sim T^{1/2}/V^2 \text{ and } Z_{1,1} \sim P^{1/2}/V^{3/2}.$$

Therefore, collision density will increase in the same direction with the *isochoric* temperature or pressure $Z_{1,1} \sim T^{0.5}$ and $Z_{1,1} \sim P^{0.5}$.

Molecule collisions represent a necessary (although insufficient) condition for the occurrence of chemical reactions in gaseous phase, as it is highlighted by the estimation of the rate constant from the collision density, in the theory of active collisions from chemical kinetics.

3.7 Molecular Diameters

Unlike mean molecular velocity and molecular number density, readily obtained from the *macroscopic* characteristics (temperature, pressure, molar mass), the mean free path depends on a *structural* characteristic: the *molecular* diameter; denoted either by σ (real diameter, independent of pressure, or temperature) either by δ (apparent diameter).

Molecular Diameter Evaluation Methods

Three methods can be used for the molecular diameter calculation, namely:

(a) Phenomenological transport coefficients method: for gases case, these coefficients depend on the mean free path from 3.8 and are relatively easy to determine, especially from viscosity measurements using relation (7.13c).
(b) Kinetic–molecular interpretation method for thermal equation of state (ES) (4.1) constants is determined especially by P, V, T compressibility measurements of noble gases at relatively high pressures.
(c) Critical volume method, using like the method at point (b), several thermal ES constants molecular significance is deducted from critical volume V_c (5.2) measurement or evaluation. By admitting Van der Waals ES (4.6) validity and considering that b (Van der Waals gas molar volume value at infinite pressure) corresponds to molecules arrangement into a cubic lattice without inter spaces, it follows that:

$$b = N_A \cdot \sigma^3 \tag{3.31a}$$

But constant b is linked to the critical volume through b = 3—relation (5.10)—from where molecule *apparent* diameter $\delta = [V_c/(3 \cdot N_A)]^{1/3}$.

A better match of results to experiments is obtained when considering that the actual placement is 4% more compact than the cubic one, by modifying dependence $\delta = f(V_c)$ to:

$$\delta = [V_c/(2.87 \cdot N_A)]^{1/3} \tag{3.31b}$$

In the simplifying hypothesis of molecules interacting only at the moment of contact, the molecular diameter is denoted by δ and is qualified as *kinetic* diameter or *apparent* diameter, due to its somewhat dependency on physical conditions—especially for temperature.

Diameter notation by $\sigma \neq F$ (P,T) is reserved for the equilibrium intermolecular distance of the Lennard–Jones potential (4.7).

Molecular Diameter Dependence on Temperature

Temperature increase, leading to molecules kinetic energy increase, narrows the distance between molecules at collision. Kinetic diameter dependence of temperature is given approximately by Sutherland's law:

$$\delta = \delta_0 \cdot \sqrt{1 + C_S/T} \tag{3.32a}$$

where δ_0 is the molecular diameter inferior limit, corresponding to an infinite temperature, while C_S is *Sutherland* constant, obtained either from gases viscosity measurements through relation (7.13c), either from critical or boiling temperatures (T_c or T_f), through one of the following relations:

$$C_S = 0.8 \cdot T_c; C_S = 1.47 \cdot T_f \tag{3.32b, c}$$

Table 3.2 presents the *apparent* molecular diameter of certain gases, obtained at 20 °C through the three above-discussed methods—columns "visc.," "compr.," and "crit."—together with the mean free path $\bar{l}$ (calculated as in 3.8) at 20 °C and 1 atm.

Methods a, b, and c for δ molecular diameter evaluation are leading to relatively concordant values, differing with at most 10%.

It is found that molecular diameters of most gaseous substances have close values: 2–5 Å (1 Å = 10^{-10} m); with molecular mass increase, the molecular diameter will be also increasing, but much less.

Table 3.2 Apparent kinetic diameters and mean free path

	M, g/mol	10^{10} δ, m				
Substance		visc.	compr.	crit.	$10^7 \cdot \bar{l}$, m	C_S, K
H_2	2.01	2.92	2.87	3.34	1.06	89
D_2	4.03	2.95	2.87		1.04	
Ne	20.1	2.86	2.78	2.90	1.10	56
Cl_2	70.9	4.11	4.15		0.53	330
N_2	28.0	3.68	3.70	3.74	0.67	104
CO	28.0	3.59	3.76	3.75	0.70	101
CO_2	44.0	4.00	4.49	3.82	0.56	254
C_2H_6	30.0	4.42	3.95	4.37	0.46	
n-C_4H_{10}*	58.0	5.00	4.97		0.28	
Hg*	200.6	2.90			1.07	

Liquid substances at 20 °C and 1 atm; $\bar{l}$ is calculated from the one of saturated vapors at 20 °C by multiplication with vapor pressure, expressed in atm.

3.8 Mean Free Path

A molecule traveled distance mean $\bar{l}$ between two consecutive collisions is denoted as the *mean free path* or "mean free route" of the respective molecule. In usual conditions, its value in gases is of 10^{-7} m order.

Mean free path calculation is made by dividing mean molecule velocity with z, its collisions density with other gas molecules. Thus, in the case of a pure substance gas:

$$\bar{l}_1 = \bar{v}_1/z_{1,1} \tag{3.33a}$$

from where it results, after introducing z_{11} from relation (3.25b) and by simplifying the velocity and by eliminating index 1 for molecule type:

$$\bar{l} = 1/\left(\sqrt{2} \cdot \pi \cdot \delta^2 \cdot \mu\right) \tag{3.33b}$$

or, with μ from formula (3.22b):

$$\bar{l} = R \cdot T/\left(\sqrt{2} \cdot \pi \cdot N_A \cdot P \cdot \delta^2\right) \tag{3.33c}$$

From relation (3.33c), which is valid for temperatures and pressures of not very low value, it results that $\bar{l}$ increases at constant temperature for isothermal pressure decrease $\left(\bar{l} \sim 1/P\right)$ and for isobaric temperature increase $\left(\bar{l} \sim T\right)$.

Free Path Dependence on Temperature

A more precise dependence $\bar{l}(T)$ is obtained from Sutherland's law (3.32a) for molecular diameters case. Relation (3.33c) is modified to:

$$\bar{l} = (T/)^2/\left[\sqrt{2} \cdot \pi \cdot N_A \cdot (\delta_0)^2 \cdot P \cdot (T + C_S)\right] \tag{3.33d}$$

where $\bar{l}$ increases more readily than T at low temperatures relative to C_s, and only at elevated temperatures ($T \gg C_S$) it results again $\bar{l} \sim T$.

For a gas mixture case (having indexes j = 1, 2, 3, ..., C of mixture components), relation (3.33a) must be replaced by:

$$\bar{l}_i = \bar{v}_i/z_i \tag{3.34a}$$

where $\bar{l}_i$ is the mean free path of a type i molecule, while z_i—frequency of a molecule collisions with any other molecule (identical or of different type). With $\bar{v}_i$ and z_i given by relations (3.12b) and (3.29b), respectively, it results that:

$$\bar{l}_i = 4/\left[\pi \cdot \sum_{j=1}^{j=C} \mu_j \cdot (\delta_i + \delta_j)^2 \cdot \sqrt{1 + m_i/m_j}\right]$$

With molecular densities μ_j from relation (3.21c) and by expressing molecular mass m through molar mass ($m_i = M_i/N_A$, $m_j = M_j/N_A$) the following is obtained:

$$\bar{l}_i = 4 \cdot R \cdot T/\left[\pi \cdot N_A \cdot P \cdot \sum_{j=1}^{j=C} X_j \cdot (\delta_i + \delta_j)^2 \cdot \sqrt{1 + m_i/m_j}\right] \tag{3.34b}$$

from where—with $C = 1$, $X_j = 1$, $\sigma_j = \sigma$, $M_j = M$— relation (3.33b) is found.

It can be observed from expression (3.34b) that the mean free path is directly proportional to temperature T and inversely proportional to pressure P. By replacing R·T/P with n/v (n is the quantity of substance, expressed in moles) it results:

$$\bar{l}_i = 4 \cdot V/\left[\pi \cdot N \cdot P \cdot \sum_{j=1}^{j=C} X_j \cdot (\delta_i + \delta_j)^2 \cdot \sqrt{1 + m_i/m_j}\right] \tag{3.34c}$$

where N is the total number of gas molecules, $N = N_A \cdot n$. Therefore, at constant volume v of a closed system without chemical reactions, the mean free path does not depend on thermo-mechanic physical quantities (P, T) state but only on composition—the set of molar fractions X_j—and on molecules characteristics (m, δ).

In the *Knudsen* domain—a field of very low pressures (under 10^{-4} torr), molecule collisions are so rare that one molecule can cross several times the distance between vessel walls without colliding other molecules, but only being reflected at each collision with the wall.

Free Path in Knudsen Regime

Within Knudsen domain of pressures, the *Knudsen* mean free path, $\bar{l}_K$, is valid, which is proportional to d_{car} characteristic dimension of the gas containing space. By denoting with f_K this proportionality *Knudsen coefficient* the following expression is obtained:

$$\bar{l}_K = f_K \cdot d_{car} \tag{3.8a2}$$

The characteristic dimension is the *main* dimensions (d_1, d_2, d_3) harmonic mean value, where $d_1 \leq d_1 \leq d_2 \leq d_3$, of the gas containing space:

$$d_{car} \equiv 3/(1/d_1 + 1/d_2 + 1/d_3) \tag{3.35a}$$

The *main* dimensions represent the distances between the walls bounding this space in three perpendicular directions. Therefore, for a gas *layer*, of thickness d_1 much lower than its width d_2 and its length d_3 the following is obtained:

$$d_1 \ll min\,(d_2; d_3) \Rightarrow d_{car} = 3 \cdot d_1 \tag{3.35b}$$

For a tubular space (a capillary), if the tube is assimilated to a cylinder of length L_{cil} and diameter d_{cil}, a length independent characteristic dimension is obtained, which has its diameter equal to the one of the capillaries:

$$d_1 = d_2 = d_{cil} \ll L_{cil} \Rightarrow d_{car} = 1.5 \cdot d_{cil} \tag{3.35c}$$

Knudsen factor f_K from relation (3.8a2) is known with an incertitude that can reach 20%, because it depends not only on the gas containing space form, but also on walls elasticity degree. Its approximate value is obtained from surface collisions number, z_s, given by formula (3.17a) which, taking into account relation (3.12b), can be put under the following form: $Z_S = \mu \cdot \overline{v}/4$. For a vessel of volume v_v and surface s_v, these collisions frequency $z' = z_s \cdot s_v$ therefore become: $z' = \mu \cdot \overline{v}/(4 \cdot s_v)$, while vessel total molecules number is $N' = \mu \cdot v_v$, so that wall molecule collisions frequency $\nu_p = z'/N'$ becomes: $\nu_p = \mu \cdot \overline{v}/(4 \cdot v_v)$. As the mean free path in Knudsen conditions is $\overline{l}_K = \overline{v}/\nu_p$, it results: $\overline{l}_K = 4 \cdot v_v/s_v$ so that, with $d_{car} = 6 \cdot s_v/v_v$ (strictly valid for a cube of d_{car} facet, or a sphere of d_{car} diameter) it results that $f_K = 2/3$, while relation (3.8a2) becomes:

$$\overline{l}_K \cong (2/3) \cdot d_{car} \tag{3.35d}$$

Free Path in Intermediate Pressures Domain

In the general case, the mean free path can be expressed as the weighted harmonic mean of its values for normal pressures $\overline{l}_n$ given by relations (3.33a,3.33b,3.33c,3.33d)—and for Knudsen domain $\overline{l}_K$ calculated from relations (3.35a,3.35b,3.35c,3.35d):

$$\overline{l} = (w_n + w_K)/\left(w_n/\overline{l}_n + w_K/\overline{l}_K\right) \tag{3.36a}$$

where the mediation shares w_n and w_K are inversions of phenomenological *transport coefficients* K_g of the kinetic–molecular treatment of transport from 7.2, g being the specific index of the transported property (η, λ, D for momentum, energy, and mass transport, respectively):

$$\forall g \in (\eta, \lambda, D) \rightarrow w_n = 1/K_{g,n} \leftrightarrow w_K = 1/K_{g,K} \tag{3.36b}$$

3.9 Triple Collisions

Triple collisions are collisions between three molecules, here supposed to be of the same type. *Density of triple collisions*—number Z_{tr} of such collisions in the unit of time and of volume—can be assessed starting from density μ_{db} of two neighboring gas molecules group.

Indeed, triple collisions can be conceived as meetings of a simple molecule with such a "double molecule," which is rather a *pseudomolecule*, because the chemical interactions between the two components of the double molecule are negligible. Considering that a mobile molecule A passing through the distance $\bar{l}$ is in "double" state over a portion equal to δ, since on the $\bar{l}$ route meets on average only one molecule, it results that molecule A is found in the state of double molecule a fraction of $\delta/\bar{l}$ from its existence, as in Fig. 3.6, where: A designates a simple mobile molecule, and B designates the double pseudomolecules, fixed.

It is deduced that the ratio between the density μ_{db} of the double molecules and μ_1 density of the simple molecules is equal to the one between the δ_1 diameter of the simple molecule and its mean free path $\bar{l}_1$.

Thus:

$$\mu_{db} = \mu_1 \cdot \delta_1/\bar{l}_1 \tag{3.37a}$$

In most collisions by a simple molecule, the collided particle is another *simple* (and not a double) molecule, the $\bar{l}$ value will be the one given by relation (3.33b) for simple identical molecules of type 1, $\bar{l}_1 = 1/\left[\sqrt{2}\cdot\pi\cdot(\delta_1)^2\cdot\mu_1\right]$, so that expression (3.37a) becomes:

$$\mu_{db} = 2^{1/2}\cdot\pi\cdot(\delta_1)^3\cdot(\mu_1)^2 \tag{3.8a3}$$

Density $Z_{1,db}$ of collisions between a simple molecule and a double one is, according to relation (3.28d), where index "2″ is replaced by index "db":

$$Z_{1,db} = \mu_1\cdot\mu_{db}\cdot(\delta_1+\delta_{db})^2\cdot\sqrt{\pi\cdot k_B\cdot T\cdot(1/m_1+1/m_{db})/2}$$

But $m_{db}/m_1 = 2$ iar $\delta_{db} = 2^{1/3}\cdot\delta_1$, because molecule's volume is proportional both with its mass and the cub of diameter. It therefore results:

$$Z_{1,db} = \mu_1\cdot\mu_{db}\cdot(\delta_1)^2\cdot\sqrt{\pi\cdot k_B\cdot T\cdot(1/m_1)/2}\cdot\left(1+2^{1/3}\right)^2\cdot 3^{1/2}/8 \tag{3.8b3}$$

Fig. 3.6 Triple collision

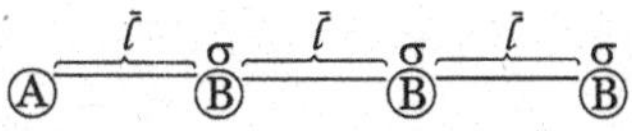

Relative Frequency of Double and Triple Collisions

Because the density of "normal" collisions—the ones between two simple molecules—is $Z_{1,1} = 2 \cdot \delta_1^2 \cdot \mu_1^2 \cdot \sqrt{\pi \cdot k_B \cdot T/m_1}$ from relation (3.28b), it results that R_T ratio between triple collisions density and "normal" collisions density:

$$R_T \equiv Z_{1,db}/Z_{1,1} \tag{3.37b}$$

becomes, considering relation (3.8b3): $R_T = (\mu_{db}/\mu_1) \cdot (1 + 2^{1/3})^2 \cdot 3^{1/2}/16$ or, with μ_d given by formula (3.8a3): $R_T = \sqrt{6} \cdot \pi \cdot \left[(1 + \sqrt[3]{2})/4\right]^2 (\sigma_1)^3 \cdot \mu_1$ where it can be replaced $\mu_1 = N_A \cdot P/(R \cdot T)$ from relation (3.21b). Therefore:

$$R_T = 2.456 \cdot (\delta_1)^3 \cdot N_A \cdot P/(R \cdot T) \tag{3.37c}$$

which is equivalent to $R_T = 4.69\, v_{mol}/v$ where $v_{mol} = N \cdot (\pi \cdot \delta^3/6)$ is all gas molecules eigenvolume; N is the total molecules number, and $\pi \cdot \delta^3/6$ is the apparent eigenvolume—that is, the volume of a δ diameter sphere—of one molecule (simple, of type 1). $R_T \ll 1$ in usual conditions, but pressure increase leads to R_T ratio increase, between two collision types: for an average dimensions molecule at 25 ° C and 220 atm: $R_T = 1/9$ —one collision of 10 is triple, but at such an elevated compression degree that the ideal gas model can no longer be considered valid.

3.10 Eleven Worked Examples

Example 3.1
The differential distribution function, Γ_x, of x property is: $8x + 6/x^2$ at $x \in [1,2]$ and 0 for $x \notin [1,2]$. The distribution functions for x are required:

(1) the integral one (non-normalized), G_x;
(2) the differential normalized one, Φ_x and
(3) normalized integrated distribution function, F_x.

Solution
For J(x) being the indefinite integral $J = \int_{x_{min}}^{x} \Gamma_x \cdot dx$. With Γ_x (x) from the example statement:

(a) $J_a = 0$ for $x \le x_{min} = 1$
(b) $J_b(x) = \int_1^x (8 \cdot x + 6 \cdot x^{-2}) \cdot dx == 4 \cdot x^2 + 2 - 6/x$ for $x \in [1, 2]$
(c) $J_c = \int_{x_{min}}^{x_{max}} \Gamma_x \cdot dx = J_b(x = x_{max} = 2) = 4 \cdot 2^2 + 2 - 6/2 = 15$ when $x \ge 2$

Particles total number is N = $\lim_{x \to \infty} J$, therefore N = 15.
From definitions (3.1b, 3.2a, 3.2b) of G, Φ, F distribution functions, it results:

1. $G_X = J$ or : $G_x = \begin{cases} 0 & x \le 1 \\ 4 \cdot x^2 + 2 - 6/x, & 1 \le x \le 2 \\ 15 & x \ge 2 \end{cases}$
2. $\Phi_X = \Gamma_x/N = \begin{cases} 0 & x \notin [1,2] \\ 8 \cdot x/15 + 2/(5 \cdot x^2), & x \in [1,2] \end{cases}$
3. $F_X = G_x/N$ or $F_x = \begin{cases} 0 & x \le 1 \\ (4 \cdot x^2 + 2 - 6/x)/15, & 1 \le x \le 2 \\ 1 & x \ge 1 \end{cases}$

Example 3.2
Calculate the integral distribution function $G_{x,y}$, knowing the concomitant x and y quantities distribution densities: $\Gamma_{x,y} = x^2 + 2 \cdot x \cdot y$ for the rectangle $x \in [0,2] \leftrightarrow y \in [-1,3]$ and $\Gamma_{x,y} = 0$ outside the specified field.

Solution
It is designated by I the semidefinite integral obtained at x = const.: $I = \int_{-\infty}^{y} \Gamma_{x,y} \cdot dy$

. By integrating the following are obtained:

(a) $I_a = 0$, for $y \le -1$
(b)
$I_{b(y)} = \int_1^{\infty} (x^2 + 2 \cdot x \cdot y) \cdot dy = x^2 \cdot y + x \cdot y^2 \cdot |_{y=1}^{y} = x^2 \cdot (y + 1) + x \cdot (y^2 - 1)$
for $-1 \le y \le 3$
(c) $I_c = I_b(y_{max}) = I_b(3) = x^2 \cdot (3 + 1) + x \cdot (3^2 - 1) = 4 \cdot x^2 \cdot 8 \cdot x$ for $y \ge 3$

Similarly to one variable integration, into relation (3.4b), of the three variables normalized distribution function, one can integrate $\Gamma_{x,y}$—the two variables non-normalized distribution function: $G_{x,y} = \int_{x_{min}}^{x} \int_{y_{min}}^{y} \Gamma_{x,y} \cdot dx \cdot dy$ namely $G_{x,y} = \int_{x_{min}}^{x} I(x) \cdot dx = \int_0^x I(x) \cdot dx$. It results that:

(a) $G_{x,\,y} = 0$ for $y \le -1 \cup x \le 0$
(b1) $G_{x,y} = \int_0^x [x^2 \cdot (y + 1) + x \cdot (y^2 - 1)] \cdot dx = \frac{x^3 \cdot (y+1)}{3} + \frac{x^2 \cdot (y^2 - 1)}{2}$ from where, $G_{x,y} = x^3 \cdot (y + 1)/3 + x^2 \cdot (y^2 + 1)/2$ for $0 \le x \le 2 \cap -1 \le y \le 3$.
(b2) For $y \in [-1, 3]$ and $x \ge 2$: variable x from b1 result is replaced by value 2 (i.e., x_{max}): $G_{x,y} = 2^3 \cdot (y + 1)/3 + 2^2 \cdot (y^2 - 1)/2$ or: $G_{x,y} = 2 \cdot (y + 1) \cdot (y - 1/3)$
For $y \ge 3$, $G_{x,y} = \int_{x_{min}}^{x} I_0 \cdot dx = \int_0^x (4 \cdot x^2 + 8 \cdot x) \cdot dx$ so that:
(c1) $G_{x,y} = 4 \cdot x^3 + 4 \cdot x^{2/3}$ if $x \in [0, 2]$l $y \ge 3$.

(c2) $G_{x,y} = \lim_{x \to 2} (4 \cdot x^3 + 4x^2/3)$ or $G_{x,y} = 3$

The results are centralized in the following table:

x→ y↓	≤ 0	[0; 2]	≥ 2
≤ −1	0	0	0
[−1;3]	0	$x^2 \cdot (y+1) \cdot (x/3 + y/2 - 1/2)$	$2 \cdot (y+1) \cdot (y + 1/3)$
≥ 3	0	$4 \cdot x^2 \cdot (x + 1/3)$	80/3

Example 3.3

Calculate the arithmetic mean $\bar{r}$ of two ideal gas molecules *relative* velocity, at equilibrium at temperature T; m is molecule mass.

Solution

Let $\overleftarrow{c_a}$ (u_a, v_a, w_a) and $\overleftarrow{c_b}$ (u_b, v_b, w_b) be the two molecules' (**a** and **b**) velocities. The first molecule relative velocity vector **r** towards the second one is defined as: $r = \vec{c_a} - \vec{c_b}$. Relative velocity projections on the three axes, r_x, r_y, and r_z, are:

$$r_x = u_a - u_b; r_y = v_a - v_b; r_z = w_a - w_b \tag{3.10a2}$$

Since one molecule velocity component distribution, for example, u_a projection distribution, does not depend on the other two components v_a and w_a, and the same variables independence is valid for u_b, v_b and w_b, so that relative velocities projections distributions can be separately obtained.

For example, for r_x relation (3.6) is the starting point—$\Phi_{v_x} = (2 \cdot \pi \cdot k_B \cdot T/m)^{-1/2} \cdot e^{-m \cdot (v_x)^2/(2 \cdot k_B \cdot T)}$. Density Φ_{u_a,u_b} of variables u_a and u_b concomitant distribution becomes, since Φ_{u_a} does not depend on u_b, and Φ_{u_b} does not depend on u_a: $\Phi_{u_a,u_b} = \Phi_{u_a} \cdot \Phi_{u_b}$ or

$$\Phi_{u_a,u_b} = (2 \cdot \pi \cdot k_B \cdot T/m)^{-1} \cdot e^{-m \cdot [(u_a)^2 + (u_b)^2]/(2 \cdot k_B \cdot T)} \tag{3.10b2}$$

The independent variables pair (u_a, u_b) changes with a new pair of physical quantities, namely difference r_x and sum **s** of the old variables is realized: $r_x = u_a - u_b$; $s = u_a + u_b$, from where: $u_a = (r_x + s)/2$ and $u_b = (-r_x + s)/2$.

The partial derivatives from Jacobian $J(u_a, u_b \| s, r_x)$, defined as in 3.1, are: $(\partial_{u_a}/\partial s)_{r_x} = 1/2$;

$$(\partial_{u_a}/\partial_{r_x})_S = 1/2; (\partial_{u_b}/\partial s)_{r_x} = 1/2; (\partial_{u_b}/\partial_{r_x})_S = -1/2$$

Therefore the Jacobian of variables change is $J = \begin{vmatrix} 1/2 & 1/2 \\ -1/2 & 1/2 \end{vmatrix} = 1/2.$

The new variables differential distribution function is, according to relation (3.4d): $\Phi_{r_x,s} = \Phi_{u_a,u_b} \cdot J(u_a, u_b \parallel s, r_x)$ or, taking into account (3.10b2): $\Phi_{r_x,S} = (8 \cdot \pi \cdot k_B \cdot T/m)^{-1} \cdot e^{-m \cdot [(u_1)^2 + (u_2)^2]/(2 \cdot k_B \cdot T)}$.

As on the other hand the sum of squares becomes, when expressed using the new variables: $(u_a)^2 + (u_b)^2 = (r_x + s)/2)^2 + (r_x - s)/2)^2] = [(r_x)^2 + s^2]/2$, for $\Phi_{r_x,S} = f(r_x, s)$ the following is obtained:

$$\Phi_{r_x,s} = (8 \cdot \pi \cdot k_B \cdot T/m)^{-1} \cdot e^{-m \cdot [(r_x)^2 + s^2]/(4 \cdot k_B \cdot T)} \tag{3.10c2}$$

The number of independent variables can be reduced from 2 to 1 through the procedure used in relation (3.8)—integral over all possible s values: $\Phi_{r_x} = \int_{S=-\infty}^{S=\infty} \Phi_{r_x,s} \cdot ds$, which leads to:

$$\Phi_{r_x} = (4 \cdot \pi \cdot k_B \cdot T/m)^{-1} \cdot e^{-m \cdot (r_x)^2/(4 \cdot k_B \cdot T)} \cdot E \tag{3.10d2}$$

where E is the integral $E = \int_{S=0}^{S=\infty} e^{-m \cdot s^2/(4 \cdot k_B \cdot T)} \cdot ds$, calculable with relation (A.2b), namely $\int_0^{\infty} x^{2 \cdot k} \cdot e^{-\alpha \cdot x^2} \cdot dx = \frac{(2 \cdot k)!}{k!} \cdot (4 \cdot \alpha)^{-(k+1/2)} \cdot \sqrt{\pi}$, with $k = 0$ and $\alpha = m/(4 \cdot k_B \cdot T)$. It results that $E = 2 \cdot (\pi \cdot k_B \cdot T/m)^{1/2}$, and relation (3.10d2) becomes:

$$\Phi_{r_x} = 0.25 \cdot (\pi \cdot k_B \cdot T/m)^{-1/2} \cdot e^{-m \cdot (r_x)^2/(4 \cdot k_B \cdot T)} \tag{3.10e2}$$

Similar expressions, where r_x is replaced by r_y, respectively r_z, are obtained for density distributions Φ_{r_y}, Φ_{r_z}. Because these three functions are independent (Φ_{r_x} does not depend on r_y and r_z, etc.), the three projections differential concomitant distribution function is therefore the product $\Phi_{r_x,r_y,r_z} = \Phi_{r_x} \cdot \Phi_{r_y} \cdot \Phi_{r_z}$ namely, since $(r_x)^2 + (r_y)^2 + (r_z)^2 = r^2$

$$\Phi_{r_x,r_y,r_z} = [m/(8 \cdot \pi \cdot k_B \cdot T)]^{3/2} \cdot exp\left[m \cdot r^2/(4 \cdot k_B \cdot T)\right] \tag{3.10f2}$$

In order to obtain the distribution of relative velocity r module, a translation is made, as in 3.2, from Cartesian coordinates $[r_x, r_y, r_z]$ to polar ones—$[r, \theta, \beta]$—with the limits $(0; 2 \cdot \pi)$ for β and $(0; \pi)$ for $\boldsymbol{\theta}$.

Translation relations are $r_x = r \cdot sin\theta \cdot cos\beta$, $r_y = r \cdot sin\theta\ sin\beta$, $r_z = r \cdot cos\theta$, from where: Jacobean $D(r_x, r_y, r_z)/D(r, \theta, \beta) = r^2 \cdot sin\,\theta$.

It results:

$$\Phi_{r,\theta,\beta} = \Phi_{r_x,r_y,r_z} \cdot D(r_x, r_y, r_z)/D(r, \theta, \beta)$$

or

$$\varphi_{r,\theta,\beta} = r^2 \cdot (\mathit{sin}\ \theta) \cdot \left(\frac{m}{8 \cdot \pi \cdot k_B \cdot T}\right)^{3/2} \cdot \mathit{exp}\left(-\frac{m \cdot r^2}{4 \cdot k_B \cdot T}\right)$$

and then:

$$\Phi_r = \iint \Phi_{r,\theta,\beta} \cdot d\theta \cdot d\beta$$

$$= \left[\int_0^{\pi} \sin\theta \cdot d\theta\right] \cdot \left[\int_0^{2\pi} d\beta\right] r^2 \cdot \left(\frac{m}{8 \cdot \pi \cdot k_B \cdot T}\right)^{3/2} \cdot \mathit{exp}\left(-\frac{m \cdot r^2}{4 \cdot k_B \cdot T}\right)$$

from where, since the two definite integrals from the square brackets are equal respectively to 2 and $2\cdot\pi$:

$$\Phi_r = (k_B \cdot T/m)^{-3/2} \cdot (32 \cdot \pi)^{-1/2} \cdot r^2 \cdot \mathit{exp}\left[-m \cdot r^2/(4 \cdot k_B \cdot T)\right] \quad (3.10g2)$$

According to relation(3.11b): $\bar{r} = \int_{r=0}^{r=\infty} r \cdot \Phi_r \cdot dr$ or, by replacing Φ_r from (3.10g2):

$$\bar{r} = (32 \cdot \pi)^{-1/2} \cdot (k_B \cdot T/m)^{-3/2} \cdot I \quad (3.10h2)$$

where I is the following integral:

$$\int_{r=0}^{r=\infty} r^2 \cdot \left\{\mathit{exp}\left[-m \cdot r^2/(4 \cdot k_B \cdot T)\right]\right\} \cdot dr \quad (3.10i2)$$

calculable by considering $k = 1$ and $\alpha = m/(4\cdot k_B\cdot T)$, with relation (A.2b), $\Phi_{2\cdot k}(\alpha) = \int_0^{\infty} x^{2\cdot k} \cdot e^{-\alpha \cdot x^2} \cdot dx = \frac{(2\cdot k)!}{k!} \cdot (4 \cdot \alpha)^{-(k+1/2)} \cdot \sqrt{\pi}$; it results: $I = 0.25\cdot(\pi/\alpha^3)^{1/2}$ sau $I = 2\cdot\pi^{1/2}\cdot(k_B\cdot T/m)^{3/2}$.

By replacing I into relation (3.10h2), it is found that: $\mathbf{\bar{r} = 4 \cdot \sqrt{(k_B \cdot T)/(\pi \cdot m)}}$.

By comparing with relation (3.12b), it is observed that: $\bar{r} = \bar{v} \cdot \sqrt{2}$.

Example 3.4
A wall of a space rocket of volume $v = 60\ m^3$ is perforated by a micrometeorite with diameter $D = 0.01$ mm. After what period of time the P air pressure within the rocket will decrease with 1%? Air molecular mass is: $M = 0.0291$ kg/mol, $t = 23$ °C, and the accommodation coefficient is $\alpha = 90\%$.

Solution
Let N and n be the number of molecules respectively of moles from the rocket, which are variables in time in the same manner as pressure P. Temperature T and

volume v of air in the rocket are obtained, after isothermal–isobaric derivation in relation to time t:

$$dN/dt = [N_A \cdot v/(R \cdot T)]dP/dt \tag{a}$$

But time derivation of rocket molecules number is also given by relation (3.22c), but with changed sign (because N from this expression does not represent molecules number remained within the vessel, but the number of the ones exiting from it): $d\,N/d\,t = -Z_s \cdot s \cdot \alpha$, where leakage area is: $s = \pi \cdot D^2/4$. Introducing the collisions frequency Z_s on the unit of area from relation (3.22b), $Z_S = P \cdot N_A/\sqrt{2 \cdot \pi \cdot M \cdot R \cdot T}$ and the leakage area $s = \pi \cdot D^2/4$ leads to: $dN/dt = -\left(P \cdot N_A \cdot D^2 \cdot \alpha\sqrt{\pi/(32 \cdot M \cdot R \cdot T)}\right)$, from where, by comparing to (a):

$$dP/dt = -\left(P \cdot N_A \cdot D^2 \cdot \alpha/v\right) \cdot \sqrt{\pi \cdot R \cdot T/(32 \cdot M)} \tag{b}$$

By denoting with P_i and P_f the initial and final pressures and with τ—the leakage time, it is obtained, after variables separation from (c) and definite integration:

$$-\int_{P_i}^{P_f} P^{-1} \cdot dP = \left(D^2 \cdot \alpha/v\right) \cdot \sqrt{\pi \cdot R \cdot T/(32 \cdot M)} \cdot \int_0^{\tau} dt \text{ or}$$

$$\tau = \left[v \cdot \sqrt{32 \cdot M/(\pi \cdot R \cdot T)}\right]/\left(D^2 \cdot \alpha\right)\Big] \cdot \ln\left(P_i/P_f\right)$$

But $P_f = (1 - \varepsilon) \cdot P_i$, where ε is the relative pressure decrease, therefore:

$$\tau = -\left[\frac{60 \cdot \ln(1 - 0.01)}{\cdot\left(10^{-5}\right)^2 \cdot 0.9}\right] \cdot \left[\frac{32 \cdot 0.0291}{3.14 \cdot 8.31 \cdot (273 + 23)}\right]^{\frac{1}{2}} = 3.37 \cdot 10^8 \text{ s}$$

Example 3.5

The mean velocity $\overline{v}_1$ of gas 1 molecules is 500 m/s and the mean velocity $\overline{v}_2\overline{v}_2$ of the same temperature for gas 2 is 800 m/s. It is required to calculate:

(1) mean relative velocities $\overline{v}_{r11}$ and $\overline{v}_{r12}$, where numerical indices show those gases whose molecules collide
(2) ratios $R_M \equiv M_2/M_1$ and $R_\varepsilon \equiv \overline{\varepsilon}_2/\overline{\varepsilon}_1$ between molecular masses and mean kinetic energies of gases

Solution

(1) From relation (3.25a): $\overline{v}_{r11} = \overline{v}_1 \cdot \sqrt{2} = 500 \cdot 1.414 = 707\ m/s$.

From relation (3.27a): $\overline{v}_{r12} = \sqrt{(\overline{v}_1)^2 + (\overline{v}_2)^2} = 500^2 + 800^2 = 944$ m/s.

(2) Since according to relations (3.12b) and (3.20), the arithmetic mean velocities (with f = 3 degrees of freedom) and the mean energies of molecules are calculated by $(\overline{v}_i)^2 = 8 \cdot R \cdot T/(\pi \cdot M_i)$ and $\overline{\varepsilon}_i = 3 \cdot k_B \cdot T/2$, it is obtained that:

$$R_M = (\overline{v}_1/\overline{v}_2)^2 = (500/800)^2 = 0.391, \text{and } R_\varepsilon = 1.$$

Example 3.6
Calculate, for ethane at 600 K and 0.4 at, the following quantities:

(1) molecules density, μ
(2) density ρ, knowing the molar mass $M = 0.03007$ kg/mol
(3) molecule mean arithmetic velocity, $\overline{v}_1$
(4) mean free path $\overline{l}$, knowing the kinetic diameter $\delta = 4.42$ Å
(5) a molecule collisions frequency, $z_{1,1}$
(6) collision density, $Z_{1,1}$
(7) mean time between two collisions, τ

Solution
By converting to the S.I. fundamental units: $P = 0.4 \cdot 1.01325 = 4.053 \cdot 10^4$ N/m^2; $M = 30.07/1000 = 0.03007$ kg/mol; $\sigma = 4.42 \cdot 10^{-10}$ m.

(1) With relation (3.21b) $\mu = N_A \cdot P/(R \cdot T)$:

$$\mu = 6.023 \cdot 10^{23} \cdot 4.053 \cdot 10^4/(8.314 \cdot 600) = 4.893 \cdot 10^{24} \text{m}^{-3}.$$

(2) By multiplying the molecular density μ from relation (3.21b) with one molecule mass M/N_A, the following is obtained for the total mass of the volume unit—that is, the density:

$$\rho = M \cdot P/(R \cdot T), \text{therefore } \rho = 0.03007 \cdot 4.053 \cdot 10^4/(8.314 \cdot 600)$$
$$= 0.2443 \text{ kg/m}^3$$

(3) From relation (3.12b) with index 1 of molecules type,

$$\overline{v}_1 = \sqrt{8 \cdot R \cdot T/(\pi \cdot M_1)} = \sqrt{8 \cdot 8,314 \cdot 600/(3,142 \cdot 0,03007)}$$

(4) With relation (3.33b), $\overline{l} = 1/\left(\sqrt{2} \cdot \pi \cdot \delta^2 \cdot \mu\right)$:

$$\bar{l} = \left(1.4142 \cdot 3.1416 \cdot 4.42^2 \cdot 10^{-20} \cdot 4.893 \cdot 10^{24}\right)^{-1} = 2.355 \cdot 10^{-7}\text{m}$$

(5) From relation (3.33a), $\bar{l}_1 = \bar{v}_1/z_{1,1}$, it results that:

$$Z_{1,1} = \bar{v}_1/\bar{l} = 669.8/\left(2.355 \cdot 10^{-7}\right) = 2.844 \cdot 10^9 \text{ s}^{-1}$$

(6) With relation (3.28a), $Z_{1,1} = z_{1,1} \cdot \mu/2$:

$$Z_{1,1} = 2.844 \cdot 10^9 \cdot 4.893 \cdot 10^{24}/2 = 6.958 \cdot 10^{33} \text{ m}^{-3} \cdot \text{s}^{-1}$$

(7) The mean time between two collisions can be obtained as the reverse of collisions frequency: $\tau = 1/z_{1,1} = 1/(2.844 \cdot 10^9) = 3.52 \cdot 10^{-10}$ s.

Example 3.7

Temperature T of a gas increases four times and pressure P increases three times. Calculate the ratios R_v, R_μ, R_l, R_z, and R_Z between final and initial values of the mean velocity $\bar{v}$, the molecular density μ, the mean free path $\bar{l}$, the collisions frequency z and respectively of intermolecular collisions densities Z.

Solution

It is denoted by $R_T = 4$ and $R_P = 3$ the for T and P.

(1) From relation (3.12b), $\bar{v} = \sqrt{8 \cdot k_B \cdot T/(\pi \cdot m)}$, it results that $\bar{v} : \sqrt{T}$, from where

$$R_V = \sqrt{R_T} = \sqrt{4} = 2$$

(2) From relation (3.21b), $\mu = N_A \cdot P/(R \cdot T)$, it is seen that $R_\mu = R_P/R_T = 3/4$.
(3) From relation (3.33b), $\bar{l} = 1/\left(\sqrt{2} \cdot \pi \cdot \delta^2 \cdot \mu\right)$, it results $R_l = 1/R_\mu = 4/3$.
(4) From relation (3.33a), $\bar{l} = \bar{v}/z$, it is obtained $R_z = R_v/R_l = 2/(4/3) = 3/2$.
(5) From relation (3.28a), $Z = z \cdot \mu/2$, it is found that $R_Z = R_z \cdot R_\mu = (3/2) \cdot (3/4) = 9/8$.

Example 3.8

Let τ_{ij} be the mean time between two successive collisions of a type i molecule with molecules of type j from the system, and τ_i—the mean time between two successive collisions of a type i molecule with any of the system's molecules. Into an ideal binary mixture of ideal gases of molar masses $M_1 = 35$ g/mol and $M_2 = 25$ g/mol, are known: $\tau_{12} = 1.5 \cdot 10^{-9}$ s, $\tau_{21} = 5 \cdot 10^{-10}$ s, and $\tau_{11} = 8.836 \cdot 10^{-10}$ s.

The following are required to be calculated: (1) τ_1; (2) gas composition into molar fractions, X_1 and X_2; (3) the R_σ ratio between the kinetic diameters σ_2 and σ_1 of the two gases molecules.

Solution

1) The mean duration between two collisions and their frequency are linked through $\tau_{ij} \cdot Z_{i,j} = 1$, so that from statement's data it is obtained that: $Z_{1,2} = 1/1.5 \cdot 10^{-9} = 6.67 \cdot 10^{8}\ s^{-1}$, $Z_{2,1} = 1/5 \cdot 10^{-10} = 2 \cdot 109\ s^{-1}$ and $Z_{1,1} = ⅛ \cdot 0.836 \cdot 10^{-10} = 1.132 \cdot 10^{9}$ s^{-1}. But from relation (3.29a): $Z_i = \sum_{j=1}^{j=C} Z_{i,j}$ it results for i = 1 and C = 2, that the total frequency of molecules collisions is:

$Z_1 = Z_{1,\,1} + Z_{1,\,2} = 1.132 \cdot 10^9 + 6.67 \cdot 10^8 = 1.799 \cdot 10^9\ s^{-1}$, from where:

$\tau_1 = 1/Z_1 = 1/1.799 \cdot 10^9 \Rightarrow \tau_1 = 5.56 \cdot 10^{10} s$

2) Relation (3.27c), $z_{1,2} = \pi \cdot (\sigma_{1,2})^2 \cdot \sqrt{\bar{v}_1^{-2} + \bar{v}_2^{-2}} \cdot \mu_2$, becomes after indines reversion: $z_{2,1} = \pi \cdot \sigma_{2,1}^2 \cdot (\bar{v}_1^{-2} + \bar{v}_2^{-2})^{1/2} \cdot \mu_1$.

By putting $\sigma_{12} = (\sigma_1 + \sigma_2)/2 = \sigma_{21}$, it is found that: $z_{2,1}/z_{1,2} = \mu_1/\mu_2$ or

$$\tau_{2,1}/\tau_{1,2} = \mu_2/\mu_1 \tag{3.10a4}$$

But $\mu_2/\mu_1 = X_2/X_1$ or, as $X_2 = 1 - X_1$: $(1 - X_1)/X_1 = 1 – 5 \cdot 10^{-10}/1.5 \cdot 10^{-9}$, therefore $X_1 = 0.75$, and $X_2 = 1–0.75$ or $X_2 = 0.25$.

3) Relation (3.27c) becomes, for τ: $1/\tau_{12} = \pi \cdot (\delta_{12})^2 \cdot \mu_2 \cdot (\bar{v}_1^{-2} + \bar{v}_2^{-2})^{1/2}$ or, by replacing $\bar{v}_1, \bar{v}_2$ using relation (3.12b), $\bar{v} = \sqrt{8 \cdot k_B \cdot T/(\pi \cdot m)}$ and by expressing δ_{12} as the semi-sum $(\delta_1 + \delta_2)/2$:

$$\tau_{1/2} = 1/\left[\mu_2 \cdot (\delta_1 + \delta_2)^2 \cdot \sqrt{0,5 \cdot \pi \cdot R \cdot T \cdot (1/M_1 + 1/M_2)}\right] \tag{3.10b4}$$

From relation (3.25b) it is obtained that: $\tau_{1,1} = 1/\left(\sqrt{2} \cdot \pi \cdot (\delta_1)^2 \cdot \bar{v}_1 \cdot \mu_1\right)$ from where, with $\bar{v}_1 = \sqrt{8 \cdot R \cdot T/(\pi \cdot M_1)}$ from relation (3.12b):

$$\tau_{1,1} = 1/\left[4 \cdot \mu_1 \cdot (\delta_1)^2 \cdot \sqrt{\pi \cdot R \cdot T/M_1}\right] \tag{3.10c4}$$

By dividing (c) to (b) it is obtained:

$\tau_{1,1}/\tau_{1,2} = (\mu_2/\mu_1) \cdot (1 + \delta_2/\delta_1)2 \cdot \sqrt{(1 + M_1/M_2)/8}$ or, with μ_2/μ_1 fom (a): $\tau_{1,1}/\tau_{1,2} = (\tau_{2,1}/\tau_{1,2}) \cdot (1 + \delta_2/\delta_1)2 \cdot \sqrt{(1 + M_1/M_2)/8}$.

By simplifying with τ_{12} the following is obtained: $(1 + \delta_2/\delta_1)^2$exponent $= (\tau_{1,1}/\tau_{2,1}) \cdot \sqrt{8/(1 + M_1/M_2)}$ or $1 + \delta_2/\delta_1 = [(8.836 \cdot 10^{-10})/(5 \cdot 10^{-10})]^{0.5}$ $[8/(1 + 36/25)]^{0.25} = 2.138$.

Therefore $R_\sigma = 2.138–1 \Rightarrow R_\sigma = 1.138$

Example 3.9

Calculate temperature T at which the total number of collisions in the unit of time Nv between n-butane molecules contained within a sphere of 0.01 m radius at pressure

$P = 1.333$ Pa equals the total number of molecules/wall collisions in the unit of time, Ns, knowing the kinetic diameter of n-butane molecule, $\delta = 5 \cdot 10^{-10}$ m.

Solution

$Z_{1,1} = 2 \cdot \delta^2 \cdot \mu^2 \cdot \sqrt{\pi \cdot k_B \cdot T/m}$ şi $Z_S = P \cdot N_A/\sqrt{2 \cdot \pi \cdot M \cdot R \cdot T}$ from relations (3.28b) (3.22b), respectively, where $Z_{1,1}$ and Zs represent the dimension of intermolecular collisions and the frequency of molecule/wall collisions on the unit of area, respectively.

From $Z_{1,1}$ and Zs the Nv and Ns are reached through $Ns = Z_s \cdot s_{sf}$ respectively $Nv = Z_{1,1} \cdot v_{sf}$, where $v_{sf} = 4 \cdot \pi \cdot r^3/3$ and $s_{sf} = 4 \cdot \pi \cdot r^2$ are the volume and area of the sphere.

Thus, $N_S = 4 \cdot \pi \cdot r^2 \cdot N_A \cdot P/\sqrt{2 \cdot \pi \cdot M \cdot R \cdot T}$.

By replacing k_B/m with R/M and μ by $N_A \cdot P/(R \cdot T)$ from relation (3.21b) into the expression for calculating Nv the following is obtained: $N_V = 8 \cdot r^3 \cdot (N_A \cdot P \cdot \delta)^2 \cdot [\pi/(R \cdot T)]^{3/2}/(3 \cdot \sqrt{M})$.

By equaling Nv = Ns, it results that: $T = \pi \cdot \sqrt{8} \cdot \delta_2 \cdot N_A \cdot P \cdot r/(3 \cdot R)$ or

$$T = 3.14 \cdot 2.82 \cdot \left(5 \cdot 10^{-10}\right)^2 \cdot 6.02 \cdot 10^{23} \cdot 1.333 \cdot 0.01/(3 \cdot 8.31) \Rightarrow T = 715\ \text{K}$$

Example 3.10

Calculate the R_l ratio between values of the mean free paths $\bar{l}_b$ and $\bar{l}_a$ at temperatures $T_b = 1080$ K respectively $T_a = 270$ K, if gas pressure remains unchanged:

(1) considering the kinetic diameter constant
(2) considering temperature variation of the kinetic diameter and knowing the critical temperature $T_{cr} = 450$ K

Solution

From relation (3.33c) for the free path: $\bar{l} = R \cdot T/\left(\sqrt{2} \cdot \pi \cdot N_A \cdot P \cdot \delta^2\right)$ and for $R_l = \bar{l}_b/\bar{l}_a$ the following expression is obtained:

$$R_l = (T_b/T_a) \cdot (\sigma_a/\sigma_b)^2 \tag{3.10a5}$$

where σ_a and σ_b are the kinetic diameters at temperatures T_a and T_b.

(1) If $\sigma \neq f(T)$, from (3.10a5) it results: $R_{l1} = (T_b/T_a) = 1080/270 \Rightarrow$ $\mathbf{R_{l1} = 4.}$
(2) The $\sigma(T)$ variation is given by relation (3.32a), $\delta = \delta_0 \cdot \sqrt{1 + C_S/T}$; after replacing in (a) and simplification with σ_0, it results:

$$R_{l2} = (T_b/T_a)^2 \cdot (C_s + T_a)/(C_s + T_b) \tag{3.10b5}$$

Sutherland's constant C_s is valuated with relation (3.32b): $C_s = 0.8 \cdot T_{cr} = 0.8 \cdot 450 = 360$ K. From (3.10b5) it is found for this case:

$$R_{lb} = (1080/270)^2 \cdot (360 + 270)/(360 + 1080) \Rightarrow \mathbf{R_{lb} = 7}$$

Example 3.11
It is required the mean free path $\bar{l}_1$ of CO_2 at pressure $P = 24$ N/m^2 and temperature $T = 290$ K within a capillary of diameter $d_{cil} = 0.002$ m containing an ideal gas mixture of CO_2, H_2, and N_2 (indices 1, 2, 3) where the molar fractions are of 20% CO_2, 30% H_2, and 50% N_2.

Equal weights of normal and Knudsen mass transport are considered.

Solution
From Table 3.2: $M_1 = 44$, $M_2 = 2.01$, and $M_3 = 28$ (expressed in g/mol); $\delta_1 = 4$; $\delta_2 = 2.92$ and $\delta_3 = 3.68$ (expressed in Å).

$\bar{l}_i = [4 \cdot R \cdot T/(\pi \cdot N_A \cdot P)]/\left[\sum_{j=0}^{j=C} X_j \cdot (\delta_i + \delta_j)^2 \cdot \sqrt{1 + M_i/M_j}\right]$ – relation (3.34b) at i = 1, C = 3, and with index n (for a "normal" regime), leads to:

$$\bar{l}_{1n} = 4 \cdot 8.31 \cdot 290/\left(3.14 \cdot 6.02 \cdot 10^{23} \cdot 24 \cdot 10^{-20} \cdot C_S\right) \quad (3.10a6)$$

where factor 10^{-20} derived from transformation of Å into m for δ^2 while:

$$S \equiv x_1 \cdot (2 \cdot \delta_1)^2 \cdot 2^{1/2} + x_2 \cdot (\delta_1 + \delta_2)^2 \cdot \sqrt{1 + M_1/M_2} + \ldots x_3 \cdot (\delta_1 + \delta_3)^2 \cdot \sqrt{1 + M_1/M_3}.$$

With the molar fractions X from the statement as well as from Table 3.2: $M_1 = 44$, $M_2 = 2.01$, and $M_3 = 28$ (expressed in g/mol) and $\delta_1 = 4$, $\delta_2 = 2.92$ and $\delta_3 = 3.68$ (expressed in Å):

$S = 0.2 \cdot (2 \cdot 4)^2 \cdot 2^{0.5} + 0.3 \cdot (4 + 2.92)^2 \cdot (1 + 44/2.01)^{0.5} + 0.5 \cdot (4 + 3.68)^2 \cdot (1 + 44/28)^{0.5} = 134$ (Å)^2. By introducing the value of S into (3.10a6) it will result, in I.S. fundamental units, the free mean path of CO_2 under normal, "molecular" regime:

$$\bar{l}_{n1} = 0.0158\,\text{m} \quad (3.10b6)$$

On the other hand, for the Knudsen regime, the mean free path $\bar{l}_K$ of molecule of composant 1 will be according to relation (3.35e), $\bar{l}_{K1} \cong (2/3) \cdot d_{car}$ with the characteristic diameter d_{car} given for a cylindrical capillary of form (3.35c), $d_{car} = 1.5 \cdot d_{cil}$, from where $\bar{l}_{K1} = d_{cil}$ or, with $d_{cil} = 0.002$ m from the statement:

$$\bar{l}_{K1} = 0.002\,\text{m} \tag{c}$$

Relation (3.36a) for intermediate pressures becomes, for component 1 and with equal weights of $w_n = w_K$, according to the statement: $\bar{l}_1 = 1/\left(1/\bar{l}_{n1} + 1/\bar{l}_{K1}\right)$.
from where, with the values from expressions (b) and (c):

$$\bar{l}_1 = 1/(1/0.0158 + 1/0.002) \Rightarrow \bar{l}_1 = \mathbf{0.00178\,m}$$

Part II
Molecular Models in Thermodynamics

Phenomenological and Molecular Thermodynamics

Phenomenological thermodynamics from the previous book (Daneş et al. 2013 [1]) offers only the *general form* of equations linking together the equilibrium thermodynamic properties.

The following distinct features are to be encountered into thermodynamic relations:

- their status of being certainties
- their fitting within an axiomatic system, any thermodynamic relation being the quantitative form of a theorem derived from a small number of basic postulates – the so-called principles of thermodynamics
- their universality
- the absence in equations of any parameter, as well as of any constant – universal or specific (constant referred to as *of material*), physical or mathematical, dimensional or not.

From all above-mentioned four quoted points of view, any equation used in technology or in non-thermodynamic scientific practice has diametrically opposed features:

- such a relationship has a high truth value but not a total one
- the different equations are largely independent, and cannot be deduced one from the other or from any fundamental axiom
- their particular character, such relationships being valid only for certain substances, groups of substances or mixtures, for certain pressure domains, temperature ranges or compositional ranges
- an almost general presence of some parameters or constants.

It is understood that thermodynamics, although providing only a binding framework for the relations used, cannot offer a model of the phenomena encountered in practice.

Through the molecular physics methods exposed in the first part of this second book of physical chemistry, and already illustrated for the ideal gas more simple case, *practical models* with numerical values of material constants, models proposing directly applicable relations for the variety of physico-chemical systems that the scientist or engineer has to do with.

These models allow both explanation and prediction of thermodynamic properties specific to the various aggregation states of the most important substances or mixtures.

The obtained results are presented in separate chapters, as following:

- Chapter 4, for real gases;
- Chapter 5, for the most important bi-phasic systems in the technical field: liquid-vapour systems, with their critical points.
- Chapter 6, for the matter found in one of the condensed states of aggregation, liquids being firstly treated, and afterwards – solids.

Chapter 4
Models in Thermodynamics of Real Gases

The *ideal* gas (lacking interactions between types of molecular motion) and the *perfect* gas (which verifies the law $P \cdot V = R\ T$) represent limiting concepts, which allow a simplified treatment of gaseous state properties under usual atmospheric conditions. At somewhat higher pressures and at low temperatures, the obtained values are so erroneous that ideal and perfect gas models become unacceptable even as primary approximations, which requires the use of *real* gas specific concepts.

Some phenomena, such as critical point existence, the condensation, and the Joule–Thomson effect of laminar relaxation, cannot even be provided for the perfect gas. Another lack of the perfect gas model consists in the identity of thermal properties of all gases, regardless of their nature: in the perfect gas equation only the *universal* constant R exists, so there are no "perfect gases" but only one perfect gas. On the contrary, valid relations for real gases contain *material* constants differing from one gas to another, which is in line with the large variety of experimentally found properties of different gases.

4.1 Equation of State (ES) and *PVT* Dependencies

Equation of state (name abbreviated by ES) is, for pure substances, a dependency linking three intensive properties of state (thermodynamic).

The ES can be either of *energetic* type (when at least one of the properties is represented by a thermodynamic potential), either of *caloric* type (comprising the entropy), and either of a *thermic* one. One or another of the infinite competing ES can be chosen in a given situation, while phenomenological thermodynamics relations are mandatory and valid in any imaginable macroscopic system, since they represent the unavoidable consequences of four fundamental thermodynamic laws.

F. E. Daneş et al., *Molecular Physical Chemistry for Engineering Applications*,
https://doi.org/10.1007/978-3-030-63896-2_4

Thermal ES is the actual ES, representing three "thermal variables" correlation, variables denoted by *x*, *y*, and *w*:

$$f(x, y, w) = 0 \tag{4.1}$$

The *thermal variables* are pressure, *P*; temperature, *T*; molar volume, *V*; and any of their combination—*P·V* product or the nondimensional *compressibility* factor, *z* and the *residual* (molar) volume, V_{rez}, which give absolute respectively relative volume deviations from perfect gas behavior:

$$V_{rez} \equiv V - V_{pf}, z \equiv V/V_{pf}$$

The perfect gas molar volume is: $V_{pf} = R \cdot T/ P$ so that

$$V_{rez} = V - R \cdot T/P \tag{4.2}$$

$$z = P \cdot V/(R \cdot T) \tag{4.3}$$

The most frequent thermal variables {*x*, *y*, *z*} triplets are in order the following:

$$f(P, V, T) = 0, \text{then} f(z, V, T) = 0, \text{followed by} f(z, P, T) = 0.$$

Until present, PVT dependencies of a few thousand gases have been proposed, with different degrees of generality. Thus, some ES are applicable in principle to the entire domain of gaseous state, possibly comprising also the critical domain and even the liquid one, while other dependencies are satisfactory only in a limited range of pressures or temperatures.

In addition to ES, which applies in principle to any gas (of course *after* the corresponding change in material constants), there are also PVT dependences valid only for a narrow range of gases (e.g., the nonpolar gases, monatomic gases, or alkanes) or even for only one important gas. In the technique, a few dozens of ESs valid only for steam have been proposed.

Graphical PVT *Dependencies*

By the way of representation, *PVT* dependencies can be analytical or graphical as shown in the diagrams given below.

The graphical dependency is the diagram of a family of curves within (*x*,*y*) coordinates, each curve corresponding to a constant value of the third thermal state variable *w*, as in the typical examples from Fig. 4.1a–f.

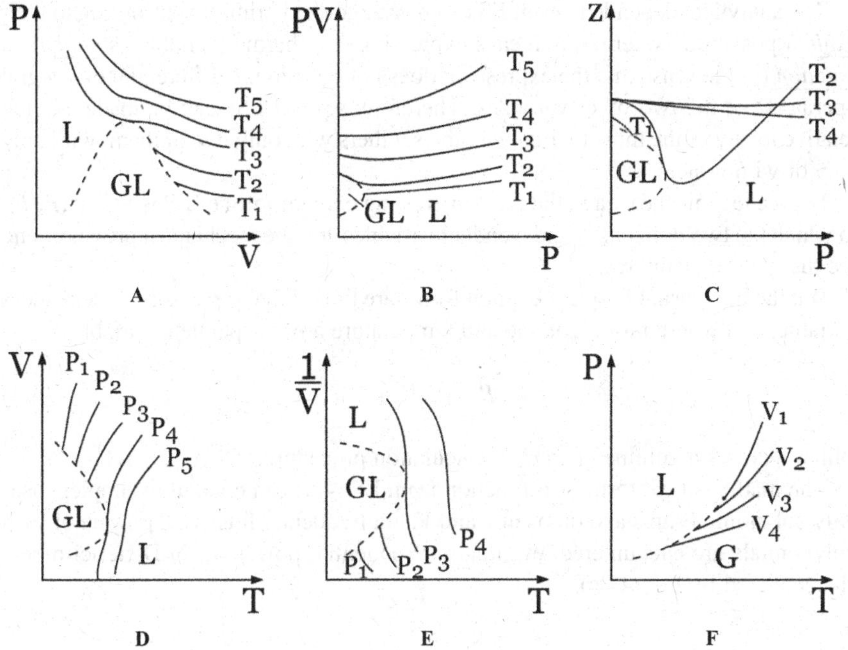

Fig. 4.1 Graphical representations for thermal ESs data. (**A**) $P = f(V)$, T constant. (**B**) $V = f(P)$, T constant. (**C**) $z \equiv f(V)$, T constant. (**D**) $V = f(T)$, P constant. (**E**) $1/V = f(T)$, P constant. (**F**) $P = f(T)$, V constant

Phase domain limits are represented in these figures by discontinuous lines: gaseous, G; liquid, L; or liquid–vapor diphasic (G-L).

The diagram curves correspond either to a mean line drawn as close as possible to an experimental point set, either to the graphical representation of an analytical dependency. In both situations, a given diagram is valid for a certain gas. However, the corresponding states theorem (Sect. 5.6) allows to obtain valid diagrams of any gas / large group of gases with good approximation.

Analytical Formulations of ES

After their origin, total empirical or absolute theoretical origin *PVT* analytical dependencies are encountered, but in most cases such deduction of equations incorporates both measurement data and modeling elements based on matter statistical mechanics, molecular physics, or kinetic theory considerations, which have been already introduced in Chaps. 1, 2, and 3, respectively, of this volume.

The analytical dependencies of ES can be categorized primarily according to their *explicit* possibilities, term that means expressing one thermal variable as *analytical function* (in the sense that the expression does not contain the differential or integral operators) of the two other variables. Therefore, several ESs can be made explicit relative to any of the three thermal variables, others with only two of them, with only one, or with none.

For the technical domain, the most interesting explicit form of ES is $V = F(P, T)$, to which the two domains of independent variables are: the mechanical pressures and the thermal temperatures.

But the most usual frequent explicit forms are those having pressure as dependent variable, and where molar volume and temperature are independent variables:

$$P = F(V, T) \tag{4.4}$$

which requires a volume *numerical* calculation procedure.

The mathematical form of F function from ES (4.4) can contain in simpler cases only polynomial sums and ratios in T and V, but frequently fractional powers of such polynomials are encountered, eventually exponential powers as in Dieterici model characterized by Eq. (4.8c).

4.2 Van der Waals (VdW) ES

The most known ES for real gases (or for fluids from a more general point of view) is the one proposed by van der Waals, serving as a way of reasoning and examples of mathematical form for all the above-mentioned models. VdW equation is of 3rd degree in V and contains two material constants: a and b, independent of P, V, and T but specific to each substance:

$$P \cdot V^3 - (b \cdot P + R \cdot T) \cdot V^2 + a \cdot V - a \cdot b = 0 \tag{4.5}$$

but can be made explicit with respect to P and T, respectively:

$$P = R \cdot T/(V - b) - a/V^2 \tag{4.6}$$

$$\mathrm{T} = \left(\mathrm{P} \cdot \mathrm{V}^2 + \mathrm{a}\right) \cdot (\mathrm{V} - \mathrm{b})/\left(\mathrm{R} \cdot \mathrm{V}^2\right)$$

Deduction of VdW ES

VdW equation was obtained on a structural basis, by replacing pressure and volume from gas perfect ES:

$$P \cdot V = R \cdot T \tag{4.2a2}$$

through corrected properties, *effective* pressure, P_{ef} and *effective* volume, V_{ef}:

$$P_{ef} \cdot V_{ef} = R \cdot T \tag{4.2b2}$$

Internal Pressure

Pressure correction is made by considering that the *external* measured pressure P is inferior to the *effective real pressure*, denoted by P_{ef}, due to the intermolecular forces of attraction, reflected by the *internal pressure*, P_i:

$$P = P_{ef} - P_i \tag{4.2c2}$$

The internal pressure's value has been considered by Van der Waals as proportional to the frequency of molecular collisions, $P_i \sim Z_{11}$. Since according to relation (3.28b), $Z_{11} \sim \mu^2$, where μ is the molecular number density inversely proportional to molar volume, and it results that $P_i \sim V^{-2}$, or:

$$P_{ef} = P + a/V^2 \tag{4.2d2}$$

Covolume

Van der Waals proposes to subtract the volume inaccessible to thermal agitation from molar volume value (V), which is termed as *covolume*. It is denoted by b.

Using a kinetic-molecular method, it is shown that volume correction b from VdW ES is four times higher than one gas mole molecule real volume when considered spherical.

When two molecules of radius "r" collide, their centers can approach one to the other up to the distance $\delta = 2 \cdot r$; the distance is denoted by σ if it is obtained from spherical intermolecular potential, as in Sect. 4.7 (Fig. 4.2).

The inaccessible volume during collision is the sphere of σ radius, of volume $(4/3) \cdot \pi \, \sigma^3$. Since two molecules participate at a given collision, to each one of them we can associate an inaccessible volume equal to half of the exclusion volume.

If it is denoted by $v_m = (4/3) \cdot \pi \cdot r^3$, the volume occupied by a molecule, thermal agitation inaccessible volume for one mole it is found to be: $4 \cdot v_m \cdot N_A = 4 \cdot V_{pr} = b$, where V_{pr} represents one gas mole molecules *eigenvolume*. Therefore:

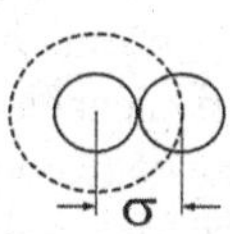

Fig. 4.2 Covolume

$$V_{ef} = V - b \quad (4.2e2)$$

where it is denoted by b, the fourfold of eigenvolume:

$$b = 4 \cdot V_{pr} \quad (4.2f2)$$

By replacing V_{ef} from (4.2e2) and P_{ef} from (4.2d2) into (4.2b2), relation (4.6) results.

Values of VdW Equation's Constants

Table 4.1. presents values for constants, a and b, for several gases.

The ES constants, a and b, are usually evaluated from the critical point (Sect. 5.3) coordinates T_c, P_c, and V_c, but can be also calculated from isochores slope:

$$a = -V \cdot [\partial(P \cdot V/T)/\partial(1/T)]_V; b = [\partial(P \cdot V - R \cdot T)/\partial P]_V \quad (4.2g2)$$

VdW Equation's Constants Incremental Calculation

Precise measurements show that neither a nor b calculated through (4.2g2) relations, and do not remain absolutely invariable: a decreases when temperature (and pressure) increases while b increases when T increases and P decreases. Thus, a and b values from tables are inaccurate, especially when located far from the critical point.

Both a and b are *constituent* properties, whose value can give information on the molecular composition and structure of a gas, and *additive* properties that can be calculated as dependencies on several *increments* sum (constants additive contributions), which are specific to each atom or bond type, for example:

$$a = 0.0503 \cdot (\Sigma_i \mu_i \cdot k_{a,i})^2 \quad (4.7a)$$

$$b = 0.0224 \cdot \Sigma_i \mu_i \cdot k_{b,i} \quad (4.7b)$$

after van Laar, μ_i being molecule's type i atoms number, and $k_{a,i}$ (Table 4.2) and $k_{b,i}$ (Table 4.3) are Van Laar increments of i type atom to a and b properties, expressed through the units of Table 4.1.

When the central atom is completely "wrapped" by other atoms (C into CH_4, S into SF_6), its contribution to a property will be 0, instead of the tabulated value.

Atoms having variable k_b increments are: **H**: 5.9 (common) or 3.4 (bonded on trivalent elements) and 1.4 (bonded on tetravalent elements); **C**: 10 (2,3 coordination) or 7.5 (4 coordination); **N**: 8.5 (1,2 coordination) and 6 (3,4 coordination); **O**: 7 (simple bonds) and 5 (double bond); **Cl**: 11.5 (inorganic compounds) and 11 (organic

Table 4.1 Values for VdW equation constants

Gas	*a*, L^2·atm/mol^2	*b*, mL/mol	Gas	*a*, L^2·atm/mol^2	*b*, mL/mol
He	0.034	23.7	SO_2	6.73	56.4
Hg	8.11	17.0	H_2S	4.44	42.9
H_2	0.245	26.6	H_2O	5.08	30.5
N_2	1.393	39.2	$GeCl_4$	22.7	148.6
O_2	1.363	31.9	PH_4Cl	4.12	45.5
Cl_2	6.53	56.3	CH_4	2.26	42.8
HCl	3.68	40.8	C_6H_6	18.05	115.4
CO	1.489	39.9	C_2H_5OH	12.05	84.1
CO_2	3.60	42.7	$(C_2H_5)_2O$	17.42	134.5

Table 4.2 Van Laar type k_a increments into relation (4.7a)

He	**Hg**	**C**	**N**	**P**	**O**	**S**	**F**	**Cl**
1	12.3	3.1	2.9	6.6	2.7	6.3	2.9	5.4
H[a](C)	**H[a](N)**	**H[a](others)**	**Period**	**II[b]**	**III[b]**	**IV[b]**	**V[b]**	**VI[b]**
3.2	1.6	1.1		3	5	7	9	11

[a]H increment depends on the nature of bonded atoms of H, N, C, or others
[b]unquoted elements, from II to VI Periodic Table periods

Table 4.3 k_b increments for relation (4.7b)

Atom	Li	Na	K	Rb	Cs	Si	Ge	Sn
k_b	14.5	27	48	58	71	15.5	21	26.5
Atom	Pb	P	As	Sb	Bi	S	Se	Te
k_b	32	14	18	25	30.5	12.5	18	23.5
Atom	F	Br	I	Cu	Ag	Au	Ti	Zr
k_b	5.5	16.5	22	11	15	15	18	23.5

compounds). The other k_b increments depend only on the atom nature and are presented in Table 4.3.

The accuracy of any additive–constitutive method is the better as it is applied to a narrower class of substances. For example, *a* and *b* mean calculation error from van Laar procedure, applicable to any substance, is of 15% and 10%, respectively, while other increment systems, proposed by *Thodos*, reduce the corresponding errors to 5% and 2%, respectively, but it is applicable only to organic substances.

4.3 Diversity of the ESs

The ESs—, thermal, or of other nature—are proposed for any state of matter, either uni- or multicomponent, mono- or polyphasic, solid or liquid (Chap. 6), either found in special conditions as are the intense electric or radiation fields, the positioning within micropores or nanoelements, etc. But ESs for gases are the most frequently used (being more accurate) and which constitute the object of this chapter. Further, Chap. 5 will discuss—from the molecular point of view—those ESs allowing liquid/vapor equilibrium, especially in monocomponent systems.

Material Constants

The ES right member for the case of pure substances made explicit in relation to pressure that contains as arguments (besides the possible mathematical or physical universal constants, like R): molar volume, V; temperature, T, and several *material constants*, properties that remain invariant for a given substance but change their value when passing from one substance to another.

A first measure of ESs mathematical complexity is the material constants number, N_{cm}, from expression $P = f(V,T)$: having 0 value for perfect gas equation and at least 1 (but commonly 2, 3, or 4) for the real gas case. Each ES is more appropriate for certain gases, certain PVT domains, or certain calculation types of technical thermodynamics.

The accuracy of PVT dependency increases with the used material constants number; but on the other hand, the property of N_{cm} involves a multiplication of constant values difficult determination with a satisfactory accuracy: for a certain accuracy, the experimental effort is proportional to the square of material constants number from equation and the calculation time—with N_c raised to fourth power. Consequently, ESs having many constants can be encountered only at gases of vapors of great practical importance (e.g., 37 constants into one of the steam ESs) and are of no use at all for the comportment stipulation of other substances.

Molecular thermodynamics of Chaps. 1, 2, and 3 allows several PVT dependencies establishment comprising simultaneously few constants, valid for a great number of gases, like the typical ESs from the following table.

Examples of ESs for Gases

Moreover, only completely defined ESs have been chosen—without unspecified functions as those from Sect. 4.4. Perfect gas ES and van der Waals must be added as representative to the tabulated relations.

Among the material constants can be besides listed also gas macroscopic measurable features, like the critical parameters (P_c, V_c, T_c), Boyle temperature from Sect. 4.5, or boiling temperature. Thus, d constant from Peng & Robinson II ES (4.8m) is in fact fluid critical temperature.

Several ESs are defined through an assembly of multiple equalities instead of a single one. For example, Peng & Robinson II ES (4.8m) from Table 4.4 consists of two equalities, each one being valid in a given domain of their parameters of state (here, the temperature domain T):

Table 4.4 Representative ESs

N_{cm} [*1]	Name	Function f from expression $P = f(T,V)$	Relation
2	Redlich & Kwong	$R \cdot T/(V-b) - a/\left[V \cdot (V+b) \cdot \sqrt{T}\right]$	4.8a
	Berthelot	$R{\cdot}T/(V-b) - a/(T{\cdot}V^2)$	4.8b
	Dieterici	$[R{\cdot}T/(V-b)]{\cdot}\exp[-a/(R{\cdot}T{\cdot}V)]$	4.8c
	Peng & Robinson I	$R{\cdot}T/(V-b) - a/(V^2 + 2{\cdot}b{\cdot}V - b^2)$	4.8d
	Scott	$R{\cdot}T{\cdot}(V+b)/[V{\cdot}(V-b)] - a/V^2$	4.8e
3	Clausius	$R{\cdot}T/(V-b) - a/[T{\cdot}(V+c)^2]$	4.8f
	Wohl	$R{\cdot}T/(V-b) - a/[V{\cdot}(V-b)] - c/(T^{4/3}{\cdot}V^3)$	4.8g
	Van Laar	$R{\cdot}T{\cdot}(V+c)/[V{\cdot}(V-b)] - a/V^2$	4.8h
	Twu	$R{\cdot}T/(V-b) - a/[V^2 + c\,(V - 2{\cdot}b) + b^2]$	4.8i
4	Martin	$R{\cdot}T/(V-b) - (a - cT)/(V+d)^2$	4.8j
	Himpan I	$R{\cdot}T/(V-b) - a/[(T{\cdot}V - c){\cdot}(V-d)]$	4.8k
	Himpan II	$R{\cdot}T/(V-b) - a/[T{\cdot}(V-c){\cdot}(V-d)]$	4.8l
	Peng & Robinson II	$R \cdot T/(V-b) - \left[a + c \cdot \left(\sqrt{\dot{d}} - \sqrt{T}\right)\right]^2/\left(V^2 + \ldots + 2 \cdot b \cdot V - b^2\right)$	4.8m
	Behar	$R{\cdot}T/(V-b) - a{\cdot}(V^2 + c{\cdot}V + d)/[V^3{\cdot}(V+b)]$	4.8n
	Keyes	$R{\cdot}T/[V - b{\cdot}\exp(-a/V)] - c/(V+d)^2$	4.8o

[*1]a: N_{cm} = number of material constants

(a, b, c, d) from Table 4.4, designate positive material constants

$$T \le d \Rightarrow P = R \cdot T/(V-b) - \left[a + c \cdot \left(\sqrt{d} - \sqrt{T}\right)\right]^2 / (V^2 + 2 \cdot b \cdot V - b^2)$$

$$T \ge d \Rightarrow P = R \cdot T/(V-b) - a/\left(V^2 + 2 \cdot b \cdot V - b^2\right)$$

Isochore curves $P(T)$ from these two relations are obtained at $T = d$ (critical point), with the same slope but with different second order derivatives $(\partial^2 P/\partial T^2)_V$.

ESs with Numerous Material Constants

The bigger is the N_{cm} number of ES constants, the more difficult is its use (values of constants are tabulated for a smaller number of gases), however becoming more accurate.

Thus, the following equations, proposed by Beattie & Bridgeman (4.8p) and Geană (4.8r), have five constants (a, b, c, d, k), all strictly positive:

$$P = \left[R \cdot T \cdot \left(V - k/T^3\right) \cdot \left(V^2 + c \cdot V - d\right) - a \cdot V \cdot (V-b)\right]/V^4 \qquad (4.8p)$$

$$P = R \cdot T/(V-b) - a \cdot T^{-k} / \left[(V-c)^2 + d\right] \qquad (4.8r)$$

For the case of equations with even more equations—8 as in relation (4.28) of Benedict, Webb and Rubin, or 9 as in Martin and Hou equation—volume calculation accuracy is of 1%, into a wide (P, T) domain.

Applications of ESs

The knowledge of ESs allows the deduction of a series of specific properties of technical importance specific to real gases, as:

- Volume, for industrial plants or pipes dimensioning
- Caloric capacities, H enthalpy (required for thermal balance calculation), and entropy isothermal dependency on pressure, for example:

$$\Delta^P H_T = H(T,P) - H(T,0) = \int_0^P \left[V - T \cdot (\partial V/\partial T)_P\right] \cdot dP \qquad (4.9)$$

- Fugacity coefficient φ (required for phase equilibriums and yield of chemical reactions) through one of the variants:

$$\ln \varphi = \int_0^P [V/(R \cdot T) - 1/P] \cdot dP \tag{4.10a}$$

$$\ln \varphi = z - 1 - \ln z + \int_V^\infty (z - 1) \cdot P^{-1} \cdot dV \tag{4.10b}$$

Virial ES

For the study of molecular interactions, virial ES is preferred:

$$P \cdot V/(R \cdot T) = 1 + \sum_{j=2}^{j=\infty} B_j/V^{j-1} \tag{4.11}$$

where B_j *virial* coefficients are not material constants but temperature dependencies.

The relation of right member series can be compared to the *MacLaurin* series (v. Appendix A.5.) of integer powers of molar volume inverse $x \equiv 1/V$ at isothermal decomposition of compressibility factor $z \equiv P{\cdot}V/(R{\cdot}T)$ found in the right member of the same relation, that is, the series:

$$z = \lim_{x \to 0}\left[z + \sum_{j=1}^{j=\infty} (1/j!) \cdot \mathrm{d}^j z/\mathrm{d}x^j\right]$$

where $\lim_{x \to 0} z$ is the perfect gas z value, namely $\lim_{x \to 0} z = 1$. By equalizing term by term, the two series it results that:

$$B_j = (1/j!) \cdot \lim_{x \to 0} \mathrm{d}^j Z/\mathrm{d}x^j \;, \forall j > 1$$

The virial coefficients have a microscopical physical significance—in order to express the interactions between gas molecules $j = 2$, $j = 3$—and are calculable through the methods based on the molecular potentials from Sect. 4.7.

Any ESs can be otherwise brought to the virial form, through its decomposition according to *Taylor* series (v. Appendix A.5) of volume negative integer powers, as in the example E.4.2 point 1 from page 172 at the end of this chapter.

The virial Eq. (4.11) is not directly usable for *PVT* calculations, but *P*(*V*) isothermal data can be correlated using truncated versions to the first term or the first two terms:

$$P \cdot V/(R \cdot T) - 1 = B_2/V$$

$$P \cdot V/(R \cdot T) - 1 = B_2/V + B_3/V^2$$

since virial coefficients become material constants if temperature is fixed.

4.4 Features of Thermal ES

Attraction and Repulsion in ES

The molecular repulsive interaction under the $R\cdot T/V$ form is the only one occurring into perfect gas pressure expression, while the more realistic ESs figure out not only the repulsive forces between molecules, but also the attraction ones:

- Some real gas ESs do not explicitly separate repulsion from attraction, as in the case of empirical established dependencies or virial ESs, both discussed earlier.
- Keyes Eq. (4.8o), $P = R\cdot T/[V - b\cdot \exp(-a/V)] - c/(V + d)^2$, combine in a complex way the repulsive features—*b* material constant—with attraction interactions reflected by negative values arguments: $-a/V$ of exponential, as the last term, $c/(V + d)^2$.
- Other ESs are *multiplicative*: *P*, *z*, or $P\cdot V$ are the products of a repulsive factor, an attractive one, namely $R\cdot T/(V - b)$, and $\exp[-a/(R\cdot T\cdot V)]$, respectively, from $P = [R\cdot T/(V - b)]\cdot \exp[-a/(R\cdot T\cdot V)]$ – Dieterici Eq. (4.8c).
- Most of PVT dependencies with molecular-statistical justification are *additive*—pressure, compressibility factor, or PVT product are two terms sums: one *repulsive* positive term corresponding to reciprocal *rejection* between molecules in opposition to a negative attractive term representing the intermolecular *attraction*, as for example the $R\cdot T/(V - b)$ and $-a/V^2$ terms in the van der Waals ES.
- More complex additive dependencies can be found, with more than two terms, like Wohl ES (4.8g) with the repulsive $R\cdot T/(V - b)$ term followed by *two* repulsive terms: $-a/[V\cdot(V - b)]$ and $-c/(T^{4/3}\cdot V^3)$.

Cubic ESs

Many ESs are cubic, namely *F* function from $F(P,V,T) = 0$. ES is a 3rd degree volume polynomial—the simplest algebraic form that can describe liquid/vapor equilibrium and the critical point. This is not possible for 2nd degree volume polynomials.

In reverse sense, not all cubic ESs allow critical phenomena and fluid phase's metastability prediction. Thus, the cubic ES

$$P = (a + b\cdot T)/(c + V)^3$$

well describes solids' and liquids' comportment at much higher pressures than their vapor pressure, but it is not capable to describe fluid equilibriums nor their critical phenomena.

The cubic ESs can be represented in the general form:

$$P = R \cdot T/\beta(V,T) + \alpha(V,T)/\gamma(V,T) \tag{4.4a2}$$

where $\gamma(V, T)$ is the 2nd degree volume trinomial.

$$\gamma(V,T) = V^2 - C(T) \cdot V + D(T) \tag{4.4b2}$$

While β (V, T) is the 1st degree volume binomial

$$\beta(V,T) = V - B(T) \tag{4.4c2}$$

eventually the polynomials ratio:

$$\beta(V,T) = [V - B(T)] \cdot [V + E(T)]/[V + F(T)] \tag{4.4d2}$$

The cubic ES can be brought to (4.4a2) form, if at least one of the following conditions: $D = 2{\cdot}C{\cdot}B - B^2$ and $D = 2{\cdot}C{\cdot}E - E^2$ is verified for any value of T.

Completely or Incompletely Defined ES

The actual ES are *completely* defined; for example, the ones made explicit in relation to pressure P allowing its value calculation as a totally determined function, that is, a function where numerical values can be assigned to each argument. These arguments are, on the one hand, the two state variables (molar volume, V and temperature, T) and, on the other hand, a series of material constants a,b,c, etc. as in Table 4.4.

But also *incompletely* defined ESs are encountered—*PVT* dependencies not allowing the achievement of a state variable, although gas nature (of which its material constants for the respective ES depend) and the other two variable values are known.

Functional Parameters

The indetermination cause is the fact that a series of undefined functions are found within equation arguments, functions that are furtherly called *functional* parameters. Thus, Redlich, Kwong & Soave ES expression, namely,

$$P = R \cdot T/(V - b) - \alpha(T)/[V \cdot (V + b)] \tag{4.12a}$$

is undefined, since it contains a *functional parameter—$\alpha(T)$*.

Arguments (α, β, γ) from (a) pressure dependence on a cubic ES are therefore examples of *bivariate* functional parameters, since their expressions consist of functions unspecified by both state parameters, V and T.

Arguments (B, C, D, E, F) are univariate functional parameters of (b, c, d) relations, since they depend—also in an undetermined manner—on a single state parameter, either V or T. For example, $\alpha(T)$ from relation (4.12a) is an univariate functional parameter, depending on temperature T.

In the cubic Van der Waals ES, all three functional parameters (α, β, γ) become independent of temperature, and α will not depend on volume, that is, $\alpha(V,T) = a$, $B(T) = b$, $E(T) \equiv F(T)$, $C(T) = 0$ and $D(T) = 0$, from where $\beta(V,T) = V - b$ and $\gamma(V,T) = V^2$.

Restrictions for the ES

From *Daneş & Geană,* ES material constants (c_{DG}, d_{DG}), namely,

$$P = T/(V-b) - (a/T)/\left[(V-b)^2 + c_{DG}\cdot(V-b) + d_{DG}\right] \tag{4.12b}$$

are readily found the ones denoted (c_{HI}, d_{HI}) of *Himpan I* ES (4.8k), through relation:

$$c_{HI} = d_{HI} = b - c_{DG}/2 \pm \sqrt{(c_{DG}/2)^2 - d_{DG}}$$

so that Himpan I ES is the particular case of Daneş & Geană Eq. (4.12b) that verifies $(c_{DG}/2)^2 - d_{DG} > 0$ restriction. Alike, van der Waals equation is the $c = 0$ customization of van Laar Eq. (4.8h).

It is to be noted that a restriction—*inequality* leaves unmodified the material constants number N_{cm} (the case of Himpan I and Daneş & Geană ESs), but the introduction of a restriction—*equality* reduces N_{cm}: Van der Waals ES have only two material constants, toward the three of van Laar ES.

Modified ES

The evolution of ESs occurs through successive operations of generalization and customization, which is usually alternately performed:

Generalization of an ES is the replacement of a certain well-defined part from ES expression by a functional parameter or an univariate functional parameter replacement—depending, therefore, only on T or only on P—through another functional parameter, but which is bivariate; eventually can be the replacement of an entire univariate group of parameters by a single bivariate functional parameter.

Customization of an ES is the reverse operation: replacement of a bivariate functional parameter by one or more functional parameters of univariate type; or the transition from a functional certain parameter to a *ES part*, namely to a well-determined expression having on the possible arguments' list: *V*, *T*, and material constants.

ES Modification

An example of ES evolution is the shifting from Redlich and Kwong Eq. (4.8a), $P = R \cdot T/(V - b) - a/\left[V \cdot (V + b) \cdot \sqrt{T}\right]$ to Soave ES, and also completely defined ES, namely:

$$P = R \cdot T/(V - b) - a\left[c - (d - T)^{0.5}\right]^2/[V \cdot (V + b)]$$

This translation is made through the above-mentioned Eq. (4.12a) of Redlich, Kwong, & Soave, $P = R{\cdot}T/(V - b) - \boldsymbol{\alpha}(T)/[V{\cdot}(V + b)]$, which is an *incompletely defined* ES. It is obtained from the starting Eq. (4.8a) of Redlich & Kwong, if this *defined part* from equation Redlich & Kwong expression offers the temperature dependence on the attractive term, namely the $\left[a/\sqrt{T}\right]$ part is replaced by the functional parameter $\alpha(T)$.

Also by a two-stage calculation, the transition from $P = R{\cdot}T/(V - b) - (a/T)/(V^2 + c{\cdot}V + d)$, Daneş & Geană ES (4.12b)—equation representing itself an extension of Himpan I ES, as seen above—to Feroiu ES with eight positive constants $\{b, c, d, T_c, k_0, k_1, k_2, k_3\}$ is made. By denoting with $y \equiv \sqrt{T_C} - \sqrt{T}$, Feroiu ES therefore comprises two parts of (4.12c) curve, which are encountered in the critical point $T = T_c$ but with different slopes $(\partial P/\partial T)_V$:

$$y \le 0 \Rightarrow P = R \cdot T/(V - b) - \left(k_0 + k_1 \cdot y + k_2 \cdot y^2 + k_3 \cdot y^3\right)^2/\left(V^2 + c \cdot V + d\right) \quad (4.12c)$$

and

$$y \ge 0 \Rightarrow P = R \cdot T/(V - b) - (k_0 + k_1 \cdot y)^2/\left(V^2 + c \cdot V + d\right) \quad (4.12d)$$

Daneş & Geană ES (4.12b) transition to Feroiu equation is made through an incompletely determined ES—*Geană & Feroiu* one, containing the univariate functional parameter $\alpha(T)$ as intermediate form among the initial and final determined parts, a/T and $(k_0 + k_1{\cdot}y)^2$, respectively: $P = R{\cdot}T/(V - b) - \alpha(T)/(V^2 + c{\cdot}V + d)$.

A major change of volume-dependent part from the *classical* repulsive term, which was $\beta(V) = 1/(V-b)$, is the one of *Carnahan and Starling*, authors proposing the following formula for $\beta(V)$ functional parameter,

$$\beta(V) = \left(1 + x + x^2 - x^3\right)/(1 - x)^3, \text{where } x \equiv V/(4 \cdot b),$$

expression theoretically justified, by mechanical–statistical means, starting from the potential model from Sect. 4.7. of molecule treated as rigid sphere.

4.5 Pressure Dependence on Volume

ESs describe correctly gas behavior only to the extent that reflects the following features, shared by all real gases:

- At low pressures, the real gas gradually becomes perfect gas:

$$\lim_{P \to 0}[P \cdot V/(R \cdot T)] = 1 \tag{4.13}$$

Perfect gas asymptotes, marked by dotted lines, are: equilateral hyperboles $P = k/V$ with $k = R \cdot T$ for Fig. 4.1a isotherms, straight linear lines $(P \cdot V) = k$ with $k = R \cdot T$, for Fig. 4.1b isotherms and overlapped linear lines, $z = 1$, for all isotherms from Fig. 4.1c.

- Real gas isotherms, from Fig. 4.1b, c, in $P \cdot V = f(P)$ or $z = f(P)$ coordinates, are linear at low pressures. Their slope is linked to the *virial second coefficient* B_2. Indeed, P and $1/V$ simultaneously tend to 0 according to relation (4.13), thus:

$$\lim_{P \to 0}[\partial(P \cdot V)/\partial(1/V)]_T = B_2 \cdot R \cdot T$$

On the other hand, by differentiating (4.13):

$$\lim_{P \to 0}[\partial P/\partial(1/V)]_T = R \cdot T$$

Through term by term division the last equalities, the following is found:

$$\lim_{P \to 0}[\partial(P \cdot V)/\partial P]_T = B_2$$

so that $m_1 = B_2$ and $m_2 = B_2/(R \cdot T)$ slopes of origin, of $P \cdot V = f(P)$, respectively $z = f(P)$ isotherms from Fig. 4.1b, c, will be also finite and dependent of temperature.

- Gas minimal volume into the thermal ESs:

At elevated pressures, gas volume tends asymptotically to a constant value, denoted by V_{min}:

$$\lim_{P\to\infty} V = V_{min} \tag{4.14}$$

and which depends not only of temperature: $(\partial Vinf/\ \partial T)_P = 0$. From the ES equations (4.6) it results that for the gas VdW this limit is exactly the b material constant:

$$(V_{inf})_{VdW} = b$$

the obtained value being defined (Sect. 4.2) as independent of temperature.

Boyle Curve

Although pressure dependence on virial second coefficient, B_2, varies from a gas to the other, it is always negative at low temperatures and positive at elevated temperatures, as in Fig. 4.3.

At temperatures close to T_B, the real gas has a quasi-perfect behavior: on a large pressures range, Boyle–Mariotte law (pressure independence on $P{\cdot}V$ product at constant temperature) is well respected. Sign change of B_2 occurs at a characteristic temperature of each gas, T_B, designated Boyle point or Boyle temperature.

Boyle Temperature

Boyle temperature corresponds to abscissa axis intersection (temperatures axis) by Boyle curve from Fig. 4.3:

$$\lim_{P\to 0}[\partial(P\cdot V)/\partial P]_T = 0 \tag{4.15}$$

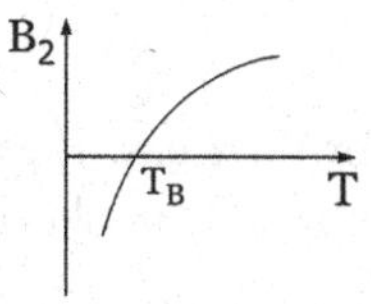

Fig. 4.3 Temperature dependence on virial second coefficient

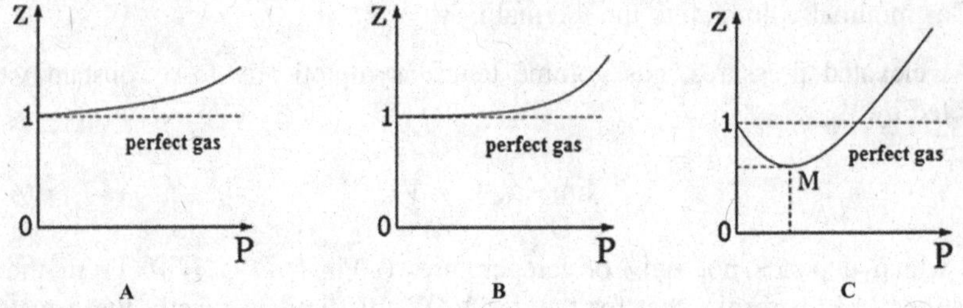

Fig. 4.4 Pressure dependence of virial second coefficient. (**A**) $T > T_B$. (**B**) $T = T_B$. (**C**) $T < T_B$

I. At temperatures T equaling or exceeding Boyle point, $z = f(P)$ or $P{\cdot}V = f(P)$ curves are monotonically increasing, as in Fig. 4.4A, B, with positive and respectively null slope.
II. At temperatures below T_B, $z(P)$ curve decreases starting from the point $z = 1$, $P = 0$, while after reaching a minimum—M point begins to grow, intersecting again the $z = 1$ horizontal, as in Fig. 4.4C.

Boyle Features of VdW Gas

For example, *Boyle curve* for VdW ES (4.6) is obtained by bringing $([\partial(P{\cdot}V)/\partial P]_T = 0$ definition to the following form:

$$V \cdot (\partial P/\partial V)_T + P = 0 \tag{4.5a2}$$

where $(\partial P/\partial V)_T$ is derived from isothermal volume derivation of VdW pressure from relation (4.6), namely:

$$(\partial P/\partial V)_T = -R \cdot T/(V-b)^2 + 2 \cdot a/V^3 \tag{4.5b2}$$

By substituting the expressions of $(\partial P/\partial V)_T$ and P from relations (4.5b2) respectively (4.6) into relation (4.5a2), Boyle curve equation in (T,V) coordinates is obtained

$$T = [a/(b \cdot R)] \cdot (1 - b/V)^2$$

for VdW gas. The analytical equations of Boyle curve expressed in (P, V), $(P{\cdot}V, V)$, or (P, T) coordinates can be obtained but not in (z, P), (z, v), or (z, T).

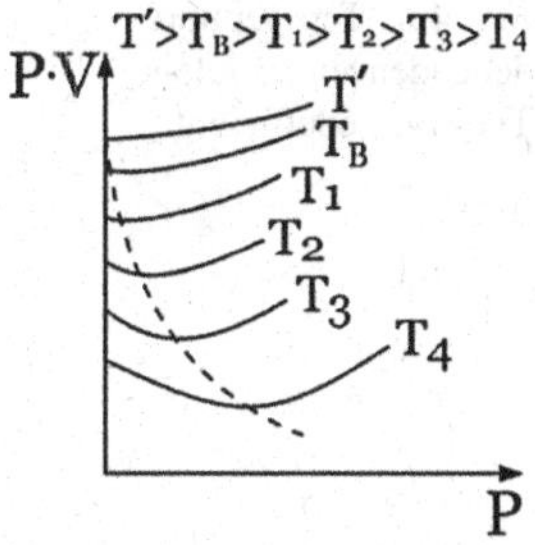

Fig. 4.5 Pressure dependence of the pressure and volume product

Boyle Temperature of VdW Gas

At temperature increases, z = $f(P)$ or $P \cdot V = f(P)$ curves minimum from Fig. 4.4 becomes less deeply and is moving toward lower pressures, as in Fig. 4.5.

The geometric locus of these minimum values—*Boyle curve*, represented by a dashed line—is calculable through:

$[\partial(P \cdot V)/\partial P]_T = 0$. VdW gas Boyle temperature is obtained from the last equality, with the condition (4.15):

$T_B = \lim_{P \to 0}[a \cdot (V - b)^2/(R \cdot b \cdot V^2)]$ or, since at $P \to 0$ volume will tend toward ∞:

$T_B = \lim_{V \to \infty}[a \cdot (V - b)^2/(R \cdot b \cdot V^2)]$ from where:

$$T_B = a/(R \cdot b) \tag{4.16}$$

4.6 Pressure Dependence on Temperature

Unlike perfect gases, real gases present a Joule–Thomson effect—a nonzero pressure dependence of temperature during the process of laminar relaxation, $\mu_{JT} \neq 0$, where μ_{JT} is the so-called Joule–Thomson coefficient (*J&T*), defined—the laminar relaxation being isenthalpic—as:

$$\mu_{JT} \equiv (\partial T/\partial P)_H$$

In phenomenological thermodynamics, it is shown that:

$$\mu_{JT} = V \cdot (\alpha \cdot T - 1)/C_P \tag{4.17}$$

so that (since V, T, and C_P are positive) Joule–Thomson coefficient has the sign of ($\alpha - 1/T$) difference between real gas coefficients of expansion α and those of ideal gas (where $\alpha = 1/T$).

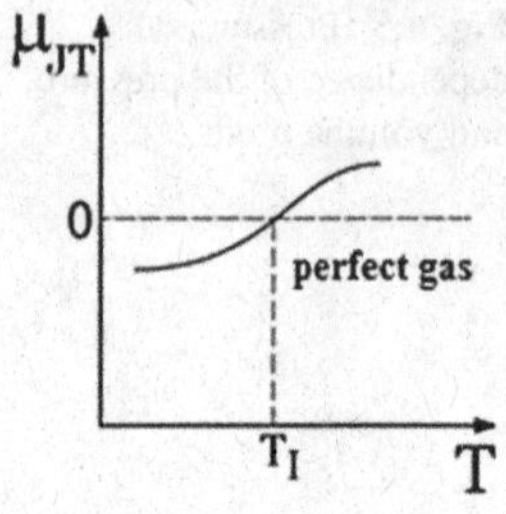

Fig. 4.6 Temperature dependence of Joule–Thomson coefficient

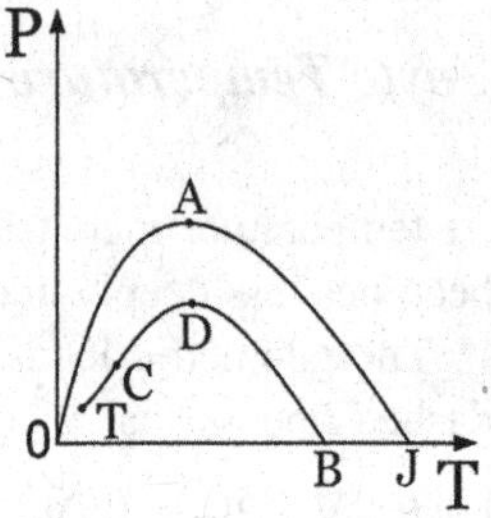

Fig. 4.7 Joule–Thomson curve

At pressures tending toward 0, Joule–Thomson effect for any real gas is found to be negative at low temperatures and positive at elevated temperatures taking according to Fig. 4.6 a zero value at inversion temperature denoted by T_I.

The physical states domain where $\mu_{JT} > 0$ is important for low temperatures technology (cryogenic) and in gas liquefaction industry: only within this domain gas cooling through adiabatic expansion is possible.

Joule–Thomson Curve

Joule–Thomson effect sign depends also on pressure: negative effect within the interior of OAJ parabola from Fig. 4.7 and positive outside this curve, designated "Joule–Thomson curve" corresponding to a null vale of the μ_{JT} coefficient namely, according to relation (4.17):

$$(\partial T/\partial V)_P = T/V \tag{4.18}$$

On the (P,T) diagram from Fig. 4.7 liquid/vapor *TC* curve also discussed in Sect. 5.4, and Boyle CDB curve, defined through the relation $V{\cdot}(\partial P/\,\partial V)_T + P = 0$ are represented; both curves are entirely situated beneath the Joule–Thomson curve and are starting from the *C* critical point.

The real gas is more compressible than the perfect gas beneath Joule–Thomson curve, $\chi_T > (\chi_T)_{id} = 1/P$, while above this curve real gas compressibility becomes inferior to $1/P$ value.

Joule–Thomson curve can be calculated for any ES from (4.18) condition. For van der Waals ES case, (4.6) ES derivation leads to: $(\partial T/\partial V)_P = (P - a/V^2 \cdot + 2 \cdot a \cdot b/V^3)/R$. After T and $(\partial T/\partial V)_P$ expressions replacement into (4.18), the following is obtained:

$$P = a \cdot (2 \cdot V - 3 \cdot b)/\left(b \cdot V^2\right)$$

and, by introducing $P(T,V)$ from here into van der Waals Eq. (4.6), it results that:

$$T = 2 \cdot a \cdot (1 - b/V)^2/(b \cdot R)$$

The inversion temperature will be the limit value of Joule–Thomson curve temperature, when $P \to 0$, and $V \to \infty$ respectively, in the last relation:

$$T_{\text{inv}} = 2 \cdot a/(b \cdot R)$$

The comparison of this value to T_B from relation (4.16) proves that for VdW gas the inversion temperature is twice higher than Boyle temperature. But experimentally it is found that $T_{\text{inv}}/T_B = 1.75 \ldots 1.8$.

Real Gas Isochores

(P, T) diagram isochores must be *almost* linear, as in Fig. 5.1f:

$$\left(\partial^2 P/\partial T^2\right)_V \cong 0 \tag{4.19}$$

This condition is for example *exactly* accomplished by van der Waals ES, for which by pressure isochoric differentiating in relation to temperature into (4.6) it results that $(\partial^2 P/\partial T^2)_V = 0$. The invalidity of this derivative implies the heat capacity temperature independence at constant volume $C_V(V,T)$—since $(\partial C_V/\partial V)_T = T \cdot (\partial^2 P/\partial T^2)_V$—in disagreement with experience.

Therefore relation (4.19) as an *equality*, although acceptable for gas density calculation, is no longer advisable for thermal effects achievement. For their calculation, only those ESs to whom $(\partial^2 P/\partial T^2)_V$ derivative is positive at vaporization curve adjacent upper area and *negative* in the rest of (P,T) domain.

Real fluid isochores are presented in Fig. 4.8, $V_1 \ldots V_5$ isochores corresponding to the liquid and V_7 to V_9 isochores to vapors and gas. The dotted line is the geometric locus of isochores inflection point, points where C_V values are maximum.

From Fig. 4.8, it is also observed that vaporization equilibrium $P(T)$ curve in the critical point as well as the critical isochore $V = V_c$ have a common tangent, which is simultaneously tangent to dotted curve.

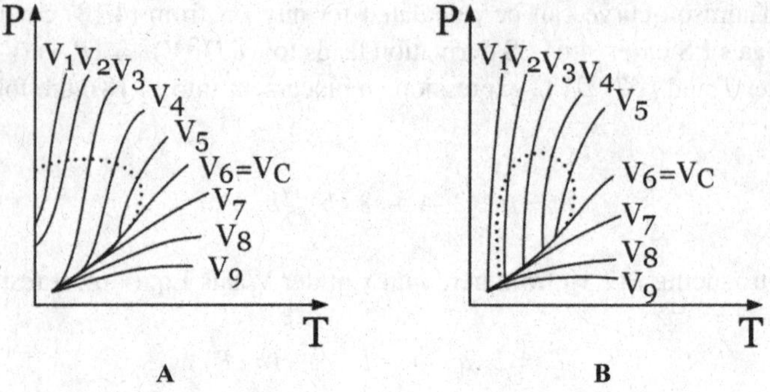

Fig. 4.8 Pressure–temperature diagram isochores. (**A**) Hydrogen and Helium. (**B**) Other gases

Fig. 4.9 Molecules pair geometry

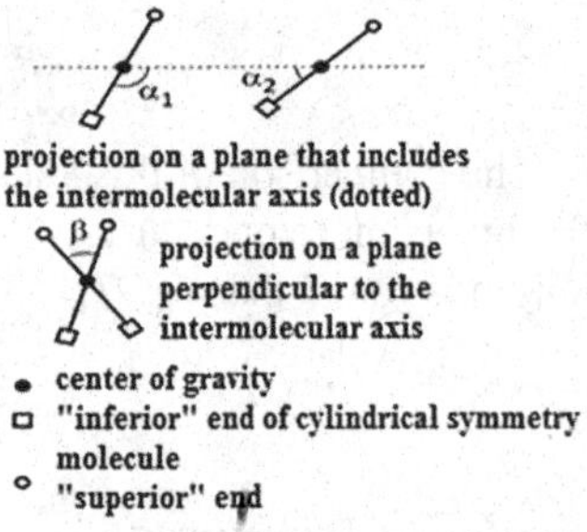

4.7 Real Gas Molecular Models

Intermolecular Potential

Real gas thermodynamic properties can be deducted from its molecular properties, if instead of considering the molecules as rigid (interaction takes place only at the direct contact time) as in the elementary kinetic theory of ideal gas, several interactions presence is admitted, continuously varying with distance between molecules and with their mutual position and characterized by the property called intermolecular potential—the difference between total energy of a system formed of two molecules positioned at an infinite distance one from the other and the total energy of the same system when molecules are positioned at a certain finite distance and in a certain mutual orientation.

Generally, the intermolecular potential u depends not only on distance d between molecules center of gravity, but also on other three geometrical parameters, as in Fig. 4.9: α_1, α_2, and β angles.

The α_1 and α_2 angles are formed by the internuclear axis with molecular axes, while β angle is the one between molecular axes projection on a perpendicular plane on the internuclear axis.

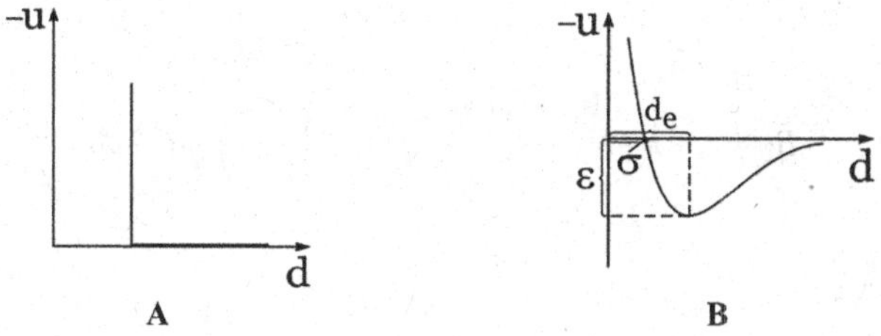

Fig. 4.10 "Spherical" intermolecular potential depending on distance. (**A**) Infinite wall. (**B**) Potential with attraction and repulsion

Spherical Potentials

In the simple case, the one of spherical symmetry molecules, the intermolecular potential depends on a single variable—distance *d* between molecules (practically, the distance between their centers of gravity): $u = u(d)$.

Simplifying is strictly valid only for monoatomic molecules—the only ones with spherical symmetry. Simplifying can be admitted with some approximation for polyatomic molecules of a not too elongated or too flattened shape, provided that these do not possess a permanent electric dipole moment (not to be "polar"). Figure 4.10b presents this potential dependence on the intermolecular distance *d*.

In the elementary collisions model from Sect. 3.6, *u*(*d*) function has the *infinite wall* shape as in Fig. 4.10a; the potential corresponds to a rigid spheres pair of radius *a*, which interacts only by direct contact:

$$u = -\infty, r \leq a \cap u = 0, r \geq a.$$

A more elaborate model introduces a *rectangular* potential well of *b* width and ε growth, where:

$$u = -\infty, r \leq a \cap u = \varepsilon, a \leq r \leq a + b \cap u = 0, r \geq a + b \tag{4.20a}$$

The more realistic model is the *attractive/repulsive* one, which is given by a continuous function on distance *d* between molecules centers, as in Fig. 4.10b: $u/\varepsilon = f(d/\sigma)$.

The *u*(*d*) dependence presents the following features:

- At long *u* distances will tend toward zero;
- At moderate distances, attraction prevails interaction energies beneath the zero-energy level, considered conventional as the one corresponding to infinite distance (energy sign is changed in *u* definition; therefore, *u* is positive in this area);
- At very low distances, repulsion prevails ($u < 0$), for $d = 0$ corresponding to an infinite energy.
- Through integration on the entire area around of spherical *u* potentials, virial second coefficient, B_2, is obtained from virial ES (4.11):

$$B_2 = 2 \cdot \pi \cdot N_A \cdot \int_{d=0}^{d=\infty} \left(e^{u/(k_B \cdot T)} - 1 \right) \cdot d^2 \cdot dd \tag{4.20b}$$

Mie Potential

One of the simplest potential forms possessing these properties is the *Mie* potential (or "double power" potential):

$$u(d) = A \cdot d^{-n} - B \cdot d^{-q} \tag{4.21}$$

with four positive material constants. The attractive part $A{\cdot}d^n$ and the repulsive one $B{\cdot}d^q$ increase when distance decreases but, since $q > n$, the increase is more rapid for the repulsive potential than for the attractive one, therefore at long distances attraction prevails, while at low distances repulsion. A and B constants can be expressed after σ—minimum distance between molecules—and after ε—maximum energy (thus minimum u) of interaction, through two conditions:

$$d = \sigma \rightarrow u = 0 \cap u_{\max} = \varepsilon. \tag{4.7a2}$$

- From the first condition (4.7a2), it results into relation (4.21): $A/\sigma^n = B/\sigma^q$ or

$$B = A/\sigma^{q-n} \tag{4.7b2}$$

- In the second condition (4.7a2), $u_{\max}$ corresponds to $du/\,dd = 0$ or $n \cdot A/(d_e)^{n+1} = q \cdot B/(d_e)^{q+1}$, where d_e is the distance at which the energy is maximum. It results that: $(d_e)^{q-n} = q \cdot B/(n \cdot A)$ with B from relation (4.7b2): $(d_e)^{q-n} = (q/n) \cdot (\sigma)^{q-n}$ or:

$$d_e = \sigma \cdot (q/n)^{1/(q-n)} \tag{4.7c2}$$

Since at $d = d_e$, the equality $u = \varepsilon$ is valid, it results from relations (4.21) and (4.7c2):

$\varepsilon = A \cdot (n/q)^{n/(q-n)}/\sigma^n - B \cdot (n/q)^{q/(q-n)}/\sigma^q$ or, by replacing B through $A{\cdot}\sigma^{q-n}$ from relation (4.7b2): $\varepsilon = A \cdot [(n/q)^{n/(q-n)} - (n/q)^{q/(q-n)}]/\sigma^n$.

Regrouping the bracket terms and decomposing after A leads to:

$$A = \varepsilon \cdot \sigma^n \cdot (q^q/n^n)^{1/(q-n)}/(q-n) \tag{4.7d2}$$

and after introducing A into relation (4.7b2), the following is obtained:

$$B = \varepsilon \cdot \sigma^q \cdot (q^q/n^n)^{1/(q-n)}/(q-n) \tag{4.7e2}$$

By replacing A and B from relations (4.7d2) and (4.7e2) into expression (4.21), it is found that:

$$u(d) = \varepsilon \cdot [(\sigma/d)^n - (\sigma/d)^q] \cdot (q^q/n^n)^{1/(q-n)}/(q-n) \tag{4.22a}$$

Mie potential depends therefore only on *two* molecular features, σ and ε, which justifies for gases the principle of corresponding states simple form from Sect. 5.6, according to which any property of any gas is uniquely determined by *two* independent parameters values.

Lennard Jones Potential

By substituting the q parameter from (4.22a) with another constant—$m \equiv q/n$ ratio, the following dependence is obtained:

$$u(d) = \varepsilon \cdot \left[\left(n^{m-2} \cdot m^2/(m-1)\right] \cdot \left[(d/\sigma)^{-n} - (d/\sigma)^{-n\cdot m}\right],$$

whose multiplicative coefficient from the right member, $n^{m-2} \cdot m^2/(m-1)$ becomes independent from the n exponent when $m = 2$; in which case:

$$u(d) = 4 \cdot \varepsilon \cdot \left[(d/\sigma)^{-n} - (d/\sigma)^{-2\cdot n}\right].$$

For nonionized gases case, exponent value $n = 6$ is justified on a quantum mechanical manner, since the energy of van der Waals attraction forces is inversely proportional to the distance to the sixth power.

Exponent q basically varies with molecule nature, but calculations have been proved to be too complicated so that Lennard Jones method was preferred—adopting the same value for any gas, that is, $q = 12$, of the repulsive exponent, which satisfies the simplifying condition $q = 2n$ from above.

Lennard Jones potential (L.J.), or *6 – 12 potential*, is thus the particular case $n = 6$, $q = 12$ of the Mie potential (4.22a) and has the following form:

$$u(d) = 4 \cdot \varepsilon \cdot \left[(d/\sigma)^6 - (d/\sigma)^{12} \right]. \tag{4.22b}$$

Lennard Jones potential constants, σ and ε, are evaluated from critical data, or from their lack from the boiling data, through relations

$$\begin{aligned} &\sigma = 0.841 \cdot (V_c/N_A)^{1/3}; \sigma = 2.44 \cdot [T_c/(P_c \cdot N_A)]^{1/3}; \\ &\varepsilon = 0.77 \cdot k_B \cdot T_c \end{aligned} \tag{4.22c, d, e}$$

$$\sigma = 1.106 \cdot \left(V_{fLich}/N_A\right)^{1/3}; \varepsilon = 1.15 \cdot k_B \cdot T_f \tag{4.22f, g}$$

indexes (*c*, Lich, *f*) indicating respectively the critical point, liquid and *boiling* – vaporization equilibrium at 1 atm, while *V* represents the molar volume.

4.8 ES for Real Gases Mixtures

The Complete ES

There is reduced information of experimental nature on mixtures of real gases. However, the great technical importance of gas mixtures, occurring in practice more frequently than pure gases, made necessary appearance of calculation models and methods for evaluating some mixture characteristic properties, as well as of several properties characterizing mixture individual components.

In the first category, these are especially included: the molar volume (depending on *T* and *P*; or inversely, pressure as dependence of *T* and *V*) and molar enthalpy (more precisely, finite isothermal variation $H(P,T) - H(0,T)$, and from the second, individual compounds fugacities.

Among the transport coefficients, viscosity and thermal conductivity are part of the first category (global parameters of mixture), and diffusion coefficients from the second category (individual properties of compounds).

Both global and individual thermodynamic properties can be calculated from the complete ES for the mixture, namely:

$$F(P, V, T, X_1, X_2, \ldots, X_{c-1}) = 0 \tag{4.23}$$

where *C* is the number of compounds in the mixture. From the complete ES, it is commonly determined, for example, density ρ at given composition, pressure, and temperature obtained as the *M*/*V* ratio, where $M = \sum_i M_i \cdot X_i$, and *V* results from Eq. (4.23). Pressure dependence of enthalpy is also determined as for the pure gas case, using relation (4.9): $H(P,T) - H(0,T) = \int_0^P \left[V - T \cdot (\partial V/\partial T)_{P,\overline{X}} \right] \cdot dP$, integrated at constant temperature and composition (composition characterized by the ordered molar fractions set).

The actual partial pressure, P_i, of the i compound is defined by:

$$P_i = P \cdot X_i \tag{4.24}$$

For the general case, P_i differs from Dalton partial pressure, $P_{i,D}$, from Dalton law—pressure of pure i compound, if the X_i moles of i would occupy themselves mixture's molar volume at temperature T. $P_{i,D}$ is, therefore, P value satisfying Eq. (4.23) when V is replaced by V/X_i, X_1 takes value 1 and the other molar fractions take value 0. The two pressure types coincide at the ideal perfect gas mixture.

Fugacity of Compounds in a Gas Mixture

For the case of i compound fugacity f_i, the calculation relation is:

$$f_i = P_i \cdot \varphi_i \tag{4.25}$$

where P_i is its partial pressure from the Expression (4.24) and φ_i is the fugacity coefficient of i compound from the mixture, defined by:

$$ln\, \varphi_i = \int_0^P \left[\overline{V_i}/(R \cdot T) - 1/P\right] \cdot dP, T = \text{const.}, X = \text{const.} \tag{4.26}$$

where $\overline{V_i} = (\partial V/\partial n_i)_{P,T,n_j(\forall j \neq i)}$ is the compound partial molar volume of the i compound, V being the system's volume.

By customization for pure substances, where $\overline{V_i}$ becomes V—pure i compound molar volume at mixture pressure and temperature—relation (4.10a) is found.

Usually, the complete ES is not made explicit after *volume*, therefore $\overline{V_i}$ cannot be calculated so that Eq. (4.26) is not usable. If complete ES is possible to be made explicit after pressure, then the fugacity coefficient is calculated using formula obtained from relation (4.26) through independent variable change and where $z \equiv P \cdot V/(R \cdot T)$ is the compressibility factor:

$$ln\, \varphi_i = z - 1 - lnz + \int_V^\infty \left\{[\partial(n \cdot P)/\partial n_i]_{V,T,n_j \forall j \neq i}/(R \cdot T) - 1/V\right\} \cdot dV \tag{4.27}$$

The complete mixture ES is seldomly known. In the absence of supplemental information on components interactions, one can use two procedures. The first one is the pseudocritical properties procedure from Sect. 5.8The second one is the material constants / ESs combinations procedures, described hereafter.

Material Constants for Mixtures

When the same ES is valid for all components and when material constants are known for each component, it is admitted that the mixture follows the same ES; mixture material constants are in these case dependencies of composition and of material constants of mixture components.

Combination Rules of Components Constants

The combination of *Individual* constants in order to obtain mixture constants is made following several rules that vary from one equation to the other. Therefore, for Benedict, Webb and Rubin (BWR) ES:

$$P = R \cdot T \cdot V^{-1} + \left(K_1 \cdot T - K_2 - K_3 \cdot T^{-2}\right) \cdot V^{-2} + (K_4 \cdot T - K_5) \cdot V^{-3} + K_5 \cdot K_6 \cdot V^{-6} + K_7 \cdot T^{-2} \cdot \left(1 + K_8 \cdot V^{-2}\right) \cdot V^{-3} \cdot \exp\left(-K_8 \cdot V^{-2}\right) \tag{4.28}$$

where the eight material constants $K_1 \div K_8$ are obtained using the following combination rules:

$$(K_k)^{m_k} = \sum\nolimits_{i=1}^{i=C} X_i \cdot \left[(K_k)_i\right]^{m_k} \tag{4.29}$$

where $(K_k)_i$ is constant's K_k value for the i compound; and m_k are the following: $m_1 = 1$; $m_2 = m_3 = m_8 = 1/2$; $m_4 = m_5 = m_6 = m_7 = 1/3$.

For the same and only ES, mixing rules can be different, according to the theory from which they have been obtained. Thus, for $P = R{\cdot}T/(V - b) - a/V^2$, the VdW equation—the material constant a—is obtained by the unique rule:

$$a = \left(\Sigma_i X_i \cdot \sqrt{a_i}\right)^2 \tag{4.30}$$

while for the b constant, the van Laar rule can be applied:

$$b = \Sigma_i X_i \cdot b_i \tag{4.31a}$$

or moreover, the Lorentz rule:

$$\mathrm{b}^{1/3} = \Sigma_\mathrm{i} \mathrm{X}_\mathrm{i} \cdot (\mathrm{b}_\mathrm{i})^{1/3} \tag{4.31b}$$

Combination Rules of Components Pairs

In some combination procedures, mixture K constants do not result from combinations of contributions of components in relations (4.29–4.31a and 4.31b), but through summation of contributions of components' *pairs*:

$$K^m = \sum_{i=1}^{i=C} \sum_{j=1}^{j=C} x_i \cdot x_j \cdot [(K_i)^m + (K_i)^m]/2 \tag{4.32}$$

Relation (4.32) produces the same results as (4.29), if $C = 2$.

Since material constants of a certain given composition mixture have been calculated, global thermodynamic properties will be obtained, like V, $H - H^0$, $C_p - C_p{}^0$, and $S - S^0$ by considering the mixture as a pure gas.

Similar combination rules are valid for constants from the intermolecular potential expression. For Lennard-Jones potential (4.29), the interaction constants ε_{ij} and σ_{ij} of (i, j) components' pair are as follows:

$$\sigma_{ij} = (\sigma_i + \sigma_j)/2; \varepsilon_{ij} = \sqrt{\varepsilon_i \cdot \varepsilon_j} \cdot (\sigma_i \cdot \sigma_j)^3 / (\sigma_{ij})^6 \tag{4.33}$$

The expression for interaction energy being seldomly simplified to:

$$\varepsilon_{ij} = (\varepsilon_i \cdot \varepsilon_j)^{1/2} \tag{4.34}$$

After the calculation of material constants through combination for a given composition mixture case, thermodynamic properties of one mole of mixture are obtained (V, $H - H^0$, $C_p - C_p{}^0$, $S - S^0$, etc.) by considering the mixture as a pure gas.

Mixture components properties can be also obtained—like molar potential properties or fugacities—if mixture material *constants* are considered as *dependencies* of composition.

Properties of Components in a Mixture

For example, for a binary mixture of van der Waals gases having indices (1, 2) and using constants combination rules (4.30 and 4.31a), namely $a = (\Sigma_i X_i \cdot \sqrt{a_i})^2$ and $b = x_1 \cdot b_1 + x_2 \cdot b_2$, for $(n \cdot P)$ product, from the ES $P = R \cdot T/(V - b) - a/V^2$, it results the following:

$$n \cdot P = \frac{(n_1 + n_2)^2 \cdot R \cdot T}{[(n_1 + n_2) \cdot V - n_1 \cdot b_1 - n_2 \cdot b_2]} - \frac{\left(n_1 \cdot a_1^{1/2} + n_2 \cdot a_2^{1/2}\right)^2}{(n_1 + n_2) \cdot V^2} \tag{4.8a2}$$

By applying partial derivative to equality (4.8a2) in relation to n_1 variable, it results the following:

$$[\partial(nP)/\partial n_1]_{V.T,n_2} = T_1 + T_2 + T_3 + T_4 \tag{4.8b2}$$

where the four terms of right term sum are:

$$T_1 = 2 \cdot (n_1 + n_2) \cdot R \cdot T/[(n_1 + n_2) \cdot V - n_1 \cdot b_1 - n_2 \cdot b_2]$$
$$T_2 = -(n_1 + n_2)^2 \cdot R \cdot T \cdot (V - b_1)/[(n_1 + n_2) \cdot V - n_1 \cdot b_1 - n_2 \cdot b_2]^2$$
$$\mathrm{T}_3 = -2 \cdot \sqrt{a_1} \cdot (n_1 \cdot \sqrt{a_1} + n_2 \cdot \sqrt{a_2})/\left[(n_1 + n_2) \cdot V^2\right]$$
$$\mathrm{T}_4 = (n_1 \cdot \sqrt{a_1} + n_2 \cdot \sqrt{a_2})^2/\left[(n_1 + n_2)^2 \cdot V^2\right]$$

or, when shifting to $X_1 = n_1/n$, $X_2 = n_2/n$ molar fractions:

$$\left[\frac{\partial(n \cdot P)}{\partial n_1}\right]_{V,T,n_2} = \frac{R \cdot T \cdot (V + b_1 - 2 \cdot b_1 \cdot x_1 - 2 \cdot b_2 \cdot x_2)}{(V - x_1 \cdot b_1 - x_2 \cdot b_2)^2} - \cdots$$
$$\cdots - [(2 - X_1) \cdot \sqrt{a_1} - X_2 \cdot \sqrt{a_2}] \cdot (X_1 \cdot \sqrt{a_1} + X_2 \cdot \sqrt{a_2})/V^2$$

by replacing $(X_1 \cdot b_1 + X_2 \cdot b_2)$ through b, and $(X_1 \cdot \sqrt{a_1} + X_2 \cdot \sqrt{a_2})$ through $\sqrt{a}$, it results that: $\partial(n \cdot P)/\partial n_1 = R \cdot T \cdot (V - 2 \cdot b + b_1)/(V - b)^2 - \left(2 \cdot a_1^{1/2} \cdot a^{1/2} - a\right)/V^2$.

The volume function $F(V) = (R \cdot T)^{-1} \cdot [\partial(n \cdot P)/\partial n_1]_{V,T,n_2} - V^{-1}$ beneath (4.27) integral becomes $= \frac{1}{(V-b)} - \frac{1}{V} + \frac{(b_1 - b_2) \cdot x_2}{(V-b)^2} - \frac{\left(2 \cdot a_1^{1/2} \cdot a^{1/2} - a\right)}{V^2 \cdot R \cdot T}$, from where it results the following for the defined integral $J_1(V) \equiv \int_V^\infty F(V) \cdot dV$:

$J_1(V) = (b_1 - b_2) \cdot x_2/(V - b) - (2 \cdot \sqrt{a \cdot a_1} - a)/(V \cdot R \cdot T) + \ln[V/(V - b)]$, and fugacity coefficient, according to relation (4.10b) of expression:

$\ln\varphi_1 = z - 1 - \ln z + J_1(V)$ where $z = \frac{P \cdot V}{R \cdot T}$ or $z = \frac{V}{(V-b)} - \frac{a}{V \cdot R \cdot T}$, becomes: $\ln\varphi_1 = b_1/(V-b) - 2 \cdot \sqrt{a \cdot a_1}/(R \cdot T \cdot V) - \ln\left[(1 - a \cdot V + a \cdot b)/\left(R \cdot T \cdot V^2\right)\right]$

Combination of ESs

Constants combination is possible only when the same ES is valid for each compound and when the values of all components' constants are known; otherwise, the procedure for *ESs* combination is used, which applies *mixture* special *rules*—rules representing the empirical generalization of some relations strictly valid for perfect gas ideal mixtures.

Let $V = F_V(P)$ and $P = F_P(V)$, the isothermal dependencies of volume and pressure for the gas mixture, and $V = F_{V,i}(P)$ and $P = F_{P,i}(V)$—similar dependencies for *i* component of mixture when found in pure state. The four mixing rules possible are:

1. Dalton rule, applicable if mixture volume *V* exceeds more than five times the *minimum* molar volume ($V_{\min} \equiv \lim_{P \to \infty} V$) resulted from the ES of each component:

$$F_P(V) = \Sigma_i F_{P,i}(V/X_i) \tag{4.35}$$

2. *Amagat* rule:

$$F_V(P) = \Sigma_i F_{V,i}(P \cdot X_i) \tag{4.36}$$

3. *Bartlett* rule:

$$F_P(V) = \Sigma_i X_i \cdot F_{P,i}(V) \tag{4.37}$$

4. *Lewis* rule:

$$F_V(P) = \Sigma_i X_i \cdot F_{V,i}(P) \tag{4.38}$$

At too elevated pressures (densities below 70% of critical density), the accuracy is approximately 5% for Amagat, Lewis or Bartlett rules, the last one being preferred at even more elevated densities.

For fugacities calculation, Lewis law (4.38) use is simpler, where component *i* molar partial volume in the real mixture is equal – as to the ideal mixture of gases – with the molar volume $(V^{o})_i$ of the same compound found in pure state at mixture temperature and pressure: $\overline{V}_i(T,P) = V_i^0(T,P)$. When using relations (4.10a, 4.26), for fugacity *Lewis & Randall* rule, the following relation is obtained:

$$\varphi_i = (\varphi^\circ)_i \tag{4.39}$$

and afterward a mixture gas fugacity coefficient φ_i is equal to the same gas fugacity $(\varphi^\circ)_i$ when found in pure state at mixture temperature and pressure.

4.9 Interactions among Components in a Mixture

A more precise theory does not consider mixture properties as an *additive* dependence of components' properties but considers the interactions specificity between different components. It has been experimentally shown that additivity deviations are the higher as:

I. The i component's composition is closer to the equimolar composition: $X_1 = X_2 = 1/2$ for a binary mixture, $X_1 = X_2 = X_3 = 1/3$ for a ternary mixture, etc.

II. Components structure is more different; among the differences at molecular level, the ones provoking more important deviations from additivity are in order the following:

- molecule polarity, measured through its electric dipole moment;
- molecule *polarizability*, characterizing the ease of deformation and whose measure is the molecular diameter σ; at macroscopic level, the molar volume corresponds to this property, mainly the critical form;
- intermolecular potential ε energy, whose macroscopic correspondent is the critical temperature T_c.

Interaction Formulae

For Redlich–Kwong ES (4.8a), $R \cdot T/(V - b) - a/\left[V \cdot (V + b) \cdot \sqrt{T}\right]$ different substances interaction leads to supplemental terms occurrence $k_{i,j}$ where combination rules for material constant a are as following:

$$a = \left(\sum_{i=1}^{i=C} X_i \cdot \sqrt{a_i}\right)^2 + \sum_{i=1}^{i=C-1} \sum_{j=i+1}^{j=C} X_i \cdot X_j \cdot k_{i,j} \tag{4.40}$$

Calculation of constant b, through relation (4.31a), does not provide interactions.

The corrective factors $k_{i,j}$, independent of composition or pressure and less dependent of temperature, are obtained from *PVT* data of the binary mixture formed from i and j gases. In the absence of data, they can be evaluated through:

$$k_{i,j} = -\sqrt{a_i \cdot a_j} + \left(V_{ci} \cdot V_{cj} \cdot \sqrt{T_{ci} \cdot T_{cj}}\right)^{3/2} \cdot \left[2/\left(\sqrt[3]{V_{ci}} + \sqrt[3]{V_{cj}}\right)\right]^6 \cdot \left(\frac{a_i \cdot z_{ci}}{V_{ci} \cdot (T_{ci})^{3/2}} + \frac{a_j \cdot z_{cj}}{V_{cj} \cdot (T_{cj})^{3/2}}\right) \cdot \left[\frac{2 \cdot \sqrt{T_{ci} \cdot T_{cj}}}{(T_{ci} + T_{cj})}\right]^{3/2} / (z_{ci} + z_{cj}) \quad (4.41)$$

Similarly, Bartlett mixing rule (4.37) can become more accurate through the modification of *Kričevski*:

$$F_P(V) = \sum_{i=1}^{i=C} X_i \cdot F_{Pi}(V) + \sum_{i=1}^{i=C-1} \sum_{j=i+1}^{j=C} X_i \cdot X_j \cdot k_{i,j} \cdot \left|F_{Pi}(V) - F_{Pj}(V)\right| \quad (4.42)$$

where k_{ij} are *other* constants, also independent of composition, of P and V.

4.10 Four-Worked Examples

Ex. 4.1
For H_2S case, Redlich–Kwong ES material constants are: $a = 8.848$ Kg·m^5·K$^{1/2}$/(mol^2·s^2), $b = 2.987 \cdot 10^{-4}$ m^3/mol. It is required to calculate:

1. Molar volume V' at: $T' = 400$ K, $P' = 1.2 \cdot 10^7$ N/m^2;
2. Compressibility factor z and residual volume V_{rez} under the same conditions;
3. Isothermal enthalpy dependence on pressure, $\Delta^P H_T$, accompanying gas compression from $P = 0$ to $P = P'$, for $T = T'$; and,
4. Coefficient of expansion α' comparison under point 1 conditions, with the coefficient of expansion α_{id} of the ideal gas under the same conditions.

Solution
1. Redlich–Kwong Eq. (4.8a), $P = R \cdot T/(V - b) - a/\left[V \cdot (V + b) \cdot \sqrt{T}\right]$, is not made explicit after volume V but only after P.

   With values (a, b, $P = P'$ and $T = T'$) from the statement, the ES becomes:

$$1.2 \cdot 10^7 = 8.31 \cdot 400/\left(V - 1.302 \cdot 10^{-4}\right) - 10.48/\left[V \cdot \left(V + 1.302 \cdot 10^{-4}\right) \cdot \sqrt{400}\right]$$

or, by denominators removing and by ordering after V powers:

$$1.2 \cdot 10^7 V^3 - 3.325\, V^2 - 1.804V - 6.818 \cdot 10^{-5} = 0.$$

Through change of variable:

$$V = x/10^5 + (1/3) \cdot \left(3325.6/1.2 \cdot 10^7\right) \tag{4.10a2}$$

which cancels the 2nd order term, the following is obtained:

$$F(x) = x^3 + 20.947 \cdot x - 127.72 = 0 \tag{4.10b2}$$

The function from (4.10b2) is increasing (its derivative, $3 \cdot x^2 + 20.947$, being positive for any x real value), therefore has a single real root. This is situated in the range between $x = 0$, where $F(x) = -127.72$ is negative and $x = (127.72)^{1/3} = 5.0360$, a point where $F(x) = 20.947 \cdot 5.0360 = 105.49$ is positive. The following table contains the successive approximations of solution using the secant method from Appendix A.5 of the final mathematical annex:

x	0	5.0360	3	3.7	3.68	3.693	3.6929
$F(x)$	−127.72	+105.49	−17.879	+0.4369	−0.7991	+0.00332	−0.00286

Solution with four exact figures is thus $x = 3.693$ from where, through relation (4.10a2):

$$V = 3.693/10^5 + (1/3) \cdot \left(3326/1.2 \cdot 10^7\right) \Rightarrow V = \mathbf{1.293 \cdot 10^{-4}\ m^3/mol}$$

2. From relations (4.2 and 4.3), it is found that:

   $V_{rez} = V - R \cdot T'/P' = 1.293 \cdot 10^{-4} - 8.31 \cdot 400/1.2 \cdot 10^7$ =>
   $\boldsymbol{V_{rez} = -1.478 \cdot 10^{-4}\ m^3/mol}$
   $z = P \cdot V/(R \cdot T) = 1.2 \cdot 10^7 \cdot 1.291 \cdot 10^{-4}/(8.31 \cdot 400)$ => $z = \mathbf{0.467}$

3. Relation (4.9), $\Delta^P H_T = H(T,P) - H(T,0) = \int_0^P \left[V - T \cdot (\partial V/\partial T)_P\right] \cdot dP$, gives the enthalpy isothermal dependence. But Redlich–Kwong ES (4.8a) is not made explicit after V so that $dP = (\partial P/\partial V)_T \cdot dV$ is transformed, from where: $\Delta^P H_T = \int_{V=\infty}^{V=V'} \left[V \cdot (\partial P/\partial V)_T - T \cdot (\partial V/\partial T)_P \cdot (\partial P/\partial V)_T\right] \cdot dV$, then product $(\partial V/\partial T)_P \cdot (\partial P/\partial V)_T$ from the function to be integrated is replaced by expression $-(\partial P/\partial T)_v$, the calculation relation for $\Delta^P H_T$ becoming therefore:

$$\Delta^P H_T = \int_{V=\infty}^{V=V'} \left[V \cdot (\partial P/\partial V)_T + T \cdot (\partial V/\partial T)_V\right] \cdot dV \tag{4.10c2}$$

The two partial derivatives beneath the integral are obtained from ES (4.8a), $P = R \cdot T/(V - b) - a/\left[V \cdot (V + b) \cdot \sqrt{T}\right]$. They will be:

$$(\partial P/\partial V)_T = -R\cdot T/(V-b)^2 + a\cdot b\cdot(2\cdot V+b)/\left[V^2\cdot(V+b)^2\cdot\sqrt{T}\right] \tag{4.10d2}$$

and

$$(\partial P/\partial T)_V = -R/(V-b) + a/\left[V\cdot(V+b)\cdot T\cdot\sqrt{T}\right] \tag{4.10e2}$$

and the function to be integrated from (4.10c2) becomes then rearranged:

$$-R\cdot T\cdot b/(V-b)^2 + \left(0.5\cdot a/\sqrt{T}\right)\cdot\left(V^2+5\cdot b\cdot V+2\cdot b^2\right)/[V\cdot(V+b)]^2 \tag{4.10f2}$$

Coefficient $(V^2 + 5\cdot b\cdot V + 2\cdot b^2)/ [V\cdot(V + b)]^2$ of $0.5\cdot a/ \sqrt{T}$ from (4.10f2) is decomposed into four terms directly integrable, namely:

$$2/V^2\cdot -2/(V+b)^2 + 1/(b\cdot V) - 1/[b\cdot(V+b)].$$

Integration from (4.10c2) will give:

$$\Delta^P H_T = \frac{R\cdot T\cdot b}{(V-b)} - \frac{a}{\sqrt{T}}\left[\frac{b}{V\cdot(V+b)} + \frac{\ln(1+b/V)}{2\cdot b}\right]$$
$$= \frac{8.3142\cdot 400\cdot 2987}{(12931-2987)} - \frac{8.848}{\sqrt{400}\cdot 10^{-8}}$$
$$\cdot\left[\frac{2987}{12931\cdot(12931+2987)} + \frac{\ln(1+2987/12931)}{2\cdot 2987}\right]$$
$$\Rightarrow \quad \mathbf{\Delta^P H_T = 456.6\ J/mole}$$

4. Partial derivative $(\partial V/ \partial T)_P$ calculation from coefficient of expansion definition $\alpha \equiv V^{-1}\cdot(\partial V/ \partial T)_P$ is not possible directly for Redlich–Kwong (R. & K.) ES (4.8a), because this ES is not made explicit neither after volume nor temperature.

However, using relation $(\partial x/ \partial y)_z\cdot (\partial y/ \partial z)_x\cdot(\partial z/ \partial x)_y = -1$ between three partial derivatives of a bivariate set $F(x,y,z) = 0$, then $(\partial V/ \partial T)_P$ can be expressed through derivatives $(\partial P/ \partial T)_V$ and $(\partial P/ \partial V)_T$ ratios, which can be calculated using $P(V,T)$ expression, available for this ES. The coefficient of expansion therefore becomes:

$$\alpha \equiv -(\partial P/\partial T)_V/\left[V\cdot(\partial P/\partial V)_T\right] \tag{4.10g2}$$

By introducing values of Redlich & Kwong constants in relations (4.10g2) and respectively (4.10e2), the statement values for hydrogen sulfide as well as of temperature and pressure, namely: $a = 8.848\ \mathrm{Kg\cdot m^5\cdot K^{1/2}/(mol^2\cdot s^2)}$;

$b = 2.987 \cdot 10^{-5}$ m³/mol; $T' = 400$ K; $P' = 1.2 \cdot 10^7$ Pa, together with value $V' = 1.293 \cdot 10^{-4}$ m³/mol from Solution no. 1, the following partial derivatives values are obtained:

$$(\partial P/\partial V)'_T = -8.31 \cdot 10^{10} \cdot 400/(12.93 - 2.99)^2 + 8.848 \cdot 10^{10}\left[1/12.93^2 \cdot -1/(12.93 + 2.99)^2\right]/\left(2.99 \cdot 400^{1/2}\right) = -3.33 \cdot 10^{12}\text{Pa} \cdot \text{mole/m}^3$$

$$\begin{aligned}(\partial P/\partial T)'_V &= -8.31/\left[10^{-5} \cdot (12.93 - 2.99)\right] \\ &\quad + 8.848/\left[10^{-10} \cdot 12.93 \cdot (12.93 + 2.99) \cdot 400^{1.5}\right] \\ &= 1.373 \cdot 10^5\ Pa/K.\end{aligned}$$

By replacing the values of $(\partial P/\partial V)'_T$, V', and $(\partial P/\partial T)'_V$ into expression (g), the following is obtained under point 1 conditions:

$$\alpha' = 1.373 \cdot 10^5/\left(1.293 \cdot 10^{-4} \cdot 3.33 \cdot 10^{12}\right) \Rightarrow \alpha' = \mathbf{0.000318\ K^{-1}}$$

The ideal gas expansibility depends only on temperature, the gas being perfect. At $T = T'$ temperature, it is found that:

$$\alpha' = 1/T' = 1/400 = 0.0025\,.$$

The real gas is therefore under these conditions of 25/3.18 ≅ 8 times less expansible than ideal gas.

Ex. 4.2

Virial coefficients have been determined for methane at 303 K, namely: $B_2 = -7 \cdot 10^{-5}$ m³/mol and $B_3 = 3.3 \cdot 10^{-9}$ m⁶/ mole². By using Dieterici equation, it is required to calculate the following:

1. ES constants (*a, b*)
2. Boyle temperature
3. Inversion curve
4. Coefficient of expansion

Solution

1. After $(R \cdot T)$ division and by substituting:

$$V = 1/x, \quad a = c \cdot R \cdot T,$$

the ES (4.8c), $P = [R \cdot T/(V - b)] \cdot \exp\,[-a/(R \cdot T \cdot V)]$, becomes: $P/(R \cdot T) = x \cdot (1 - b \cdot x)^{-1} \cdot e^{-c \cdot x}$. MacLaurin series expansion (with formulae from annex Appendix A.5.) of the last two factors after x powers leads to: $P/(R \cdot T) = x \cdot (1 + b \cdot x + b^2 \cdot x^2 + K) \cdot (1 - c \cdot x \cdot + c^2 \cdot x^2/2 + \ldots)$ or

$$P/(R\,T) = x + (b - c)\cdot x^2 + \left(b^2 - b\cdot c + c^2/2\right)\cdot x^3 + \ldots.$$

If the first three x power coefficients are equalized with the first three (1/ V) powers from member II of virial Eq. (4.11)), $P/(R\,T) = 1/V + B_2/V^2 + B_3/V^3 + \ldots$, the following two equations system of two unknowns (b, c):

$$b - c = B_2 \tag{4.10a3}$$

$$b^2 - b\cdot c + c^2/2 = B_3 \tag{4.10b3}$$

From Eq. (4.10a3) the following is obtained:

$$c = b - B_2 \tag{4.10c3}$$

By substituting c into relation (4.10c3) from Eq. (4.10b3), it is found that:

$b^2 = 2{\cdot}B_3 - (B_2)^2/2$ or, because b cannot take negative values:
$b = [2{\cdot}B_3 - (B_2)^2/2]^{1/2}$ from where, through relation (4.10c3) and returning by $a = c{\cdot}R{\cdot}T$ at material constant a from ES: $a = R{\cdot}T{\cdot}(b - B_2)$

With values T, B_2, and B_3 from the statement, it is found that:

$$b = \left[2\cdot 3.3\cdot 10^{-9} - \left(7\cdot 10^{-5}\right)^2/2\right]^{0.5} \Rightarrow \mathbf{b = 6.44\cdot 10^{-5} m^3/mole}$$

$$a = \left(6.44\cdot 10^{-5} + 7\,10^{-5}\right)\cdot 8.31\cdot 303 \Rightarrow \mathbf{a = 0.368\ kg\cdot m^5/\left(s^2\cdot mole^2\right)}$$

2. Isothermal derivation after volume of ES Dieterici (4.8c)

$$P = [R\cdot T/(V - b)]\cdot \exp[-a/(R\cdot T\cdot V)] \tag{4.10d3}$$

leads to:

$$(\partial P/\partial V)_T = (V - b)^{-1}\cdot \left[a/V^2 - R\cdot T/(V - b)\right]\cdot \exp[-a/(R\cdot T\cdot V)] \tag{4.10e3}$$

Boyle temperature is the temperature verifying condition (4.15), namely: $\lim_{P\to 0}[\partial(P{\cdot}V)/\partial V]_T = 0$. By replacing P with V as independent variable, the condition becomes: $\lim_{V\to\infty}[\partial(P{\cdot}V)/\partial V]_T = 0$ or $\lim_{V\to\infty}[P + V{\cdot}(\partial P/\partial V)_T] = 0$. By introducing P and $(\partial P/\partial V)_T$ from equalities (4.10d3) and respectively (4.10e3) into square bracket expression leads to:

$$T = T_B \Rightarrow \lim_{V\to\infty}[(b\cdot R\cdot T/(V - b) + a/V] = 0$$

from where: $T_B = \lim_{V\to\infty}[(a{\cdot}(V - b)/(b{\cdot}R{\cdot}T{\cdot}V)]$.

It results that: $T_B = a/(R{\cdot}b)$ or, with (a,b) from point 1:

$$T_B = 0.368/\left(8.31 \cdot 6.44 \cdot 10^{-5}\right) \Rightarrow \mathbf{T_B = 688\ K}$$

3. On the inversion curve, according to condition (4.18): $T - V{\cdot}(\partial T/\partial V)_P = 0$ or:

$$T \cdot (\partial P/\partial T)_V + V \cdot (\partial P/\partial V)_T = 0 \tag{4.10f3}$$

ES isochore derivative-relation (4.10c2) - relative to temperature:

$$(\partial P/\partial T)_V = [R + a/(V \cdot T)] \cdot (V - b)^{-1} \cdot \exp[-a/(R \cdot T \cdot V)] \tag{4.10g3}$$

By replacing the two partial derivatives from (4.10e3) and (4.10g3) into (4.10f3), the following is obtained:

$$[2 \cdot a/V - R \cdot T/(V - b)] \cdot (V - b)^{-1} \cdot \exp[-a/(R \cdot T \cdot V)] = 0 \text{ or :}$$

$$T = 2 \cdot a \cdot (V - b)/(R \cdot b \cdot V) \tag{4.10h3}$$

By introducing $T(V)$ given by (4.10h3) into ES, relation (4.10d3), it results that:

$$P = [2 \cdot a/(b \cdot V)] \cdot \exp[-0.5/(V/b - 1)]$$

or, with values $a = 0.368$ and $b = 6.44{\cdot}10^{-5}$ from point 1:

$$P = \left[2 \cdot 0.368/\left(6.44 \cdot 10^{-5} \cdot V\right)\right] \cdot \exp\left[-0.5/\left(V/6.44 \cdot 10^{-5} - 1\right)\right]$$

$$\Rightarrow \mathbf{P} = \left(\mathbf{1.14 \cdot 10^5/V}\right) \cdot \boldsymbol{exp}\left[\mathbf{-0.5/\left(1.553 \cdot 10^4 \cdot V - 1\right)}\right]$$

Ex. 4.3

Sulfuryl chloride molar mass SO_2Cl_2 is $M = 0.135$ kg/mol, and O and Cl atoms are arranged on an irregular tetrahedron tops, containing S in the center. It is required to calculate:

1. Van der Waals constants, starting from the molecular structure;
2. Temperature from state A, for which density is $\rho = 225$ kg/m^3 at pressure $P = 4.5$ MPa;
3. Fugacity in state A.

Solution

1. By using van Laar procedure from Sect. 4.2, the following increments sums are obtained:

   - For constant a: $2{\cdot}k_{a,O} + 2{\cdot}k_{a,Cl}$ (because atom S is "wrapped" by the other atom, it does not contribute to value a) or: $2{\cdot}2.7 + 2{\cdot}5.4 = 16.2$ (from Table 4.2).
   - for constant b: $2{\cdot}k_{b,O} + 2{\cdot}k_{b,Cl} + k_{b,S} = 2{\cdot}5 + 2{\cdot}11.5 + 12.5 = 45.5$ (from Table 4.3, by adopting for O the value corresponding to double bond, and for Cl, value from inorganic compounds).

From relations (4.7a), (4.7b) are obtained respectively:

$$a = 0.0503 \cdot \left(\sum\nolimits_i \mu_i \cdot k_{a,i}\right)^2 = 0.0503 \cdot 16.2^2 = 13.2\ \text{atm} \cdot \text{L}^2/\text{mol}^2$$

$$b = 0.00224 \cdot \sum\nolimits_i \mu_i \cdot k_{b,i} = 0.00224 \cdot 45.5 = 0.102\ \text{L/mol}.$$

By expressing into S.I. fundamental units system with 1 atm = 101325 Pa and 1 L = 0.001 m^3:

$$\Rightarrow \mathbf{a = 1.337\ kg \cdot m^5 \cdot s^{-2} \cdot mol^{-2}; b = 1.019 \cdot 10^{-4} m^3/mol}$$

2. $V = M/\rho = 0.135/225 = 6{\cdot}10^{-4}$ m^3/mol. From van der Waals ES (4.6) made explicit after temperature, $T = (P{\cdot}V^2 + a){\cdot}(V - b)/(R{\cdot}V^2)$, the following is obtained:

$$T = \left[4.5 \cdot 10^6 + 1.337/\left(6 \cdot 10^{-4}\right)^2\right].(6 - 1.02) \cdot 10^{-4}/8.31 \Rightarrow \mathbf{T = 492\ K}$$

3. Relation (4.10b) for fugacity coefficient calculation is equivalent to:

$$\ln \varphi = z - 1 - \ln z + J \tag{4.10a4}$$

where: $J \equiv \int_V^\infty [(z-1)/P] \cdot dV$ or, by substituting z from relation (4.3):

$$J = \int_V^\infty [P/(R \cdot T) - 1/V] \cdot dV \tag{4.10b4}$$

By replacing into (4.10b4) function $P = R{\cdot}T/(V - b) - a/V^2$ from ES, it results that: $J = \int_V^\infty \left[1/(V-b) - 1/V - a/\left(R \cdot T \cdot V^2\right)\right] \cdot dV$ from where:

$$J = -a/(R \cdot T \cdot V) - \ln[(V-b)/V] \tag{4.10c4}$$

On the other hand, compressibility coefficient definition (4.3), $z \equiv P{\cdot}V/(R{\cdot}T)$, becomes for van der Waals ES (VdW):

$$z = V/(V-b) - a/(R \cdot T \cdot V) \tag{4.10d4}$$

By introducing z and J from (4.10d4) and (4.10c4) equalities into (4.10a4), it is found that:

$\ln \varphi = \frac{V}{(V-b)} - \frac{a}{R \cdot T \cdot V} - 1 - \ln\left[\frac{V}{(V-b)} - \frac{a}{R \cdot T \cdot V}\right] - \frac{a}{R \cdot T \cdot V} + \ln[(V-b)/V]$ from where: $\ln \varphi = b/(V - b) - 2 \cdot a/(R \cdot T \cdot V) - \ln[1 - a \cdot (V - b)/(R \cdot T \cdot V^2)]$.

and afterwards, with (a, b) from point 1 at temperature from point 2:

$$\ln \varphi = 1.02/(6 - 1.02) - 2 \cdot 1.337/\left[8.31 \cdot 492 \cdot 6 \cdot 10^{-4}\right] - \ldots$$

$$.. - \ln\left[1 - 1.337 \cdot (6 - 1.02)/\left(8.31 \cdot 492 \cdot 6^2 \cdot 10^{-4}\right)\right] = -0.27$$

$$\text{Fugacity}\, f \equiv P \cdot \varphi = 4.5 \cdot 10^6 \cdot e^{-0.27} \Rightarrow \mathbf{f = 34.4 \cdot 10^6 Pa}$$

Ex. 4.4

For a van der Waals gaseous binary mixture, it is required to calculate the following:

1. Rapid procedures identifications (without transcendental equations or superior degree equations) that use only (*a, b*) constants values;
2. Pressure of CH_4 + H_2 mixture with 40% molar methane, of molar volume $6{\cdot}10^{-5}$ mol/ m^3, at temperature $T = 300.69$ K, through the different procedures.
3. The most probable pressure calculation, P_{est}, under conditions of point 2.

Solution

1. Among constants combination procedures, the only ones that do not use other information than values (*a,b*) are van Laar ones, with relations (4.30), (4.31a) for *a* respectively *b*, and the Lorentz one, using also relation (4.30) for a calculation but with *b* from relation (4.31b).

Among the four pure components ESs combination procedures, Amagat and Lewis procedures combining *volume* are in this case too heavy, because van der Waals ES is of third degree in V molar volume.

Therefore, remain acceptable only the two procedures based on components *pressures* combination, namely Dalton and Bartlett rules, which use relations (4.35) respectively (4.37).

Possible procedures are therefore: Van Laar, Lorentz, Dalton, Bartlett

2. The individual material constants for methane (component no. 1) and for hydrogen (no. 2) are removed from Table 4.1, namely:

$a_1 = 2.26\,L^2{\cdot}atm/mol^2$, $b_1 = 42.8$ mL/mol, $a_2 = 0.245\,L^2{\cdot}atm/mol^2$, $_2 = 26.6$ mL/mol or, when converting into S.I. fundamental units for 1 atm = 101325 Pa:

$$a_1 = 0.229\ \text{kg} \cdot \text{m}^5/\left(\text{s}^2 \cdot \text{mol}^2\right); a_2 = 0.0248\ \text{kg} \cdot \text{m}^5/\left(\text{s}^2 \cdot \text{mol}^2\right);$$

$$b_1 = 4.28 \cdot 10^{-5}\text{m}^3/\text{mol}; b_2 = 2.66 \cdot 10^{-5}\text{m}^3/\text{mol}$$

Mixture molar fractions are: $X_1 = 0.4$ from the statement and $X_2 = 1 - X_1 = 0.6$.

van Laar pressure: With relations (4.30, 4.31a): $a_{VL} = \left(X_1 \cdot \sqrt{a_1} + X_2 \cdot \sqrt{a_2}\right)^2$ $= \left(0.4 \cdot \sqrt{0.229} + 0.6 \cdot \sqrt{0.0248}\right)^2 = 0.0817\ \text{kg} \cdot \text{m}^5/\left(\text{s}^2 \cdot \text{mol}^2\right)$ and $b_{VL} = X_1{\cdot}b_1 + X_2{\cdot}b_2 = (0.4{\cdot}4.28 + 0.6{\cdot}2.66){\cdot}10^{-5} = 3.31{\cdot}10^{-5}$ m^3/ mol

From van der Waals ES (4.6) $P = R{\cdot}T/(V - b) - a/V^2$ is calculated with van Laar procedure, with R = 8.314 J/(mol·K) and with values T, V from the statement: $P_{VL} = 8.314{\cdot}300.7/[(6-3.31){\cdot}10^{-5}] - 0.0817/(6{\cdot}10^{-5})^2$

$$\Rightarrow \mathbf{P_{VL} = 6.69 \cdot 10^7 Pa}$$

Lorentz pressure: Value of a from relation (4.30) is the Van Laar procedure one: $a_{Lo} = 0.0817$ kg·m^5/ (s^2·mol^2), and with relation (4.31b):

$$b_{Lo} = \left[X_1 \cdot (b_1)^{1/3} + X_2 \cdot (b_2)^{1/3}\right]^3 = \left[0.4 \cdot (4.28)^{1/3} + 0.6 \cdot (2.66)^{1/3}\right]^3 \cdot 10^{-5}$$
$$= 3.27 \cdot 10^{-5} m^3/mol.$$

From van der Waals ES, the following is obtained:

$$P_{Lo} = R \cdot T/(V_{Lo} - b) - a/(V_{Lo})^2 = 8.3142 \cdot 300.69/\left[(6 - 3.27) \cdot 10^{-5}\right] -$$
$$\ldots - 0.0817/(6 \cdot 10^{-5})^2 \Rightarrow \mathbf{P_{Lo} = 6.89 \cdot 10^7 Pa}$$

Dalton pressure: According to definition (4.14), *minimum* molar volumes of pure state components (V_{min}) are equal at VdW gas with respective b constants, $4.28 \cdot 10^{-5}$ and $2.66 \cdot 10^{-5}$ m^3/ mol. But mixture molar volume, $6 \cdot 10^{-5}$ m^3/mol, does *not* exceed at least five times the higher value of V_{min} values; therefore, Dalton procedure is not applicable in the conditions stated at point 2.

Bartlett pressure:

- Methane from relation (4.6): $P_{Ba,1} = R{\cdot}T/\ (V - b_1) - a_1/V^2 = 8.3142{\cdot}300.69/ [(6{-}4.28){\cdot}10^{-5}] - 0.229/\ (6{\cdot}10^{-5})^2 = 8.17{\cdot}\ 10^7$ Pa
- Pure hydrogen, with the same relation: $P_{Ba,2} = R{\cdot}T/\ (V - b_2) - a_2/\ V^2$ or

$$P_{Ba,2} = 8.3142 \cdot 300.69/\left[(6 - 2.66) \cdot 10^{-5}\right] - 0.0248/(6 \cdot 10^{-5})^2 = 6.80 \cdot 10^7 \text{ Pa}$$

- For the mixture, from Bartlett relation (4.37):

$$P_{Ba} = X_1 \cdot P_{Ba,1} + X_2 \cdot P_{Ba,2} \ \ or$$

$$P_{Ba} = (0.4 \cdot 8.17 + 0.6 \cdot 6.80) \cdot 10^7 \Rightarrow \mathbf{P_{Ba} = 7.35 \cdot 10^7 Pa}$$

3. Calculated values at point 2 by three procedures—Van Laar, Lorentz, and Bartlett—seem all equally justified, only that the first procedures are based on 50% on the same relation, namely at one of the two mixtures constants calculation: a, by relation (4.30).

Weights' assignment during mean calculation: For a relative weight $w_{Ba} = 4$ of the Bartlett value, to each of van Laar and Lorentz procedures should be given a weight of 2 if they would be completely identical, and to a weight of 4 if they would be completely different. Since they differ in proportion of 50%, their weights which seem justified are of $w_{VL} = w_{Lo} = (2 + 4)/2 = 3$.

The weighted mean of pressure is thus:

$$P = (w_{VL} \cdot P_{VL} + w_{LO} \cdot P_{LO} + w_{Ba} \cdot P_{Ba})/(w_{VL} + w_{LO} + w_{Ba}) \text{ or}$$

$$P = (3 \cdot 6.69 + 3 \cdot 6.89 + 4 \cdot 7.35) \cdot 10^7/(3 + 3 + 4) = 7.01 \cdot 10^7 Pa.$$

Since procedures values scattering is relatively important (namely, of $7.35 \cdot 10^7 - 6.69 \cdot 10^7 = 0.66 \cdot 10^7$), the estimate mean value is in fact less accurate—for example it will be rounded to ⇒ $\mathbf{P_{est} = 7.0 \cdot 10^7}$ **Pa.**

Chapter 5
Liquid-Vapor Equilibrium Models – Critical Point, Corresponding States, and Reduced Properties

5.1 Phase Equilibrium of Pure Substances

Vapors in Molecular Physics

The term "steam" derives from the technical field, where *vapors* (at plural) is the name given to a gaseous phase which can be relatively easily obtained – namely, without major pressure or temperature variations – starting from what is called a condensed phase: either from a liquid one (through evaporation or boiling process) or from a solid one (through sublimation).

On the microscopic level, there is no structural difference between the gas itself and the vapors of a substance, so that their physicochemical demarcation is purely conventional: any gas of pressure and temperature (P, T) lower than the critical pressure and temperature (P_c, T_c) is designated as vapors, namely:

$$\text{"vapors"} \equiv \text{gas}, (P < P_c) \cap (T < T_c) \tag{5.1a}$$

The other fluid state definitions conventionally delimited as being according to Fig. 7.1 from page 126 of the book on thermodynamics of the same authors are as follows (Daneş et al. 2013 [1]):

$$\text{"gas itself"} \equiv \text{gas}, (P < P_c) \cap (T > T_c) \tag{5.1b}$$

$$\text{"supercritical fluid"} \equiv \text{fluid}, (P > P_c) \cap (T > T_c) \tag{5.1c}$$

$$\text{"liquid itself"} \equiv \text{liquid}, (P < P_c) \tag{5.1d}$$

The term critical point has been already introduced there, as the terminal point of biphasic liquid-vapor equilibrium curve, namely, the end having higher temperature and pressure. But this definition remained formal, since this point's existence does not necessarily follow the thermodynamic principles.

F. E. Daneş et al., *Molecular Physical Chemistry for Engineering Applications*,
https://doi.org/10.1007/978-3-030-63896-2_5

Mono-Component System: Phase Diagrams

As shown in the phenomenological thermodynamics, the Gibbs law becomes, for the system comprising a single component (pure substance) in usual conditions – namely, without major contributions of the electrical and magnetic energies:

$$L = 3 - F$$

according to relation (13.6) from page 296 of the same book (Daneş et al. 2013 [1]), relation where F is system's number of phases at equilibrium while L represents the *variant* or *degrees of freedom number*, namely the state variables number that can be independently modified.

Since F is a positive integer, three cases of system states can be distinguished:

- Monophasic system, therefore bivariant ($F = 1$ from where $L = 2$)
- Diphasic system, which can only be univariate, $F = 2$ and $L = 1$
- Triphasic system, invariable (zero-variant), where $F = 3$ and $L = 0$

Since the maximum possible variance is 2, it follows that all equilibrium states of a mono-component system can be graphically represented on a unique plot – a system phase diagram. To any system's equilibrium state corresponds a representative point in the system phase diagram – pressure P is graphed dependently on temperature T; reciprocally, any point of arbitrary coordinates (P, T) designates a well-defined equilibrium state. Certainly, there is an exception, namely the (P, T) field, where the substance is unstable and suffering a chemical reaction (decomposition, disproportionation, dimerization or more generally polymerization, association, etc.)

Mono-Component System: The State Diagrams

Unlike phase diagram of a substance, which represents only the existing information regarding system phase (or phases) for a given *thermomechanical* set of conditions (P and T), the state diagram indicates a certain quantity value of the given substance, found in a certain phase, at values set for two intensive properties of state.

Molar volume and pressure are chosen, for example, as independent variables of the state diagram or – as in the phase diagram – the pressure and temperature pair.

Quantity is meant here as any property which is intensive (which does not depend on the system's mass, which is at the same time specific to the respective substance).

The quantity can be *thermodynamic* (volume, enthalpy, entropy, etc.) or *non-thermodynamic* (viscosity, refractive index, mechanical resistance). But a field – as it is the case for the quantities: pressure, temperature, electrical potential, magnetic intensity, unitary stress, or movement speed – is not considered as a quantity, because it can be meant regardless of the nature of the system's matter.

A phase diagram turns into a state diagram (of *P*, *T* type) through drawing – into each phase 2D field – of an *isoquantity* curve family, each indicating a value different from the one of the respective quantities.

Triple Points

Into the phase diagram, to monophasic systems corresponds a part of the graph's surface, to biphasic systems, a line segment (curve) while to triphasic systems, a point, designated as *triple point* where three biphasic equilibrium curves meet, curves separating three monophasic fields.

For example, the triple point SLG which is present in any system where all three states of aggregation present stability of eigenfields (*P*, *T*) is located at the intersection of the three biphasic equilibrium curves of sublimation (solid-vapor equilibrium), melting (liquid-vapor equilibrium), and vaporization (liquid-vapor equilibrium). Certainly, other triple points will exist in *polymorphic* systems (many stable solid phases) of type SSS, SSL, and/or SSG.

Singularity of Vaporization among Phase Transitions

Vaporization equilibrium state is, for the mono-component system, the equilibrium between the liquid and the vapor phases of that substance. Liquid-vapor equilibrium is usually treated as *liquid/gas* equilibrium, abbreviated *LG equilibrium*, since vapors are nothing but a gas placed under certain temperature and pressure conditions, as seen from definition (5.1a).

The *vaporization* curve (*LG equilibrium* curve) into (*P*, *T*) coordinates, inherent to phase diagram, differs from the other types of biphasic equilibrium from the ordinary mono-component systems: this curve starting from SLG triple point ends by a *critical* point, more completely designated as "liquid-vapor critical point."

The curves of the other biphasic equilibrium types – SS, SL, and SG where S, L, and G indices designate the solid, liquid, and gaseous states of aggregation – never have a critical point, but only triple ones. Certainly, many of the equilibrium curves – including the LG equilibrium one – are not completely traceable, since the substances decompose or equilibrium achievement is too slow (very low temperatures) or the measurement conditions (the field of very high pressures) cannot be achieved.

The critical point can exist in poly-component systems with several condensed phases, or even in mono-component systems if they have special electrical or magnetic properties, but in the *usual* mono-component systems, where system state is completely described by its mechanical and thermal quantities, the liquid-vapor critical point is the only one that exists, so that it will be briefly designated as "critical point."

Variation of Properties on the Vaporization Curve

Two values of a certain given quantity – molar enthalpy, conductivity, light or sound speed, etc. – correspond to any point of an equilibrium curve, usually different ones, each value corresponding to one of the two phases in equilibrium.

Gibbs free energy and fugacity represent the exceptions, with values equal for both phases, at the given pressure and temperature.

Certainly, that is also the difference between quantity values in the two phases; for example, enthalpy of sublimation ΔH_{sb}, defined as the difference between molar enthalpy of gaseous phase (vapors) denoted by H_G and the solid phase one denoted by H_S, will variate from a point to the other of the SG equilibrium curve.

But liquid-vapor equilibrium behavior is unique among biphasic equilibrium curves: all quantity differences regularly decrease when moving from the triple SLG point to the critical point and become zero in this later point: viscosity, enthalpy, entropy, or gas volume become equal to the liquid ones at the critical point, while interfacial tension between liquid and gas is canceled in critical conditions, as well as refractive index difference of these two phases.

Single-Phase Fluid

Liquid cannot be distinguished from its vapors at the critical point, so that liquid and gaseous states of aggregation appear as instances of the same phase: the *fluid* phase.

Besides, starting from liquid, one can reach to the vapors with which this one is in equilibrium at T' temperature and P' pressure (T' and P' being lower than the corresponding critical properties, T_c and P_c), not only through phase transitions represented by vaporization but also by a continuous series of infinitesimal changes that have as result the critical point *bypassing*.

Such a bypassing can take place, for example, through the following four sets, where each stage represents a simple process:

1. An isobaric heating (at pressure P') from T' up to a temperature T'' higher than T_c
2. An isothermal compression (at temperature T'') from P' up to a pressure P'' which exceeds P_c, the substance passing thus from the field of compressed liquid to the one of supercritical fluid

After which, the following processes are made:

3. An isobaric cooling (at the new pressure, P'') from T'' back to the initial temperature T', so that the substance passes to the gaseous field
4. An isothermal relaxation (at temperature T') of the gas at pressure P'' until the equilibrium pressure P' of vapors is thus obtained

5.2 Pure Substances' Critical Point

Critical Point in Molecular Thermodynamics

Phenomenological thermodynamics cannot explain the LG equilibrium unique comportment among all phase equilibria, and besides, it is not a measure to predict liquid-vapor critical point existence.

Molecular thermodynamics is the only one allowing the understanding of the cited facts. Indeed, both liquid and gas present a labile molecular arrangement which can be modified relatively easily, while the normal specific order of every crystalline network makes impossible the gradual passage from one phase to the other. This happens whether both phases possess such an order (polymorphic translations of SS equilibria) or whether only one of the phases is crystalline (SL and SG equilibria for melting and sublimation).

Critical point notion occupies a central part in description, prediction, and evaluation of matter's physical properties – including thermodynamic but also kinetic ones – as well as for the achievement of many technical processes from matter-transforming industries, like supercritical carbon dioxide extraction of natural compounds in food industry, perfumery industry, and textile industry or for waste recovery for energy valorization.

Excepting entropy, volume, and temperature coefficient of pressure $(\partial \ln P/\partial T)_V$, the other characteristic properties of different substances do not take the expected "normal" positive values at critical point; some of them become null (interfacial and surface tension), others being infinite (thermal conductivity, compressibility, expansibility, isobaric heat capacity C_p, difference between C_p and C_v, possibly viscosity) or taking controversial values. Thus C_v – isochoric heat capacity – is either null, finite positive according to some critical point physical-structural theories, and infinite according to others (which become dominant in the last years, without being unanimously accepted), and reliable methods for its measurement at the critical point have not been proposed yet.

Experimental difficulties are encountered otherwise for most property determinations, in the critical point area. It is indeed observed that, as the critical point is approached, the inherent spatial amplitude of structural fluctuations – which usually take place at the submicroscopic level – will increase exponentially, similar to quantity value oscillations.

Besides, for many of substances having bigger molecules, the critical point is not significant, due to an irreversible decomposition suffered by the substance during heating, and for polymers the vapor phase does not exist.

Critical Exponents

Critical phenomena physical theory has not yet reached applicable technical results, the mathematical formulation leading until nowadays to <u>*nonanalytical dependencies*</u> –

at functions that cannot be decomposed according to Taylor series of distance integer powers (on a state diagram) from the critical point. This point constitutes a mathematical singularity, being manifested through discontinuities, of not well-established types, of the physical properties, or of their derivatives.

Peri-critical Domain and Critical Exponents

The preliminary condition: in a restricted region of the LG phase diagram, the *peri-critical domain* adjacent to the critical point (temperature and pressure differences below 0.1% or 1% of T_c and P_c), the properties of state vary according to special rules, differing from the ES describing well the P, T field.

These rules are – in most of vaporization models – rules of *power* type, according to which in a point of the (V, T) state diagram the Y value of the considered quantity (eventually $Y - Y_c$ deviation value from Y_c finite value that this quantity takes at the critical point) is proportional to a certain power of the distance between the representative point to the critical point. The distance is measured either on volume axis or on the temperature one, and the respective power, designated as the *critical exponent* ε, is a *universal* constant, depending on the physical quantity Y but not on the considered substance. The "power" rule has therefore the following form:

$$Y = K \cdot (D - D_c)^{\varepsilon} + L \tag{5.2}$$

where Y is the estimated quantity value, in the point having the coordinates (V, T_c) or (V_c, T), D represents depending on the case either temperature or molar volume (see Table 5.1), and D_c is D quantity value at critical point, while K and L are *material* constants.

The most important "power" rules are presented in Table 5.1.

Relations (5.3) from Table 5.1 cannot be extrapolated beyond the peri-critical domain unless with the price of several essential changes: both multiplicative coefficients of $(D - D_c)$ distance, namely K_1 to K_4 factors from the table, and critical

Table 5.1 Power type rules for peri-critical domain

Conditions	"Power" rule	Critical exponent			Rel. No.
		Notation	Value		
			Theory	VdW	
$V = V_c$; $T \geq T_c$	$C_v = K_1/(T - T_c)^{\alpha}$	α	≅ 0.11	0	(5.3a)
LG equilibrium $T \leq T_c$	$V_G - V_L = K_2 \cdot (T_c - T)^{\beta}$	β	≅ 0.32	1/ 2	(5.3b)
$V = V_c$	$\chi_T = K_3/(T_c - T)^{\gamma}$	γ	≅ 4.8	3	(5.3c)
$T = T_c$	$(P - P_c)/K_4 = [sgn\,(V_c - V)] \cdot [abs(V_c - V)]^{\delta}$	δ	≅ 1.2	1	(5.3d)

exponents will take values different from the ones valid in the peri-critical domain, whether they are empirically determined or are calculated from different ESs.

From Table 5.1 it can be seen, on VdW ES example, that critical exponents *outside* the peri-critical domain are very far from the values predicted through statistical physics theory for the states found in the vicinity of the critical point.

Experimental Determination of Critical Quantities

Specific experimental methods relatively complex allow the measurement of three critical quantities: critical temperature, pressure, and density (volumetric mass - m/v ratio), denoted by T_c, P_c, and ρ_c, respectively. The critical point can be reached through liquid-vapor biphasic closed system compression, in conditions allowing both volume modification and heating. Critical point achievement is signaled by example through optical methods, based on inferior liquid layer and superior vapor layer meniscus disappearance, or on the sudden drop of image transparency (critical opalescence).

Critical Quantities Examples

Experimental error is, at usual or medium temperatures and pressures, of 0.5%, 1%, and 2%, respectively, for these properties, but can be much higher at temperatures of thousands of degree orders. On the other hand, critical density is less significant, from a structural point of view, than the critical (molar) volume V_c, calculated by dividing the critical molar mass, M_c, at critical density:

$$V_c = M_c/\rho_c.$$

In many cases, the molecule suffers a reversible chemical transformation during the process through which the transition is made from the usual conservation conditions to critical pressure and temperature. Such transformations can be constituted of dissociations, which reduce the molar mass, or dimerization (or other polymerizations) which increase M_c towards the molar mass M valid in usual conditions. Therefore, elemental sulfur, of which the molar mass for the monoatomic molecule is approximately 0.03206 kg/mol, contains at the critical point 2% molecules of monoatomic S, 57% of S_2 dimers, and 41% tetramer molecules, from where a critical molar mass M_c of $(0.02 \cdot 1 + 0.57 \cdot 2 + 0.41 \cdot 4) = 2.8$ is higher than the usual molar mass M. The degree of association establishment requires complex methods (quantitative IR spectroscopy) and is often avoided, leading to a more important undetermination of critical volume data.

Critical Quantities Values

Data of Table 5.2 illustrate the fact that kinetic quantities differ very much from one substance to the other and that variability changes with critical quantity nature: extremely pronounced for critical temperature or pressure (even only among tabulated data, T_c varies from 3 to 8000 *K*, while P_c – from 1 to 3000 bar), the variation range does not exceed an order of magnitude at critical volume V_c, for which the extreme values from the table are 40 and 360 cm^3/mol. In truth, the bulky molecules are always less stable, i.e., decomposing through heating before reaching the critical point.

An even more restricted variation is found for the three discussed critical quantities: dimensionless parameter z_c, designated *critical compressibility factor,* and defined – by analogy with definition (4.3) of the compressibility factor, $z = P{\cdot}V/(R{\cdot}T)$ – by the expression:

$$z_c = P_c \cdot V_c/(R \cdot T_c) \tag{5.4}$$

It can be seen that z_c values rarely go out of the range 0.2–0.3.

Table 5.2 Quantities of LG critical point substances

Substance	T_c, *K*	P_c, bar	V_c, cm^3/ mole
C_2H_4 ethylene	282.5	50.8	127 ± 3
Al aluminum	7960	3500	61
C_2H_5OH ethanol	515 ± 1	63.5	167
C_2H_6 ethane	305.4	49.6	146
C_6H_{14} n-hexane	508	30.3	368
C_6H_6 benzene	560	49.2	260
CCl_4 carbon tetrachloride	564 ± 8	45.6	276
CH_3OH methanol	513	79.5	118
CH_4 methane	191	46.4	99
Cl_2 chlorine	417	78 ± 2	124
CO carbon monoxide	133	35.0	93
CO_2 carbon dioxide	304.2	73.8	95
H_2 hydrogen	33.2	13.0	65.0
H_2O water	647.2	220.9 ± 0.3	58
H_2S hydrogen sulfide	373.6	90	98
HCl hydrochloric acid	324.6	83	87
He helium	5.19	2.28	57.4
^{3}He helium «3»	3.30	1.18	72.5
Hg mercury	1750 ± 10	1720 ± 20	43
K potassium	2220	160	209
N_2 azote	126.2	33.9	90
NH_3 ammonia	405.5	113	72
NO nitrogen monoxide	180	65	58

Dependence of the Critical Point on the Nature of the Substance

Like other physical features, critical quantities are relatively sensitive to isotopes' nature of which atom nuclei are formed. This isotope effect becomes significant only for very small molecules – with under five nucleons – of the *quantum* gases like monatomic helium, where critical pressure of normal isotope ^4He is three-quarters higher than the three-nucleon isotope, ^3He, similar to the diatomic molecule of hydrogen, where D_2 variant comprising two deuterium atoms (D is the ^2H isotope, with one extra nucleon from the usual isotope ^1H) has T_c and P_c values with 20–30% higher than H_2 from Table 5.2.

Regarding substance chemical nature, it has a decisive influence on the critical properties. Therefore:

- Metals have pressures and temperatures particularly elevated, this effect being weakened for alkaline metals.
- T_c and P_c are much lower for noble gases, for metalloids with diatomic molecules – *H*, *N*, *O*, and halogens – or for molecules comprising at least one of these metalloids, including hydrocarbons.
- Intermediate values for critical pressure and temperature are encountered at divalent and trivalent metalloids, like Se, As, B, etc., partially due to high temperature association (S_2 and S_4 as has been seen for sulfur, P_4 at phosphorus), which is also valid for their combinations like sulfides.
- Finally, ionic compounds like NaCl or CaO and the intermetallic ones decompose during vaporization.

One of the isoquantity curves passing in (P, T) coordinates through the critical point, for example, *isenthalpic line*, *isentropic line*, and *isochore* – curves on which enthalpy, entropy, and volume molar values H, S, or V are maintained – since on this equilibrium curve differences ΔH_v, ΔS_v, and ΔV_v between thermodynamic functions of vapors and liquid are weakened as temperature increases, becoming zero when reaching the critical point.

Equilibrium curve and critical isochore line collinearity will be illustrated in Sect. 5.5 through a comparison of relations (5.39b, 5.39c) for the case of gas respecting VdW ES.

5.3 ES and the Critical Point

From ES to Critical Point

ES knowledge allows the deduction of the following:

- molar variation ΔYv at vaporization, representing the difference $\Delta Yv \equiv Y_G - Y_L$, on the liquid-vapor equilibrium curve, of several molar properties Y of the two phases that are in contact, among which volume, enthalpy and entropy:

from page 14 of the book on thermodynamics of the same authors (Daneş et al. 2013 [1]):

- critical point coordinates (P_c, V_c, T_c)
- $P(T)$ dependence represented by the liquid-vapor pressure curve

In order to give the two features, namely, critical point and liquid-vapor phase equilibrium existence, the form of ES must satisfy a set of special conditions as the following:

I. At temperatures exceeding the critical temperature, $P(V)$ isotherms must be monotonically decreasing:

$$(\partial P/(\partial V)_T < 0, \forall T > T_c \tag{5.5a}$$

II. <u>*Critical isotherm*</u>, $P = f(V)$ at $T = T_c$, must present in the critical point (i.e., at $V = V_c$) an inflection point with the horizontal tangent and must be decreasing outside the critical point:

$$T = T_c \cap V \neq V_c \Rightarrow (\partial P/\partial V)_T < 0 \tag{5.5b}$$

Therefore, at the critical point, $T = T_c \cap V = V_c$, an *even* number of successive derivatives must be null, and the following derivative must be negative:

$$(\partial^n P/\partial V^n)_T = 0, \quad \forall n \leq 2 \cdot k \cap \left(\partial^{2 \cdot k+1} P/\partial V^{2 \cdot k+1}\right)_T < 0; \; k \in Z_+ \tag{5.6}$$

where Z_+ designates the positive integer set of numbers.

Relations (5.6), equivalent to the existence of an inflection point where the tangent is horizontal, become for k = 1:

$$(\partial P/\partial V)_T = 0 \cap \left(\partial^2 P/\partial V^2\right)_T = 0 \cap \left(\partial^3 P/\partial V^3\right)_T < 0. \tag{5.7}$$

III. At subcritical temperatures, the ES must have at least three real and positive V roots, for any pair of values (T, P).

The fulfilling of these three conditions will be furtherly discussed in the following paragraphs.

VdW ES Critical Quantities

Thus, for VdW isotherm (4.6), $P = R \cdot T/(V - b) - a/V^2$, the first two partial derivatives of pressure in relation to volume are the following:

$$(\partial P/\partial V)_T = -R \cdot T/(V-b)^2 + 2 \cdot a/V^3 \tag{5.8a}$$

$$\left(\partial^2 P/\partial V^2\right)_T = 2 \cdot R \cdot T/(V-b)^3 - 6 \cdot a/V^4 \tag{5.8b}$$

It is found from (5.7) that in the critical point, where $(P, V, T) = (P_c, V_c, T_c)$:

$$P_c = R \cdot T_c/(V_c - b) - a/(V_c)^2 \tag{5.9a}$$

$$0 = -R \cdot T_c/(V_c - b)^2 + 2 \cdot a/(V_c)^3 \tag{5.9b}$$

$$0 = 2 \cdot R \cdot T_c/(V_c - b)^3 - 6 \cdot a/(V_c)^4 \tag{5.9c}$$

Elimination of T_c between the last two relations leads to $V_c - b = 2 \cdot V_c/3$ or:

$$V_c = 3 \cdot b \tag{5.10}$$

By introducing V_c from (5.10) into relation (5.9b), it results that:

$$T_c = 8 \cdot a/(27 \cdot R \cdot b) \tag{5.11a}$$

And by replacing T_c from (5.11a) and V_c from (5.10) into (5.9a), the following is obtained:

$$P_c = a/\left(27 \cdot b^2\right) \tag{5.11b}$$

Relations (5.10, 5.11a and 5.11b) allow the calculation of critical point coordinates (V_c, T_c, P_c), from quantities (a, b) of VdW ES.

By comparing the VdW values for Boyle point from relation (4.16), $T_B = a/(R \cdot b)$, with the expression (5.11a) of critical temperature, $T_c = 8 \cdot a/(27 \cdot R \cdot b)$, it leads to $T_B/T_c = 3.4$, a value relatively far from the average real ratio, which is 2.5.

By replacing P_c, V_c, and T_c from (5.10, 5.11a and 5.11b) into definition (5.4) for the critical compressibility factor z_c, it results the following for the Van der Waals gas:

$$z_{c,VdW} = 0.375 \tag{5.12}$$

The experimental z_c values are lower: from 0.20 up to 0.31, depending on the gas.

The critical values can be calculated from ES not only for (P, V, T, z) but also for other properties. For example, from $P = R \cdot T/(V - b) - a/V^2$ – Van der Waals ES (4.6) –calculating firstly $(\partial P/\partial T)_V = R/(V - b)$ derivative gets the *pressure temperature coefficient*, defined through relation (2.4) from page 14 of the book on thermodynamics of the same authors (Daneş et al., 2013 [1]):

$$\beta \equiv (1/P) \cdot (\partial P/\partial T)_V \tag{5.3a2}$$

After introducing into definition (a) the expressions for P and $\partial P/\partial T$, the following relation is obtained:

$$\beta = R \cdot V^2/(R \cdot V^2 \cdot T - a \cdot V + a \cdot b)$$

which becomes in critical conditions:

$$\beta_c = R \cdot (V_c)^2/\left[R \cdot (V_c)^2 \cdot T_c - a \cdot V_c + a \cdot b\right]$$

from where, with $V_c = 3 \cdot b$ and $T_c = 8 \cdot a/(27 \cdot R \cdot b)$ from relations (5.10) and (5.11a), respectively:

$$\beta_c = 27 \cdot R \cdot b/(2 \cdot a) \tag{5.13}$$

The above-described procedure for critical quantity determination for Van der Waals gas is applicable for any ES that can be made explicit in relation to pressure, as in the examples from the next paragraph.

Critical Quantities for Other ES

Redlich and Kwong

For example, from Redlich-Kwong ES, $P = R \cdot T/(V - b) - a/\left[V \cdot (V + b) \cdot \sqrt{T}\right]$ (4.8a), through successive differentiation, it results that:

$$(\partial P/\partial V)_T = -R \cdot T/(V - b)^2 + a \cdot (2 \cdot V + b)/\left[V^2 \cdot (V + b)^2 \sqrt{T}\right]$$

$$\left(\partial^2 P/\partial V^2\right)_T = 2 \cdot R \cdot T/(V - b)^3 - 2 \cdot a \cdot (3 \cdot V^2 + 3 \cdot V \cdot b + b^2)/\left[V^3 \cdot (V + b)^3 \sqrt{T}\right].$$

A system of three equations of unknowns V_c, T_c, and P_c is obtained when imposing that $P = P_c$, $(\partial P/\partial V)_T = 0$, $(\partial^2 P/\partial V^2)_T = 0$ for $T = T_c$ and $V = V_c$, a system of three equations with V_c, T_c, and P_c unknowns is obtained, respectively:

$$P_c = R \cdot T_c/(V_c - b) - a/\left[V_c \cdot (V_c + b) \cdot \sqrt{T}\right],$$

$$R \cdot (T_c)^{1,5} \cdot [V_c \cdot (V_c + b)]^2 = a \cdot (V_c - b)^2 \cdot (2 \cdot V_c + b)$$

$$R \cdot (T_c)^{1,5} \cdot [V_c \cdot (V_c + b)]^3 = a \cdot (V_c - b)^3 \cdot \left[3 \cdot (V_c)^2 + 3 \cdot V_c \cdot b + b^2\right]$$

System solution is found through the already used procedure for critical VdW quantity determination – T_c variable elimination from equations obtained through isothermal derivatives $(\partial P/\partial V)_T$ and $(\partial^2 P/\partial V^2)_T$ nullification.

Therefore, the following solution is obtained:

$$V_c = b/\left(2^{1/3} - 1\right) \tag{5.14a}$$

$$T_c = \left(2^{2/3} - 1\right)^2 / (R \cdot b/a)^{2/3} \tag{5.14b}$$

$$P_c = \left[\left(2^{1/3} - 1\right)^2 / \left(2^{1/3} + 1\right)\right] \cdot \left(R \cdot a^2/b^5\right)^{1/3} \tag{5.14c}$$

or

$$V_c = 3.847 \cdot b, T_c = 0.3450/(R \cdot b/a)^{2/3}, P_c = 0.02989 \cdot \left(R \cdot a^2/b^5\right)^{1/3}$$

By replacing Redlich-Kwong expressions (5.14a, 5.14b and 5.14c) into relation (5.4), the following is obtained:

$$z_c = 1/3 \tag{5.15}$$

the calculated value being closer to the experimental data for the critical compressibility coefficient than the 3/8 VdW value.

Critical Quantities of ESs with more than Two Constants

Clausius

For Clausius ES with three constants, $P = R \cdot T/(V - b) - a/[T \cdot (V + c)^2]$, from relation (4.8f), the indicated procedure leads to:

$$V_c = 3 \cdot b + 2 \cdot c; T_c = \sqrt{8 \cdot a \cdot (b + c)/(27 \cdot R)};$$
$$P_c = \sqrt{R \cdot a}/[6 \cdot (b + c)]^{3/2} \tag{5.16}$$

z_c being defined by relation (5.4) will no longer be a number like in the case of ES with two material constants, but becomes a dependence on the a, b, and c quantities, namely:

$$z_c = (3 \cdot b + 2 \cdot c)/[8 \cdot (b + c)] \tag{5.17}$$

and can take values between 0.25 (for b/c = 0) and 0.375 (for c/b = 0).

Martin

For $P = R{\cdot}T/(V-b)-(a-c{\cdot}T)/(V+d)^2$, Martin ES with four material constants from relation (4.8j); this method application leads to:

$$V_c = 3 \cdot b + 2 \cdot d; T_c = 8 \cdot a/[8 \cdot c + 27 \cdot R/(b + d)];$$
$$P_c = [R \cdot a/(b + d)]/[8 \cdot c + 27 \cdot R/(b + d)] \tag{5.18}$$

and the critical compressibility factor z_c – calculated through its definition (5.4) – $z_c \equiv P_c{\cdot}V_c/(R{\cdot}T_c)$ – will be given by:

$$z_c = (3 \cdot b + 2 \cdot d)/[8 \cdot (b + d)] \tag{5.19}$$

ranging also between 0.25 and 0.375 for the value of b/d ratio.

From Critical Point to ES

In the case of equations with less than four material constants, these constants can be determined from the critical point coordinates.

Thus, for Clausius Eq. (4.8f), with three constants, considering relations (5.16) as being a system with three equations of a, b, and c unknowns, through system solutioning, it is found that:

$$\begin{aligned} &a = R^2 \cdot (3 \cdot T_c/4)^3/P_c; b = 3 \cdot R \cdot T_c/(8 \cdot P_c) - V_c; \\ &c = V_c - R \cdot T_c/(4 \cdot P_c) \end{aligned} \tag{5.20}$$

For a ES with two material constants, namely, a and b, the three equations expressing P_c, V_c, and T_c as dependencies of a and b form an incompatible system.

The solution consists of taking into account only these three equations. For example, for the Van der Waals gas, where V_c, T_c, and P_c are expressed through relations (5.10, 5.11a, 5.11b), the following expressions are obtained:

I. When considering only relations (5.10, 5.11a):

$$a = 9 \cdot R \cdot T_c \cdot V_c/8; b = V_c/3 \tag{5.21a}$$

II. Starting from Eqs. (5.10, 5.11b):

$$a = P_c \cdot (V_c)^2; b = V_c/3 \tag{5.21b}$$

III. By combining the equalities (5.11a, 5.11b):

$$a = 27 \cdot (R \cdot T_c \cdot /8)^2 / P_c; b = \cdot (R \cdot T_c / (8 \cdot P_c) \qquad (5.21c)$$

Usually procedure III is used – by expressing a and b by T_c and P_c – since critical volume experimental determination is less accurate than the one of critical pressure or temperature. Therefore, for Redlich-Kwong Eq. (4.8a), it is similarly obtained from relations (5.14b, 5.14c):

$$a = \left[\left(2^{1/3} + 1\right)/2\right]^3 \cdot R^2 \cdot (T_c)^{5/2} / P_c; b = \cdot \left(2^{1/3} + 1\right)^3 \cdot R \cdot T_c / P_c \qquad (5.22)$$

5.4 ES and Liquid-Vapor Equilibrium

Liquid-Vapor Equilibrium in the Pressure/Volume Graph

When representing the ESs in *P*(*V*) coordinates, the presence of several isotherms is observed, which have maximum and minimum values, at temperatures below critical temperature; thus, the isotherm from Fig. 5.1a corresponds to the existence of three solutions (three V values) for any P value situated between extremum C point (minimum) and E point (maximum) coordinates.

The CDE isotherm part is physically unachievable, since on this part $(\partial P/\partial V)_T > 0$, which corresponds to a *negative* compressibility.

The achievable parts – with *positive* compressibility – of ABC and EFH isotherms correspond to liquid and vapors, respectively, since the molar volume is lower on the first part than on the second one.

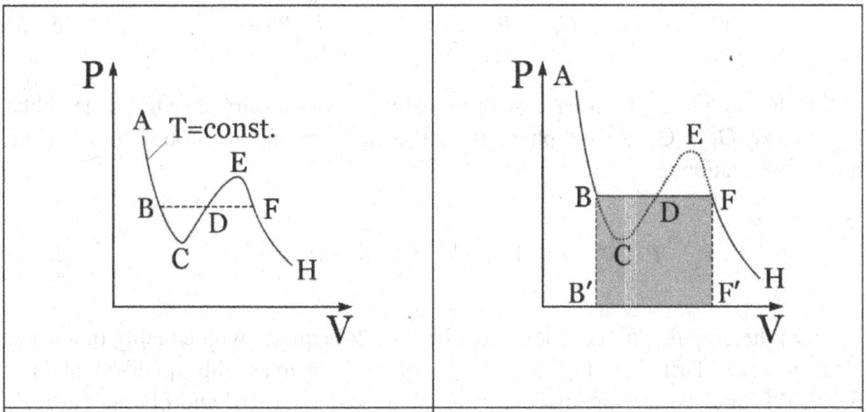

A. Subcritical isotherm	B. Equilibrium isotherm

Fig. 5.1 L-G equilibrium in P(V) coordinates

Stable, Metastable, and Unstable Monophasic States

When the two phases – liquid and vapors, marked by indices *L* and *G* – are at thermodynamic equilibrium, their pressure and temperature are identical; therefore, their representative points B (liquid) and F (vapors) are found at the intersection of a common horizontal line – BDF – with isotherm decreasing parts, AC and EH, respectively. The isotherm real allure is consequently the one corresponding to the full path of Fig. 5.1b.

The straight part BF corresponds to biphasic liquid-vapor mixture, mixture of whose molar volume V_{am} variates with vapor molar fraction y_G:

$$V_{am} = V_B \cdot (1 - y_G) + V_F \cdot y_G.$$

The AB and FH parts correspond to the system consisting of a single *absolute stable* phase: absolute stable liquid from thermodynamic point of view and absolute stable vapors, respectively. The dotted parts BC and FE correspond each to a single *metastable* phase: metastable liquid (overheated) and metastable vapors (subcooled), and the CE part marked by line-dot is not physically realizable, being an absolute *unstable* fluid phase.

From the ES one can calculate the coordinates of points B and F, respectively their vapor pressure P at temperature T and equilibrium liquid and vapors molar volumes, V_{Le} (point B) and V_{Ge} (point F), also at temperature T, can be made starting from the ES. Since $dG = -S{\cdot}dT + V{\cdot}dP$ it results through isothermal integration between points B (liquid) and F (gas):
$G_F - G_B = \int_B^F V \cdot dP$ where, since $V{\cdot}dP = d(P{\cdot}V) - P{\cdot}dV$, the integral can be replaced by $(P \cdot V)_F - (P \cdot V)_B - \int_B^F P \cdot dV$.

From here it results, since $P_F = P_B = P$, that:

$$G_F - G_B = P \cdot (V_F - V_B) - \int_B^F P \cdot dV \tag{5.23}$$

But the two phases found in isothermal-isobaric equilibrium have the same Gibbs free energy, $G_F = G_B$, so that phase analytical equilibrium condition is given by the following relation:

$$P \cdot (V_G - V_L) = \int_{V_L}^{V_G} P(V,T) \cdot dV, T = \text{cst} \tag{5.24}$$

From the graphic point of view, condition (5.24) equates with equality of shaded areas BCD and DEF from Fig. 5.1b: by denoting these areas with S_1 and S_2, and the B'BCDFF' area (where the discontinuously marked parts BB' and FF' are vertical) with S_3, relation (5.24) equals with $S_1 + S_3 = S_2 + S_3$, from where: $S_1 = S_2$.

Calculation of PVT Equilibrium Features for VdW Fluid

Isotherm belonging to points B (liquid) and F (vapors) for VdW Eq. (4.6) $P = R{\cdot}T/(V - b) - a/V^2$, respectively, corresponds to the following equations:

$$P = R \cdot T/(V_L - b) - a/(V_L)^2 \tag{5.25a}$$

$$P = R \cdot T/(V_G - b) - a/(V_G)^2 \tag{5.25b}$$

and the equilibrium condition (5.24) becomes:

$$P \cdot (V_G - V_L) = \int_{V=V_L}^{V=V_G} \left[R \cdot T/(V - b) - a/V^2\right] \cdot dV,$$

which leads, after integration, to:

$$[P + a/(V_G \cdot V_L)] \cdot (V_G - V_L) = R \cdot T \cdot ln\,[(V_G - b)/(V_L - b] \tag{5.25c}$$

Relations (5.25) form a three-equation system whose solutioning allows determination of unknowns P, V_G, and V_L as a dependence on T. Usually, the system has no analytical solution, but only a numerical one. In the particular case of Van der Waals gas, a parametric solution can be obtained, namely:

By eliminating P between relations (5.25a, 5.25b), the following expression is obtained for T:

$$T = a \cdot (V_G + V_L) \cdot (V_G - b) \cdot (V_L - b)/\left[R \cdot (V_G \cdot V_L)^2\right] \tag{5.26a}$$

By replacing T in one of relations (5.25a, 5.25b), it results that P becomes equal to:

$$P = a \cdot [V_G \cdot V_L - b \cdot (V_G + V_L)]/(V_G \cdot V_L)^2 \tag{5.26b}$$

By replacing T and P from (5.26a) and (5.26b), respectively, into (5.25c), it leads to:

$$\frac{(V_G - V_L) \cdot [2 \cdot V_G \cdot V_L - b(V_G + V_L)]}{(V_G + V_L) \cdot (V_G - b) \cdot (V_L - b)} = ln\,\frac{V_G - b}{V_L - b} \tag{5.27}$$

Through change of variables:

$$V_G = b \cdot (1 + y \cdot x); V_L = b \cdot (1 + y/x) \tag{5. 28a}$$

Relation (5.26a) becomes: $T_B = a/(R{\cdot}b)$, or, by converting to explicit formula on *y*, it becomes:

$$y = \left[\left(x^2 - \cdot 1/x^2 - 4 \cdot \ln \mathrm{x}\right)/2\right]/\left[(x + \cdot 1/x) \cdot (\ln \mathrm{x}) - (x - 1/x)\right] \qquad (5.\ 28\mathrm{b})$$

Thus, x is the unique parameter of which V_G, V_L, *P*, and *T* depend.
By replacing y(x) from (5.28b) into (5.28a), it results V_G(x) and V_L(x):

$$V_G = \left[b \cdot (x - 1/x) \cdot \left(x^2 - 1 - 2 \cdot \ln x\right)/2\right]/\left[(x + 1/x) \cdot (\ln x) - (x - 1/x)\right] \qquad (5.4\mathrm{b}2)$$

$$V_L = \left[b \cdot (x - 1/x) \cdot \left(1/x^2 - 1 + 2 \cdot \ln x\right)/2\right]/\left[(x + 1/x) \cdot (\ln x) - (x - 1/x)\right] \qquad (5.4\mathrm{c}2)$$

Finally, by introducing V_G and V_L from (b) and (c), respectively, into (5.26a) and (5.26b), the *T*(x) and *P*(x) dependencies, respectively, are obtained:

$$T = 2 \cdot \mathrm{a} \cdot (\mathrm{x} - 1/\mathrm{x}) \cdot [(\mathrm{x} + 1/\mathrm{x}) \cdot (\ln \mathrm{x}) - (\mathrm{x} - 1/\mathrm{x})] \cdot [\mathrm{F}_T/(\mathrm{R} \cdot \mathrm{b} \cdot \mathrm{F}*)]^2 \qquad (5.29\mathrm{a})$$

$$P = 4 \cdot a \cdot \left[(x - \cdot 1/x)^2 - (2 \cdot \ln\ x)^2\right] \cdot [F_P/(R \cdot b \cdot F*]^2 \qquad (5.29\mathrm{b})$$

where F^*, F_T, and F_P represent the functions:

$$F* = \left(x^2 - \cdot 1 - 2 \cdot \ln\ x\right) \cdot \left(1/x^2 - \cdot 1 + 2 \cdot \ln\ x\right),$$

$$F_T = x^2 - \cdot 1/x^2 - 4 \cdot \ln\ x; F_P = -1 + \left[\left(x^2 + 1\right)/\left(x^2 - 1\right)\right] \cdot \ln\ x$$

Thus, the properties (V_G, V_L, *T*, *P*) for the same state of equilibrium, are obtained from relations (5.29a, 5.29b) by introducing an arbitrary value (> 1) of x parameter.

5.5 Stability of the Liquid-Vapor Equilibrium

Molar volume values' graphical representation for the two phases (*L*, *G*) at equilibrium, in relation to pressure or temperature, is called *binodal* curve of liquid-vapor equilibrium. The general aspect of AKB binodal curve is presented as a thick line in Fig. 5.2 A, B.

The *spinodal* curve is graphically represented also in Fig. 5.2 – geometric locus of points separating the mono-component metastable phase areas from the biphasic field. Both binodal and spinodal curves are comprised of two branches: liquid (AK, CK) and vapors (BK, DK). The four branches meet at the critical point, where they

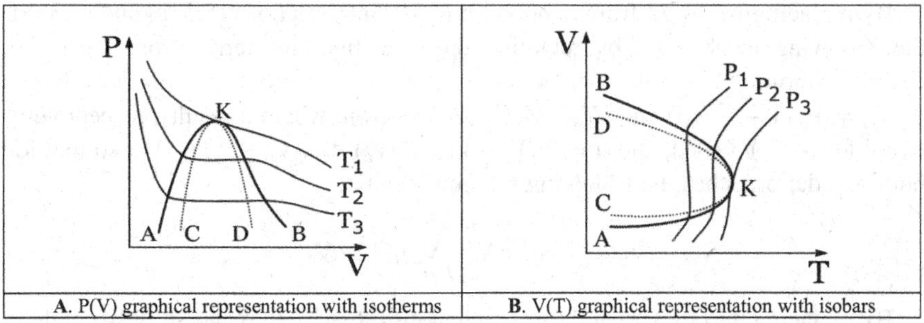

Fig. 5.2 Binodal and spinodal curves

have a common tangent: critical isotherm, horizontal, from PV graphical representation, and critical isobar, vertical, from VT graphical representation, respectively.

Binodal Curve

In vicinity of critical point, the binodal curve of real gases has a symmetrical parabolic allure, both in the PV and the VT graphical representation:

$$abs(V - V_c) \ll V_c \Rightarrow P = P_c - K_P \cdot [abs(V - V_c)]^m \tag{5.30a}$$

$$abs(V - V_c) \ll V_c \Rightarrow T = T_c - K_T \cdot [abs(V - V_c)]^m, \tag{5.30b}$$

where K_P and K_c are positive material constants and m > 1. From the experimental data, for the majority of real gases, $m \cong 3$. The binodal curve of Van der Waals gas has however m = 2. Indeed, by effectuating the change of variable $x = e^u$ where $0 < u \ll 1$, it is obtained, by replacing x into (5.28b) and by series expanding e^u relative to u powers: $y = 2 \cdot (1 + u^2/10 - u^4/350 + ...)$. By replacing y into (5.28a), it results:

$$V_G = b \cdot (3 + 2 \cdot u + 6 \cdot u^2 + \ldots)/(5 + \ldots) \tag{5.31a}$$

$$V_L = b \cdot (3 - 2 \cdot u + 6 \cdot u^2 + \ldots)/(5 + \ldots) \tag{5.31b}$$

And after replacing V_G and V_L into (5.26a, 5.26b), the following expressions are obtained:

$$T = 8 \cdot a \cdot (1 - u^2/9 + \ldots)/(27 \cdot R \cdot b) \tag{5.31c}$$

$$P = a \cdot (1 - 4 \cdot u^2/9 + \ldots)/(27 \cdot b^2) \tag{5.31d}$$

By replacing $b = V_c/3$ from expression (5.10) into relations (5.31a) and (5.31b), the following is obtained by retaining only the first two terms from the series decomposition:

$V_G = V_c\,(1 + 2\cdot u/3)$ and $V_L = V_c\cdot(1-2\cdot u/3)$, from where it results, respectively: $u = 3\cdot(V_G - V_c)/(2\cdot V_c)$, and $u = 3\cdot(V_c - V_L)/(2\cdot V_c)$. But $V_L \le V_c \le V_G$, so that for both binodal branches the following relation is valid:

$$u = [3 \cdot abs(V - V_c)]/(2 \cdot V_c) \tag{5.32}$$

By replacing into (5.31c) and (5.31d) the ratios $8\cdot a/(27\cdot R\cdot b)$ and $a/(27\cdot b^2)$ with T_c and with P_c, respectively – according to relations (5.11a, 5.11b) – the following is obtained:

$$T = T_c - (u/3)^2 \cdot T_c; P = P_c - (2 \cdot u/3)^2 \cdot P_c$$

and after replacing u from (5.32):

$$P = P_c - \left[P_c/(V_c)^2\right] (V - V_c)^2; T = T_c - \left[T_c/(2 \cdot V_c)^2\right] (V - V_c)^2 \tag{5.33}$$

Therefore, relations (5.30) are obtained, with $K_P = P_c/\,(V_c)^2$, $K_T = T_c/\,(2\cdot V_c)^2$, and $m = 2$.

Let α be the logarithmic slope of vapor pressure curve, defined by:

$$\alpha \equiv (\partial\, ln\, P/\partial\, ln\, T)_{LG} \tag{5.34}$$

This slope value at the critical point, α_c, identic with Riedel criterion from Sect. 5.8, is an important feature of LG equilibrium, defined by:

$$\alpha_c = \lim_{T \to T_c} \alpha \tag{5.35a}$$

By replacing $d\, ln\, P = d\, P/\, P$ and $d\, lnT = d\, T/\, T$, it results that:

$$\alpha_c = (T_c/P_c) \cdot \lim_{V \to V_c} [(P_c - P)/(T_c - T)] \tag{5.35b}$$

But from relations (5.33) it is found that $P_c - P = P_c\cdot(V/\,V_c - 1)^2$ and $T_c - T = (T_c/4)\cdot(V/\,V_c - 1)^2$, from where $(P_c - P)/\,(T_c - T) = 4\cdot P_c/\,T_c$, so that after the replacement into (5.35b) the following expression is obtained for the Van der Waals gas:

$$VdW : \alpha_c = 4 \tag{5.36}$$

Actually, α_c takes experimental values from 5 to 9 for different gases.

Spinodal Curve

Spinodal equation is obtained from condition $(\partial P/\partial V)_T = 0$.

For the Van der Waals gas, for which $(\partial P/\partial V)_T = -R\cdot T/(V-b)^2 + 2\cdot a/V^3$, according to relation (5.8a), it results the following for the spinodal in (*V*, *T*) coordinates:

$$T = 2 \cdot a \cdot (V - b)^2/(R \cdot V^3) \tag{5.37a}$$

and spinodal equation in (*P*, *V*) coordinates is obtained when replacing $T = (P + a/V^2)\cdot(V - b)/R$ from the ES (4.6) into (5.37a):

$$P = a \cdot (V - 2 \cdot b)/V^3 \tag{5.37b}$$

By applying the differential to relations (5.37b) and (5.37a), respectively, it is obtained that $(\partial P/\partial V)_T = 2\cdot a\cdot(3\cdot b - V)/V^4$ and $(\partial T/\partial V)_P = 2\cdot a\cdot(3\cdot b - V)\cdot(V-b)/(R\cdot V^4)$, from where it results that for both coordinate types, *P*(*V*) and *T*(*V*), the spinodal presents a maximum with an horizontal tangent at $V = 3\cdot b$, a maximum which corresponds therefore to the critical point.

Spinodal and Binodal Curves within the Peri-critical Domain

The *T*(*V*) or *P*(*V*) spinodal allure within the critical domain is identical to the respective binodals' allure – Eqs. (5.30a, 5.30b). Indeed, by replacing into (5.37a) the a and b constants from (5.21a) and into (5.37b) the a and b constants from (5.21b), it results, respectively, that $T = T_c\cdot V_c\cdot(3\cdot V - V_c)^2/(4\cdot V^3)$ and $P = P_c\cdot(V_c)^2\cdot(3\cdot V - 2\cdot V_c)/V^3$, or by subtracting *T* and *P* from T_c and P_c, respectively:

$$1 - T/T_c = (V - V_c/4) \cdot (V_c - V)^2/V^3;$$
$$1 - P/P_c = (V + 2 \cdot V_c) \cdot (V_c - V)^2/V^3$$

It is obtained that in the critical point: $\lim_{T\to T_c, V\to V_c}\left[\frac{(T_c-T)}{(V_c-V)^2}\right] = 3\cdot\frac{T_c}{4\cdot(V_c)^2}$, $\lim_{P\to P_c, V\to V_c}\left[\frac{(P_c-P)}{(V_c-V)^2}\right] = 3\cdot\frac{P_c}{4\cdot(V_c)^2}$. Therefore, spinodal Eqs. *T*(*V*) and *P*(*V*) in the vicinity of the critical point are:

$$T = T_c - 3 \cdot T_c \cdot (V_c/V - 1)^2/4; \quad P = P_c - 3 \cdot P_c \cdot (V_c/V - 1)^2 \tag{5.38}$$

having the same form as the binodal (5.33), with the same exponent $m = 2$ of the quantity $abs(V_c - V)$, but with coefficients $K_P = 3\cdot P_c/(V_c)^2$ and $K_T = 3\cdot T_c/(2\cdot V_c)^2$ three times higher than the ones for the case of binodal.

From relations (5.37a, 5.37b) can be obtained under parametric form (with parameter V) the spinodal equation in $P(T)$ coordinates, the spinodal slope being in this case:

$$\left(\frac{dP}{dT}\right)_V = -\left(\frac{dP}{dV}\right)_T : \left(\frac{dT}{dV}\right)_P = \frac{\left[2 \cdot a \cdot (3 \cdot b - V)/V^4\right]}{\left[2 \cdot a(V-b) \cdot (3 \cdot b - V)/(R \cdot V^4)\right]} = \frac{R}{(V-b)}$$

The slope $(dP/dT) = R/(V - b)$ is obtained at the critical point or with $V_c = 3 \cdot b$:

$$(dP/dT)_{cr} = R/(2 \cdot b) \ (spinodal\ curve) \tag{5.39a}$$

Binodal slope is obtained from (5.34) $(dP/dT)_{cr} = \alpha_c \cdot P_c/T_c$ or by replacing $\alpha_c = 4$, $P_c = a/(27 \cdot b^2)$, and $T_c = 8 \cdot a/(27 \cdot R \cdot b)$ from (5.36, 5.11b, 5.11a):

$$(dP/dT)_{cr} = R/(2 \cdot b) \ (binodal\ curve) \tag{5.39b}$$

Finally, for the case of critical isochore ($V = V_{cr}$) of the real Van de Waals gas, through ES (4.6) differentiation $P = R \cdot T/(V - b) - a/V^2$, it results that: $dP/dT)_{cr} = R/(2 \cdot b)$ or at the critical point ($V_{cr} = 3 \cdot b$):

$$(dP/dT)_{cr} = R/(2 \cdot b) \ (izochore) \tag{5.39c}$$

From comparison of relations (5.39a, 5.39b, 5.39c), it can be seen that the critical isochore, the binodal, and the spinodal are tangent in the critical point, i.e., they have the same slope.

5.6 Corresponding States

Reduced Properties

The (simple) principle of the *corresponding states* is in fact the following approximate *rule*: "The value of any reduced quantity (X_{red}) is univocally determined by other two reduced properties values (Y_{red}, Z_{red})":

$$X_{red} = F_{x,y,z}(Y_{red}, Z_{red}) \tag{5.40}$$

where (X_{red}, Y_{red}, Z_{red}) are three *reduced* properties characterizing the mono-component system and $F_{x,y,z}$ is a universal function, independent of system composition, but depending both on choice of properties X, Y, and Z and on their reduction mode. Properties X, Y, and Z involved in this theorem can be of any physical nature, provided that they are properties of state and that they are intensive properties (independent of the substance quantity). For example, they can occur in the corresponding states theorem as P, V, T, S, α, $G - G_{(0\ \mathrm{K},\ 1\ \mathrm{atm})}$, viscosity, refractive index, extinction coefficient, etc.

A quantity *reduction* is realized by its reporting to the same quantity value in the *reference* state ("ref" index):

$$X_{\mathrm{red}} \equiv X/X_{\mathrm{ref}};\, Y_{\mathrm{red}} \equiv Y/Y_{\mathrm{ref}};\, Z_{\mathrm{red}} \equiv Z/X_{\mathrm{ref}} \tag{5.41a}$$

so that the reduced properties X_{red}, Y_{red}, and Z_{red} are dimensionless. The X, Y, and Z values refer to a certain given system state, and X_{ref}, Y_{ref}, and Z_{ref} refer to each *reference* state, which can differ from one quantity to the other but must be a "corresponding" state for the different substances. For example, by choosing as reference states: point Boyle (B index) for Y, triple solid-liquid-vapor for Z (tr index), and inversion curve point for X where pressure is maximum (IPM index), then the corresponding states law (5.40), where $Y = T$ and $Z = D$ (self-diffusion coefficient of vapors) becomes:

$$X = X_{\mathrm{IPM}} \cdot F(T/T_{\mathrm{B}}, D/D_{\mathrm{tr}})$$

where X is any quantity of gas and the universal function F depends only on the quantity X, but not also on the substance nature. The *reduced* quantity must be differentiated from the *relative* quantity, defined as the difference (and not as the ratio) between the quantity current value and its reference value.

Reduction Method through Critical Quantities

For gases properties (and liquid, at temperatures not very close to the melting temperature) study the most convenient choice of Y and Z variables which consists in pressure and temperature choice:

$$X_{\mathrm{red}} = F(T_{\mathrm{red}}, P_{\mathrm{red}}) \tag{5.41b}$$

It is practical to choose, both for X quantity reduction and for temperature and pressure reduction, the same reference state – the critical state, denoted by index c. The reduced properties using this procedure are denoted by the index r:

$$X_{\mathrm{r}} \equiv X/X_{\mathrm{c}};\, T_{\mathrm{r}} \equiv T/T_{\mathrm{c}};\, P_{\mathrm{r}} \equiv P/P_{\mathrm{c}} \tag{5.42}$$

And the corresponding states binding becomes:

$$X_r = F(T_r, P_r) \tag{5.43}$$

Principle of Corresponding States (PCS)

The principle of corresponding states is not an exact law and cannot be proved; it is a consequence of the similar manner in which different substances molecules interact, regardless their nature.

It is understood that, since the similarity is not perfect, the principle of corresponding states has only an approximate validity, its accuracy depending, among others, on the choice of independent variables as well as on their reduction mode (the choice of reference states for X, for Y, and for Z).

The corresponding state's rule, as set forth, refers meantime to single phase thermodynamic systems found at equilibrium.

It is understood that the number of independent variables of which X_{red} (as well as X) depends on cannot be *inferior* to 2, since single phase system where composition changes take place is a bivariate system. From thermodynamics it does not however result that the number of these independent quantities could be *superior* to 2. Indeed, the number of reduced independent variables increases to 3, 4, etc. (see the corresponding states *extended* theorem, at Sect. 5.7) which leads to an accuracy improvement, for properties of state calculation.

The corresponding state's law, applied for $\{X, Y, Z\} = \{P, V, T\}$, leads to the conclusion of a universal equation existence, if reduced coordinates are used; this universal equation is:

$$\Psi(P_{red}, V_{red}, T_{red}) = 0 \tag{5.44a}$$

while for critical point reduction, considering notations (5.42):

$$\Psi(P_r, V_r, T_r) = 0 \tag{5.44b}$$

The *universal* ES in *reduced coordinates* (5.44a) is consistent with any of the two material constants ES. Each of these equations can be brought to the form (5.44b) by replacing a and b material constants depending on the reduced pressure and temperature.

Reduced ES

For example, in the case of VdW ES (4.6), $P = R{\cdot}T/(V - \mathrm{b}) - \mathrm{a}/V^2$, the replacement of P, V, and T variables with the reduced variables:

$$P_r \equiv P/P_c; V_r \equiv V/V_c; T_r \equiv T/T_c \tag{5.45}$$

According to relations (5.42) with (*P, V, T*) instead of (*X, Y, Z*) leads to: $P_r{\cdot}P_c = R{\cdot}T_r{\cdot}T_c/\ (V_r{\cdot}V_c - \mathrm{b}) - \mathrm{a}/\ (V_r{\cdot}V_c)^2$ where from relations (5.10, 5.11a, 5.11b) the terms $V_c = 3{\cdot}\mathrm{b}$, $T_c = 8{\cdot}\mathrm{a}/(27{\cdot}R{\cdot}\mathrm{b})$, and $P_c = \mathrm{a}/(27{\cdot}\mathrm{b}^2)$ are replaced by:

$$\left[P_r + 3/(V_r)^2\right] \cdot (3 \cdot V_r - 1) - 8 \cdot T_r = 0 \tag{5.46a}$$

It is observed that *reduced* Van der Waals Eq. (5.46a) is an universal equation, without material constants, having thus the form (5.44b).

From the ES $P = R \cdot T/(V - b) - a/\left[V \cdot (V + b)\sqrt{T}\right]$ of Redlich-Kwong (4.8a), passing from (*P, V, T*) to reduced variables and introducing critical properties from relations (5.14), namely:

$V_c = b/(2^{1/3} - 1)$; $T_c = (2^{2/3} - 1)^2/(R \cdot b/a)^{2/3}$ and

$P_c = [(2^{1/3} - 1)^2/(2^{1/3} + 1)] \cdot (R \cdot a^2/b^5)^{1/3}$, it is obtained similarly:

$$\begin{aligned} P_r &= 3 \cdot T_r/\left(V_r - 2^{1/3} - 1\right) \\ &\quad - 1/\left[\left(2^{1/3} - 1\right) \cdot (T_r)^{1/2} \cdot (V_r) \cdot \left(V_r + 2^{1/3} - 1\right)\right] \end{aligned} \tag{5.46b}$$

Heat Capacities from Reduced ES

The reduced ESs allow also calculation of any thermodynamic quantity in relation to pressure; its critical pressure and temperature are known, so that the reduced thermodynamic parameters P_r and T_r can be calculated. For example, from the $(C_V)°$ value which corresponds to zero pressure, the C_V value can be obtained for any pressure conditions when starting from thermodynamic relation (6.27b) of page 104 of the book on thermodynamics of the same authors (Daneş et al. 2013 [1]):

$$C_V = (C_V)^{\circ} - 0.75 \cdot (V_c \cdot P_c/T_c) \cdot (T_r)^{-3/2} \cdot \int_{\infty}^{V_r} \left\{V_r \cdot \left[\left(V_r/\left(2^{1/3} - 1\right)\right) + 1\right]\right\} \cdot dV$$

where V, T, and P are replaced by the respective reduced coordinates (viz., reduced *properties*). The following expression is obtained:

$$C_V = (C_V)^{\circ} + (V_c \cdot P_c/T_c) \cdot T_r \cdot \int_{\infty}^{V_r} \left(\partial^2 P/\partial T_r^2\right)_{V_r} \cdot dV_r \tag{5.47a}$$

where the integration is made at constant reduced temperature.

For the Van der Waals gas, it results from relation (5.46a) that:

$P_r = 8{\cdot}T_r/{\cdot}(3{\cdot}V_r - 1) - 3/\ (V_r)^2$, from where, by derivation, it results that:

$\partial P_r/\partial T_r = 1/(3 \cdot V_r - 1)$, and then $\partial^2 P_r/\partial T_r^2 = 0$; from relation (5.47a) it is found that $C_v = (C_v)°$; therefore C_v does not depend on pressure.

For the Redlich-Kwong gas, from the reduced ES (5.46b), it results in two times derivation in relation to T_r, maintaining constant the reduced volume $\partial^2 P_r/\partial (T_r)^2 = -4/[3 \cdot (2^{1/3} - 1) \cdot (T_r)^{5/2} \cdot (V_r) \cdot (V_r + 2^{1/3} - 1)]$, so that relation (5.47a) becomes:

$$C_v = (C_v)^{\circ} - 0.75 \cdot (V_c \cdot P_c/T_c) \cdot (T_r)^{-3/2} \cdot \int_{\infty}^{V_r} \{V_r \cdot \left[\left(V_r/\left(2^{1/3} - 1\right)\right) + 1\right]\}^{-1} \cdot dV_r$$

or after integration:

$$C_v = (C_v)^o - (3/4) \cdot (P_c \cdot V_c/T_c) \cdot (T_r)^{-3/2} \cdot ln\left[1 + \left(2^{1/3} - 1\right)/V_r\right]$$

and since for Redlich-Kwong gas: $z = 1/3$ according to relation $Z_c = P_c \cdot V_c/T_c$ (5.15) and from definition (5.4), it results that $(P_c \cdot V_c/T_c) = R \cdot z_c$:

$$C_v = (C_v)^o - (R/4) \cdot (T_r)^{-3/2} \cdot ln\left[1 + \left(2^{1/3} - 1\right)/V_r\right] \tag{5.47b}$$

Ideality Deviation Calculation through Reduced ES

Molar enthalpy isothermal dependence on pressure, given by expression: $(\partial H/\partial P)_T = V - T \cdot (\partial V/\partial T)_P$, becomes in reduced coordinates:

$$(\partial H/\partial P_r)_{T_r} = P_c \cdot V_c \cdot [V_r - T_r \cdot (\partial V_r/\partial T_r)]_{P_r}$$

from where it results, for the difference between enthalpy H at pressure P and enthalpy H^0 at the same temperature but at pressure tending towards zero:

$$H - H^{\circ} = P_c \cdot V_c \cdot \int_0^{P_r} \left[V_r - T_r \cdot (\partial V_r/\partial T_r)_{P_r}\right] \cdot dP_r \tag{5.48a}$$

the integration being isothermally effectuated.

For the case of ESs that can be made explicit in relation to pressure, but not in relation to volume, it is convenient to replace the integration independent variable P_r with V_r. Relation (5.48a) is therefore transformed into:

$$H - H^{\circ} = P_c \cdot V_c \cdot \left\{P_r \cdot V_r - T_r/z_c - \int_{\infty}^{V_r} \left[P_r - T_r \cdot (\partial P_r/\partial T_r)_{V_r}\right] \cdot dV_r\right\} \tag{5.48b}$$

Relation (5.48b) for Van der Waals gas leads to:

$$H - H^o = R \cdot T_c \cdot [T_r \cdot (3 - V_r)/(3 \cdot V_r - 1) - 6/V_r] \tag{5.48c}$$

Similarly, the reduced variables can be used for calculation of difference between real gas entropy and ideal gas one, for the same temperature and pressure:

$$S - S_{id} = R \cdot \int_0^{P_r} \left[1/P_r - z_C \cdot (\partial V_r/\partial T_r)_{P_r}\right] \cdot dP_r \tag{5.49a}$$

as well as real gas fugacity coefficient calculation φ:

$$ln\, \varphi = \int_0^{P_r} (z_c \cdot V_r/T_r - 1/P_r] \cdot dP_r \tag{5.49b}$$

Experimental data comparison of V, $H - H^0$, $S - S_{id}$ and φ values calculated from the different reduced ESs, directly (V) or through relations (5.48a), (5.49a) and (5.49b), shows that calculation accuracy is reduced, errors of 10–20% being frequent at elevated pressures or in the critical point area.

5.7 Physicochemical Similarity

Hougen-Watson Diagram

The thermodynamic quantities experimentally measured in the same conditions (equal P_r and T_r) fairly well coincide for different gases. This allowed elaboration of several *universal* state diagrams, among which the most known is the *Hougen-Watson* diagram.

This consists in a $z = f(P_r)$ diagram, on which T_r = constant curves are drawn, the diagram being represented in Fig. 5.3 and which corresponds to an universal ES $F(P_r, V_r, T_r) = 0$ without material constants, an equation which is not representable into an analytical form.

Based on Hougen-Watson diagram, $z(P_r, T_r)$ have been realized, using relations (5.47a, 5.47b, 5.48a, 5.48b, 5.48c, 5.49a, 5.49b), similar diagrams representing, depending on T_r and P_r, dimensionless properties C_p/R, $(S - S_{id})/R$, $(H - H^o)/(R \cdot T_c)$, and φ.

There is a 5% calculation accuracy when using Hougen-Watson diagrams.

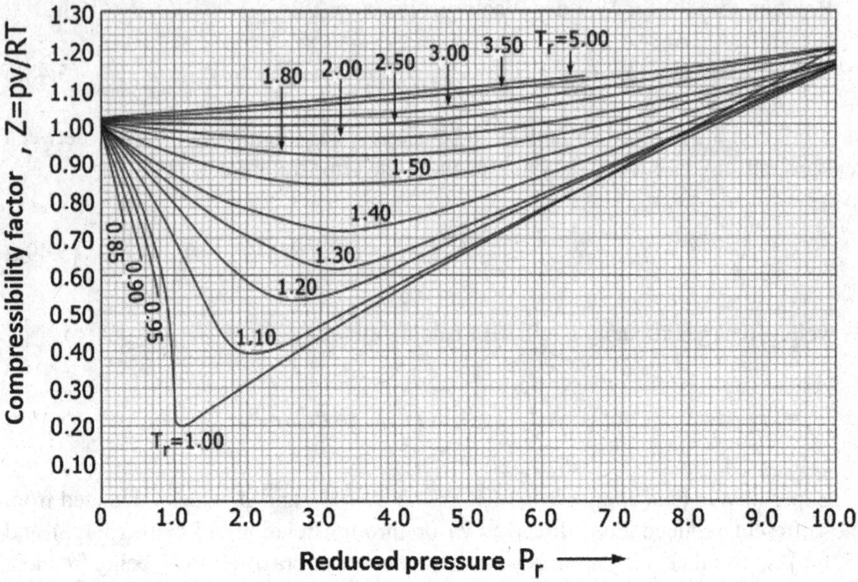

Fig. 5.3 Curves of fixed T_r, on the universal state diagram $z = f(P_r)$

The relatively high errors observed at He and H_2 quantum gases are empirically removed by replacing P_c and T_c from the denominator of definitions $P_r \equiv P/\,P_c$ şi $T_r \equiv T/\,T_c$ with modified properties: $P_c + 8$ bar and $T_c + 8\ K$:

$$H_2, He : P_r = P/(P_c + 8) \ and \ T_r = T/(T_c + 8) \tag{5.50}$$

Extended PCS

The accuracy obtained at calculations made using the principle of corresponding states is insufficient for many of the technical purposes. On the other hand, it is observed that two similar substances errors are always close as sign and size. It can be deduced that calculations results will be more accurate if relation (5.40): $X_{red} = F(Y_{red}, Z_{red})$ is replaced by a dependence on more variables – for example, by *three* independent variables instead of *two*:

$$X_{red} = F(Y_{red}, Z_{red}, W_{red}) \tag{5.51a}$$

the statement of this *extended* principle of corresponding states (improperly designated *theorem*) is "the value of any reduced quantity is univocally determined by the value of other N reduced properties, where $N \geq 3$." For $N = 2$ the *simple* theorem of corresponding states from Sect. 5.6 is obtained.

For real gas study, Y_{red} and Z_{red} properties are usually temperature and pressure, reduced by division to the critical values corresponding to relations (5.42), as in the simple theorem of corresponding states. Besides, W_{red} is often replaced by another dimensionless quantity sufficiently characteristic for different substances. Therefore, the extended theorem of corresponding states existent for three variables is:

$$X/X_c = F(P_r, T_r, K_{sim}) \tag{5.51b}$$

where X_c is quantity X value at critical point and K_{sim} is a *physicochemical similarity* criterion – a dimensionless quantity taking close values for two substances, if they are similar from the physicochemical point of view. Unlike reduced properties, K_{sim} takes the same value for any state it would be fond into.

The K_{sim} criterion must be not only *relevant* (correlation coefficient between K_{sim} and X_r to be close to 1 for as many properties and substances, on an as wide as possible reduced temperature and pressure fields) but also *discriminating* (the difference between K_{sim} values of different substances to be as higher as possible in relation to the measurement error) and *easily* measurable.

An optimal K_{sim} quantity does not exist which satisfies all the three points of view, which is why different physicochemical similarity criteria are used.

Physicochemical Similarity Criteria

The most known such criteria are as follows:

- Van der Waals criterion, denoted by K_{Wa} or [Wa], defined by:

$$[Wa] = 1/z_c = R \cdot T_c/P_c V_c \tag{5.52a}$$

- Riedel criterion, K_{Ri} or [Ri], and also denoted by α_c:

$$[Ri] = \alpha_c \equiv \lim_{T \to T_c} (dP_r/dT_r)_{LG} = (T_c/P_c) \cdot \lim_{Tc \to 1} (dP/dT)_{LG} \tag{5.52b}$$

where LG index refers to liquid-vapor equilibrium curve:

- Guldberg-Guy criterion, K_{Gu} or [Gu], defined by the ratio below:

$$[Gu] = T_f/T_c \tag{5.52c}$$

- Trouton criterion, K_{Tr} or [Tr]:

$$[Tr] = \lambda_f/(R \cdot T_f) \tag{5.52d}$$

- Mathias criterion, K_{Ma} or [Mt]:

$$[Mt] \equiv -(T_c/\rho_c) \cdot (d\rho_m/dT) \tag{5.52e}$$

where "mean density" ρ_m is the mean value of liquid and vapor densities at equilibrium at temperature T, $\rho_m \equiv (\rho_L + \rho_G)/2$; on a wide temperature range, ρ_m is a linear dependency function of T, according to <u>*Cailletet and Mathias law of rectilinear diameter*</u> which has a statistical justification.

- Pitzer criterion, K_{Pi} or [Pi], designated also as an acentric factor (denoted by ω):

$$[Pi] \equiv -1 + \log_{10}(P_c/P_{Pi}) \tag{5.52f}$$

- Filippov criterion, K_{Fi} or [*Fi*]:

$$[Fi] \equiv 100 \cdot P_{Fi}/P_c \tag{5.52 g}$$

pressures of Pitzer P_{Pi} and of Filippov P_{Fi} being the vapor pressures corresponding to low temperatures of $T_r = 0.7$ and $T_r = 0.625$, respectively.

The abovementioned physicochemical similarity criteria are calculable from the thermodynamic properties. However, molecular criteria can be also used, like the reduced dipole moment μ_{red}:

$$\mu_{red} = K * \cdot \mu \cdot (V_c \cdot T_c)^{1/2} \tag{5.53}$$

with K* as universal constant product and μ as electric dipole moment.

Altenburg criterion, q, can be also used – exponent of intermolecular distance d reverse d from the *repulsive* part, $B \cdot d^{-q}$, of the Mie potential from relation (4.21), u (d) = $A \cdot d^{-n} - B \cdot d^{-q}$, where $A \cdot d^{-n}$ is the intermolecular potential *attractive* par and $0 < n < q$.

The extended principle of corresponding states allows, even in the tri-parametric version (5.51b), more accurate correlations of thermodynamic quantities than the ones obtained from the simple principle, a bi-parametric principle (5.43). Better results can be obtained in the case of a four-quantity form, like:

$$X_r = F(T_r, P_r, K_1, K_2) \tag{5.54}$$

where K_1 and K_2 are two different physicochemical similarity criteria.

Material Constants Calculation from ES

The molecular constants – minimum distance σ and maximum interaction energy ε from Sect. 4.7. – can be determined either from structural measurements or, with approximation, from the ESs. For the case of Van der Waals gas, ε and σ constants from the Lennard-Jones potential dependence on intermolecular distance d can be obtained from ES (4.6), $P = R{\cdot}T/(V-\text{b}) - \text{a}/V^2$, from whose material constants (a, b) the following expressions can be found:

$$\sigma = 1.014 \cdot (b/N_A)^{1/3}; \varepsilon = 8 \cdot a/(27 \cdot b \cdot N_A) \tag{5.55}$$

Intermolecular potential constants (ε, σ) can be also evaluated from characteristic microscopical quantity values of the given gas. Usually, ε constant is calculable from the characteristic temperatures, while σ – from characteristic volume, the $\varepsilon/(k_B{\cdot}T)$ and $N_A{\cdot}\sigma^3/V$ groups being dimensionless. Using indices c, f, t, and L for "critical, boiling, melting, and liquid," the following relations have been proposed in the various thermodynamic-statistical models:

$$\varepsilon/(k_B \cdot T_c) = 0.751; \varepsilon/\left(k_B \cdot T_f\right) = 1.17; \varepsilon/(k_B \cdot T_t) = 1.92 \tag{5.56}$$

$$N_A \cdot \sigma^3/V_c = 0.348; N_A \cdot \sigma^3/V_{Lf} = 0.955; N_A \cdot \sigma^3/V_{Lt} = 1.10 \tag{5.57}$$

In the *extended* corresponding states theorem, ε and σ depend, according to Stiel and Thodos, not only on T_c and V_c, respectively, but also on the critical compressibility factor z_c, namely:

$$\varepsilon = 65.3 \cdot k_B \cdot T_c \cdot (z_c)^{3.6} \tag{5.58}$$

$$\sigma = 0.1576 \cdot (V_c)^{1/3}/(z_c)^{1.2} \tag{5.59}$$

5.8 Critical Point of Mixtures

Gas mixture's behavior in the critical field is complex, since liquid and vapor compositions at equilibrium with liquid are different in the case of mixtures (at mono-component systems, of identical compositions) (Fig. 5.4).

But the results are unsatisfactory if the reduction is made through division at mixture critical coordinates – the ones corresponding to critical point C. Instead of all these, reduction is effectuated using at denominator the *pseudocritical* properties, T_{pc}, P_{pc}, and V_{pc}, established as in the next paragraph. The reduced properties (P_r, V_r, T_r) are defined in this case by reporting (P, V, T) thermodynamic variables to the corresponding pseudocritical properties, distinguished by index "pc":

$$P_r \equiv P/P_{pc}; V_r \equiv V/V_{pc}; T_r \equiv T/T_{pc} \tag{5.60}$$

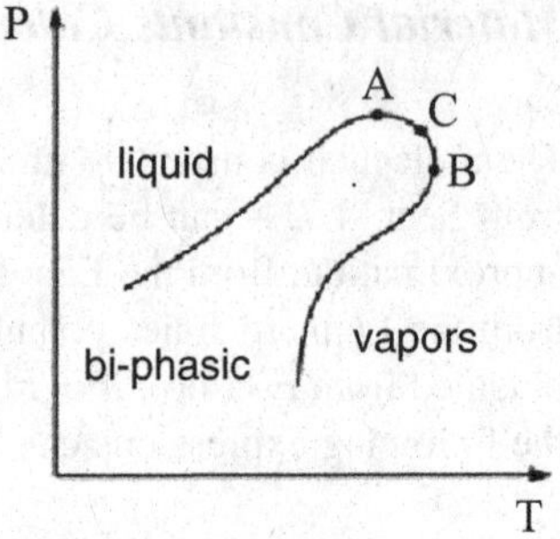

Fig. 5.4 Biphasic P/T field for the binary system

Particularly C, mixtures critical point the point where, for a given composition, liquid physical properties coincide with the ones of vapors does not correspond neither to the maximum temperature of vapor existence (B) nor to the maximum pressure of liquid existence (A), as in Fig. 5.4. It has been experimentally observed that mixtures satisfy the principle of corresponding states, with the same degree of approximation as the pure substances.

Any mixture of given composition is treated by this method as a pure gas, whose thermodynamic or transport properties can be obtained using the corresponding state theorem or using the general procedures of 5.3. It is understood that this method can directly obtain only mixture's *global* properties (like density, enthalpy and entropy, viscosity, thermal or electrical conductivity).

Individual properties calculation, like their fugacities or interdiffusion coefficients from the mixture of three or more of their compounds, is possible when using the partial-molar properties method from Sect. 8.4. of the book on thermodynamics of the same authors (Daneş et al. 2013 [1]), which however is not accurate enough.

Pseudocritical Properties

The *pseudocritical* properties depend on the compounds' critical properties and on the mixture composition. For the *simple* principle of corresponding states, the independent variables P_r and T_r and the pseudocritical properties P_{pc} and T_{pc} required for reduction, are calculated from composition through different procedures, namely:

I. Kay procedure, with:

$$T_{pc} = \Sigma_i x_i \cdot T_{c,i}; P_{pc} = \Sigma_i x_i \cdot P_{c,i} \tag{5.61}$$

simple, but with results who agree only for nonpolar substance mixture case;

II. Joffe procedure, where:

$$S' = \sum_{i=1}^{i=C} \sum_{j=1}^{j=C} x_i \cdot x_j \left[(T_{c,i}/P_{c,i})^{1/3} + (T_{c,j}/P_{c,j})^{1/3} \right]^3 /8$$

$$S'' = \sum_{i=1}^{i=C} T_{c,i}/\sqrt{P_{c,i}}; T_{pc} = (S'')^2/S'; P_{pc} = (S''/S')^2 \tag{5.62}$$

III. Prausnitz procedure, for which:

$$T_{pc} = \Sigma_i x_i \cdot T_{c,i} : P_{pc} = R \cdot T_{pc} \cdot \Sigma_i x_i \cdot z_{c,i}/\Sigma_i x_i \cdot V_{c,i}; \tag{5.63}$$

In the case of the *extended* principle of corresponding states, P_r, T_r and ω, or P_r, T_r, and z_c can be used as independent variables. The P_r and T_r properties are calculated as in the Prausnitz procedure, and constants ω or z_c for a mixture are calculated through the weighted average of the pure components' corresponding constants ω_i and z_{ci}:

$$\omega = \sum x_i \cdot \omega_i; z_c = \sum x_i \cdot z_{ci}; V_c = \sum x_i \cdot V_{ci} \tag{5.64}$$

5.9 Six Worked Examples

Ex. 5.1
For the Van Laar ES with three material constants, it is required to determine:

1) Equation constants, relative to the critical quantities
2) Equation in reduced form

Solution

1. Van Laar equation (4.8h), $P = R \cdot T \cdot (V + c)/[V \cdot (V - b)] - a/V^2$, is brought, through the following decomposition into rational fractions of the functional parameter depending on V from the first term of the right member:

$$(V + c) \cdot [V \cdot (V - b)] = [(b + c)/(V - b) - c/V]/b$$

and introduction of simplifying notations:

$$f \equiv b + c \tag{5.8a2}$$

$$g \equiv a \cdot b/(R \cdot T) \tag{5.8b2}$$

$$X \equiv P \cdot b/(R \cdot T) \tag{5.8c2}$$

to the form:

$$X = f/(V - b) - c/V - g/V^2 \tag{5.8d2}$$

Conditions (5.7), $(\partial P/\partial V)_T = 0 \leftrightarrow (\partial^2 P/\partial V^2)_T = 0$, become in the critical point, when using the change of variables (5.8c2):

$$(\partial X/\partial V)_T = 0; \left(\partial^2 X/\partial V^2\right)_T = 0; V = V_c; T = T_c \tag{5.8e2}$$

From (5.8d2), the following is obtained, through isothermal derivation in relation to volume:
$(\partial X/\partial V)_T = -f/(V-b)^2 + c/V^2 + 2 \cdot g_c/V^3$
$(\partial^2 X/\partial V^2)_T = 2 \cdot f/(V-b)^3 - 2 \cdot c/V^2 - 6 \cdot g_c/V^4$, where:

$$g_c \equiv a \cdot f/(R \cdot T_c) \tag{5.8f2}$$

is the g value in the critical point, so that relation (5.8e2) therefore becomes:

$$c/(V_c)^2 + 2 \cdot g_c/(V_c)^3 = f/(V_c - b)^2 \tag{5.8g2}$$

$$c/(V_c)^3 + 3 \cdot g_c/(V_c)^4 = f/(V_c - b)^3 \tag{5.8h2}$$

Relations (5.8g2, 5.8h2) can be treated as a linear system of two equations with two unknowns (c, g_c), having b and V_c as quantities. System solution is:

$$c = f \cdot (V_c - 3 \cdot b) \cdot (V_c)^2/(V_c - b)^3 \tag{5.8i2}$$

$$g_c = f \cdot b \cdot [V_c/(V_c - b)]^3 \tag{5.8j2}$$

By noting with X_c the critical value of X parameter:

$$X_c = (b \cdot P_c)/(R \cdot T_c) \tag{5.8k2}$$

relation (5.8d2) becomes in the critical point $X_c = f/(V_c - b) - c/V_c - g_c/(V_c)^2$ or after introducing the quantities (c, g_c, X_c) from (i, j, k) expressions:

$$P_c/(R \cdot T_c) = f \cdot b/(V_c - b)^3$$

from where $f = (3 \cdot V_c - R \cdot T_c)/P_c$.

By considering $z_c = P_c \cdot V_c/(R \cdot T_c)$ – definition (5.4) of the critical compressibility factor – it will result in:

$$f = z_c \cdot (V_c - b)^3/(b \cdot V_c) \tag{5.8l2}$$

When replacing f from relation (5.8l2) into equality (5.8j2), the following expression is obtained:

$$g_c = z_c \cdot (V_c)^2 \tag{5.8m2}$$

When replacing f from relation (5.8l2) into equality (5.8i2), the following expression is obtained:

$$c = z_c \cdot V_c \cdot (V_c - 3 \cdot b)/b \tag{5.8n2}$$

Replacing afterwards (f, g_c, c) from relations (5.8l2, 5.8m2, 5.8n2) into (5.8a2) and (5.8f2) definitions, it results that:

$$a = P_c \cdot (V_c)^2 \cdot z_c/(3 \cdot z_c - 1)$$
$$b = V_c \cdot (3 \cdot z_c - 1)/z_c$$

and after introducing b expression into relation (5.8n2), the following expression is obtained:

$$c = V_c \cdot z_c(3 - 8 \cdot z_c)/(3 \cdot z_c - 1)$$

2. When using ratios to critical quantities from (5.42) and (5.45) as reduction methods: $P = P_r \cdot P_c$, $T = T_r \cdot T_c$, and $V = V_r \cdot V_c$, state variables P, V, and T can be replaced from Van Laar ES with corresponding reduced variables P_r, T_r, and V_r:

$$P_r \cdot P_c = R \cdot T_r \cdot T_c \cdot (V_r \cdot V_c + c)/[V_r \cdot V_c \cdot (V_r \cdot V_c - b)] - a/(V_r \cdot V_c)^2$$

With (a, b, c) from point 1 and then by replacing the $P_c \cdot V_c / (R \cdot T_c)$ ratio with z_c:

$$\Pr = \frac{T_r \cdot [V_r + z_c \cdot (3 - 8 \cdot z_c)/(3 \cdot z_c - 1)]}{V_r \cdot [V_r \cdot z_c - (3 \cdot z_c - 1)]} - \frac{z_c}{(3 \cdot z_c - 1) \cdot (V_r)^2}$$

It can be observed that Van Laar ES is not universal – unlike Van der Waals reduced Eq. (5.46a) – but contains a material constant, z_c, which is specific to each substance. Indeed, by eliminating V_c between expressions for c and b from point 1, it results that $c/b = (3-8 \cdot z_c)/(3-1/z_c)^2$, from where it can be seen that z_c depends on the c/b ratio, differing from one fluid to another.

Ex. 5.2
For Berthelot gas it is required to determine the following criteria values: (1) Van der Waals and (2) Riedel.

Solution
1. Berthelot ES (4.8b) represents the starting point:

$$P = R \cdot T/(V - b) - a/\left(T \cdot V^2\right) \tag{5.8a3}$$

By ES successive derivations, it is found that:

$$(\partial P/\partial V)_T = -R \cdot T/(V - b)^2 + 2 \cdot a/\left(T \cdot V^3\right);$$

$$\left(\partial^2 P/\partial V^2\right)_T = -2 \cdot R \cdot T/(V - b)^2 + 6 \cdot a/\left(T \cdot V^4\right).$$

Since at the critical point V, T, and P are replaced by V_c, P_c, and T_c and both derivatives are null, the three above relations become, respectively:

$$P_c = R \cdot T_c/(V_c - b) - a/\left[T_c \cdot (V_c)^2\right]; R \cdot T_c/(V_c - b)^2 = 2 \cdot a/\left[T_c \cdot (V_c)^3\right];$$

$$R \cdot T_c/(V_c - b)^3 = 3 \cdot a/\left[T_c \cdot (V_c)^4\right]; \tag{5.8b3}$$

It is observed that (5.8b3) constitutes a three-equation system, with P_c, V_c, and T_c unknowns. By eliminating T_c between the last two relations, it results that:

$$V_c = 3 \cdot b \tag{5.8c3}$$

When replacing V_c into the second or third equations (5.8b3), the following expression is obtained:

$$T_c = [8 \cdot a/(27 \cdot R \cdot b)]^{1/2} \tag{5.8d3}$$

and by introducing V_c and T_c into the first equation (5.8b3), it results that:

$$P_c = \left[a \cdot R/\left(216 \cdot b^3\right)\right]^{1/2} \tag{5.8e3}$$

From relation (5.52a), [Wa] = $R \cdot T_c / (V_c \cdot P_c)$, the following is obtained, by replacing (V_c, P_c, T_c):

$$[Wa] = R \cdot [8 \cdot a/(27 \cdot R \cdot b)]^{1/2}]/\{\left[a \cdot R/\left(216 \cdot b^3\right)\right] \cdot 3 \cdot b\}^{1/2} \Rightarrow [\mathbf{Wa}] = \mathbf{8/3}$$

2. From (a) the following is obtained, by isochoric derivation in relation to T:

   $(\partial P/\partial T)_V = R/(V - b) + a/(T \cdot V)^2$ from where at the critical point:

$$(\partial P/\partial T)_{V,c} = R/(V_c - b) + a/(T_c \cdot V_c)^2$$

or with expressions (5.8c3) and (5.8d3) for V_c and T_c, respectively, the following expression is obtained:

$$(\partial P/\partial T)_{V,c} = 7 \cdot R/(8 \cdot b) \tag{5.8f3}$$

On the other hand at the critical point, the isochore curve and vapor pressure curve have the same slope, so that relation (5.52b), $[Ri] = (T_c/P_c) \cdot \lim_{Tc \to 1} (dP/dT)_{LG}$, becomes [Ri] = $(T_c/P_c) \cdot (\partial P/\partial T)_{V,c}$ or by replacing T_c, P_c, and $(\partial P/\partial T)_{V,c}$ from (5.8d3), (5.8e3), and (5.8f3), respectively:

$$[Ri] = [8 \cdot a/(27 \cdot R \cdot b)]^{1/2} \cdot \left[a \cdot R/\left(216 \cdot b^3\right)\right]^{-1/2} \cdot [7 \cdot R/(8 \cdot b)] \Rightarrow [\mathbf{Ri}] = \mathbf{7}$$

Ex. 5.3

The reduced volume and temperature of a mono-component and monophasic Van der Waals fluid have the following values: $V_r^* = 5/3$ and $T_r^* = 0.96$. It is required:

1) To establish if the respective state corresponds to an unstable fluid, stable liquid, or stable gas
2) To determine fluid stability type in the given state: metastability or absolute (thermodynamic) stability

Solution

1. Van der Waals reduced ES (5.46a):

$$P_r = 8T_r/(3V_r - 1) - 3/(V_r)^2 \qquad (5.8a4)$$

becomes, with values V_r* T_r* from the statement:

$$P_r = 8 \cdot 0.96/ \cdot (3 \cdot 5/3 - 1) - 3/ \cdot (5/3)^2, or\ P_r{*} = 0.84.$$

It is now considered (5.8a4) as an equation with unknown V_r and with quantities $P_r = P_r$* and $T_r = T_r$*. By regrouping the terms for the unknown decreasing powers, it results that:

$$3P_r * (V_r)^3 - (P_r * +8T_r{*}) \cdot (V_r)^2 + 9V_r - 3 = 0$$

or after replacing P_r* = 0.84 and T_r* = 0.96 and simplification with 3:

$$0.84(V_r)^3 - 2.84(V_r)^2 + 3V_r - 1 = 0 \qquad (5.8b4)$$

Equation (5.8b4) is of third degree, but one of the roots, denoted now by V_r', is already known, $V_r' = V_r{*} = 5/3$, so that the other roots can easily be determined, by dividing the left member polynomial of equation (5.8b4) with $0.6 \cdot (V_r - 5/3)$. It results that $1.4 \cdot (V_r)^2 - 2.4 \cdot V_r + 1 = 0$.

The other two roots, V_r'' and V_r''', will therefore have the following values:

$$V_r = \left[2.4 \pm \sqrt{2.4^2 - 4 \cdot 1.4}\right]/(2 \cdot 1.4).$$

or by performing the calculations:

$$V_r'' = 5/7; V_r''' = 1.$$

Therefore, the three roots of equation (5.8b4) that was obtained using the isotherm curve $T_r = 0.96$ intersection with isobar curve $P_r = 0.84$, are in increasing order: 5/7 (V''_r), 1 (V_r'''), and 5/3 (V_r').

According to the discussion from Sect. 5.4, the middle root – here V_r''' – corresponds to an unstable absolute monophasic system, and the other two roots correspond to a stable single-phase state of the system (states of which stability it is not yet known if it is absolute or relative – metastability).

The smallest root corresponds to the stable liquid state, $V_{Lr} = V_r'' = 5/7$, and the highest root – to the stable gaseous state: $V_{Gr} = V_r' = 5/3$. Therefore:

$$\Rightarrow \textbf{gaseous fluid (vapors), stable}$$

2. The type of equilibrium is given by the vaporization Gibbs free energy sign $\Delta G_{vap} \equiv G_G - G_L$ – the difference between gas enthalpy, G_G, and liquid enthalpy, G_L, since phases are compared at equal temperature and pressure.

At $\Delta G_{vap} < 0$ the vapors are absolutely stable and the liquid – metastable, while $\Delta G_{vap} > 0$ implies metastable vapors, the absolutely stable phase being the liquid one.

Relation (5.23), $G_F - G_B = P \cdot (V_F - V_B) - \int_B^F P(V,T) \cdot dV$ becomes, using the notations from here, $\Delta G_{vap} = P \cdot (V_G - V_L) - \int_{V=V_L}^{V=V_G} P(V,T) \cdot dV$ or in relation to the reduced variables:

$$\Delta G_{vap}/(P_c \cdot V_c) = P_r * \cdot (V_{G,r} - V_{L,r}) - \int_{V_r=V_{L,r}}^{V_r=V_{G,r}} P_r(V_r, T_r) \cdot dV_r$$

or with $P_r = f(V_r, T_r)$ from expression (5.8a4) and values $P_r^* = 0.84$, $V_{Lr} = 5/7$, and $V_{Gr} = 5/3$ calculated at point 1:

$$\Delta G_{vap}/(P_c \cdot V_c) = (0.84) \cdot (5/3 - 5/7) - I \qquad (5.8c4)$$

where I is the following integral (where 0.96 is the value $T_r^* = 0.96$ from the statement).

$I = \int_{V_r=5/7}^{V_r=5/3} \left[8 \cdot 0.96/(3 \cdot V_r - 1) - 3/(V_r)^2 \right] \cdot dV_r$ or after integration:

$$I = (8 \cdot 0.96/3) \cdot \{ ln\,[(5/3 - 1/3)/(5/7 - 1/3)] \} + 3(5/3 - 5/7) = 2.56 \cdot [ln\,(7/4)] + 20/7 = 4.29$$

By introducing into relation (5.8c4) the value for I, this leads to $\Delta G_{vap} = -3.49 \cdot P_c \cdot V_c$ or, since P_c and V_c are positive, $\Delta G_{vap} < 0$.

$\Rightarrow$ **vapors stability is absolute(thermodynamics)**

Ex. 5.4

For capronitrile the following are known: molar mass (0.12915 kg/mol), critical temperature (622 K), and critical pressure (32.4 bar).

It is required to determine, following the principle of corresponding states:

1) Density at 100 bar and 750 K
2) Pressure at which molar volume value is 530 cm^3/mol, at 684 K

Solution

1. $P_r = P/P_c = 100/32.2 = 3.11$; $T_r = T/T_c = 1.206$. By graphic interpolation for P_r and T_r on the universal Hougen and Watson diagram of state (Fig. 5.3), $z = 0.58$ is obtained.

   From definition (4.3) of the compressibility factor, $z = P{\cdot}V/(R{\cdot}T)$, it results that $V = R{\cdot}T{\cdot}z/P = 8.314{\cdot}750{\cdot}0.58/(100{\cdot}10^5) = 36.16{\cdot}10^{-5}$ m^3/mol.

   With the elementary mass from the statement, density becomes:

$$\rho = M/V = 0.12915/\left(36.16 \cdot 10^{-5}\right) \Rightarrow \boldsymbol{\rho} = \mathbf{357\ kg/m^3}$$

2. By replacing into expression, $z = P{\cdot}V/(R{\cdot}T)$, pressure size, $P = P_c{\cdot}P_r$, it is obtained that $z = P_r{\cdot}P_c{\cdot}V/(R{\cdot}T)$. By numerically replacing (P_c, V, R, T), it is found that $z = P_r{\cdot}32.4{\cdot}0.53/(8.314{\cdot}684)$ or:

$$z/P_r = 0.302 \tag{5.8a5}$$

   The reduced temperature is $T_r = T/T_c = 684/622$ from where:

$$T_r = 1.100 \tag{5.8b5}$$

The point of whose coordinates satisfy (5.8a5) and (5.8b5) conditions can be determined from the universal state diagram of Fig. 5.3, but only indirectly (on the diagram it is not represented the z/P_r ratio). In order to find this point, different isotherm curves $T_r = 1.1$ from Fig. 5.3 points are read, i.e., coordinates' value domains (z and P_r), and afterwards their ratio is calculated:

P_r (read)	0.9	1.0	1.2	1.4	1.6	1.8	2
z (read)	0.72	0.68	0.61	0.52	0.445	0.41	0.42
z/ P_r (calculated)	0.80	0.68	0.51	0.371	0.278	0.228	0.210

A linear interpolation of the reduced pressure in relation to z/P_r ratio values, in the z/P_r range (0.278; 0.371), leads for $z/P_r = 0.302$ to:

$P_r = 1.6 - (1.6 - 1.4) \cdot (0.302 - 0.278)/(0.371 - 0.278) \cong 1.55$ from where:

$$P = P_r \cdot P_c = 1.55 \cdot 32.4 \Rightarrow \mathbf{P_r \cong 51\ bar}$$

Ex. 5.5

For methane, a and b constants of the Dieterici ES (4.8c) are a = 0.2802 $Kg \cdot m^5/(s^2 \cdot mol^2)$ and b = $4.123 \cdot 10^{-5}$ m^3/mol, and its molar mass is 0.01604 Kg/mol.

It is required to determine the critical constants T_c and P_c and the critical density ρ_c.

Solution

For critical coordinates determination ES (4.8c) is used:

$$P = R \cdot T \cdot (V - b)^{-1} \cdot e^{-a/(R \cdot T \cdot V)} \tag{5.8a6}$$

And the first two pressure isotherm derivatives in relation to volume are:

$$(\partial P/\partial V)_T = \left[a/V^2 - R \cdot T/(V - b)\right] (V - b)^{-1} \cdot e^{-a/(R \cdot T \cdot V)} \tag{5.8b6}$$

$$\left(\frac{\partial^2 P}{\partial V^2}\right)_T = \left[\frac{a^2}{R \cdot T \cdot V^4} - \frac{2 \cdot a \cdot (2 \cdot V - b)}{V^3 \cdot (V - b)} + \frac{2 \cdot R \cdot T}{(V - b)^2}\right] \cdot (V - b)^{-1} \cdot e^{-a/(R \cdot T \cdot V)} \tag{5.8c6}$$

For the critical point, *P*, *T*, and *V* are replaced by P_c, T_c, and V_c into relations (5.8a6, 5.8b6, 5.8c6), while $(\partial P/\partial V)_T$ and $(\partial^2 P/\partial V^2)_T$ are zero:

$$P_c = R \cdot T_c \cdot (V_c - b)^{-1} \cdot exp\left[-a/(R \cdot T_c \cdot V_c)\right] \tag{5.8d6}$$

$$a/(V_c)^2 - R \cdot T_c/(V_c - b) = 0 \tag{5.8e6}$$

$$\frac{a^2}{R \cdot T_c \cdot V_c^4} - \frac{2 \cdot a \cdot (2 \cdot V_c - b)}{V_c^3 \cdot (V_c - b)} + \frac{2 \cdot R \cdot T_c}{(V_c - b)^2} = 0 \tag{5.8f6}$$

By replacing $R \cdot T_c = a \cdot (V_c - b)/(V_c)^2$ from relation (5.8e6) into expression (5.8f6), the following is obtained, after dividing by a and term regrouping:

$$V_c = 2 \cdot b \tag{5.8g6}$$

Replacement of V_c into (5.8e6) equality, $a/(4 \cdot b^2) - R \cdot T_c/(2 \cdot b - b) = 0$, leads to:

$$T_c = a/(4 \cdot R \cdot b) \tag{5.8h6}$$

By introducing (V_c, T_c) from relations (5.8g6) and (5.8h6) into expression (5.8d6), it is found that:

$$P_c = a/\left(4 \cdot b^2 \cdot e^2\right) \quad (5.8i6)$$

From $V_c = 2{\cdot}b$ and M the critical density is obtained, $\rho_c = M/ V_c$, namely:

$$\rho_c = M/(2 \cdot b) \quad (5.8j6)$$

From relations (5.8h6, 5.8i6, 5.8j6) the following expressions are obtained:

$$T_c = 0.2802/\left(4 \cdot 8.314 \cdot 4.123 \cdot 10^{-5}\right),$$

$$P_c = 0.2802 \cdot \left(2 \cdot 4.123 \cdot 10^{-5} \cdot 2.7182\right)^{-2}$$

and

$$\rho_c = 0.01604/\left(2 \cdot 4.123 \cdot 10^{-5}\right) \Rightarrow \mathbf{T_c = 204\ K; P_c = 5.58 \cdot 10^6 N/m^2};$$

$$\boldsymbol{\rho_c = 194.5\ Kg/m^3}.$$

The real values are 191 K, $4.64{\cdot}10^6$ N/m^2, And162 Kg/m^3, the error being 7%, 20%, and 20%, respectively.

Ex. 5.6
Binary gas mixture components, where $x_1 = 0.4$, respect the Clausius ES (4.8f), $P = R{\cdot}T/ (V – b) – a/ [T{\cdot}(V + c)^2]$, and have the following features:

i	formula	T_c, K	P_c, bar	V_c, cm^3/ mol
1	CH_3Cl	416.3	66.7	143
2	CH_4	191	46.4	99.5

It is required to determine mixture molar volume at 380 K and 200 bar, following the extended principle of corresponding states, using the Prausnitz method for pseudocritical quantities determination.

Solution
From definition (5.4), the following are obtained:

$$z_{c1} = P_{c1} \cdot V_{c1}/(R \cdot T_{c1}) = \left(66.7 \cdot 10^5 \cdot 143 \cdot 10^{-6}/(8.314 \cdot 416.3\right) = 0.276$$

$$z_{c2} = P_{c2} \cdot V_{c2}/(R \cdot T_{c2}) = \left(46.4 \cdot 10^5 \cdot 99.5 \cdot 10^{-6}/(8.314 \cdot 191\right) = 0.291.$$

The molar fraction $x_2 = 1 – x_1 = 1–0.4 = 0.6$.
From the Prausnitz rules (5.64) the following are obtained:

$$z_{pc} = z_{c1} \cdot x_1 + z_{c2} \cdot x_2 = 0.276 \cdot 0.4 + 0.291 \cdot 0.6 = 0.285$$

$$V_{pc} = V_{c1} \cdot x_1 + V_{c2} \cdot x_2 = 143 \cdot 0.4 + 99.5 \cdot 0.6 = 116.9\ cm^3/mol$$

and from rules (5.63) of the same author results:

$$T_{pc} = T_{c1} \cdot x_1 + T_{c2} \cdot x_2 = 416.3 \cdot 0.4 + 191 \cdot 0.6 = 281.1\ K$$

$$P_{pc} = R \cdot T_{pc} \cdot Z_{pc}/V_{pc} = 8.314 \cdot 281.1 \cdot 0.285/\left(116.9 \cdot 10^{-6}\right) = 56.9\ bar$$

Relation (5.16) for the critical quantities of the Clausius gas, $V_c = 3 \cdot b + 2 \cdot c, T_c = \sqrt{8 \cdot a \cdot (b + c)/(27 \cdot R)}$, and $P_c = \sqrt{R \cdot a}/[6 \cdot (b + c)]^{3/2}$ leads, when eliminating (b + c) between the second and third relations, to $a = 27 \cdot R^2 \cdot (T_c)^3/(64 \cdot P_c)$. By introducing a into relation for T_c, it is obtained that $c = (R \cdot T_c)/(8 \cdot P_c - b)$, from where, by replacing into the relation for V_c, $b = V_c - R \cdot T_c/(4 \cdot P_c)$. By replacing then a and b into one of the starting relations, it is obtained that $c = 3 \cdot R \cdot T_c/(8 \cdot P_c - V_c)$. By replacing a, b, and c expressions into the Clausius ES from the statement, it results that:

$$P = \frac{R \cdot T}{(V - V_c + R \cdot T_c/4 \cdot P_c)} - \frac{27 \cdot R^2 \cdot T_c^3}{64 \cdot P_c \cdot T \cdot (V - V_c + 3 \cdot R \cdot T_c/8 \cdot P_c)^2}.$$

After reducing (5.40, 5.45) the state variables, the reduced Clausius ES is found, when replacing $P = P_c \cdot P_r$, $V = V_c \cdot V_r$, and $T = T_c\ T_r$:

$$P_r = \frac{T_r}{(z_c \cdot V_r - z_c + 1/4)} - \frac{27}{64 \cdot T_r \cdot (z_c \cdot V_r - z_c + 3/8)^2} \qquad (5.8a7)$$

Using T and P reduction (5.60) by pseudocritical properties, it results that $P_r = P/P_{pc} = 200/56.2 = 3.559$ and $T_r = T/T_{pc} = 380/281.12 = 1.352$.

When introducing into (a) these values P_r, T_r, as well as $z_c = z_{pc} = 0.2848$, it results that:

$$3.5587 = \frac{1.3517}{(0.2848 \cdot V_r - 0.2848 + 0.25)} - \frac{27}{64 \cdot 1.3517} \cdot \frac{1}{(0.2848 \cdot V_r - 0.2848 + 0.375)^2}$$

which is rearranged to $3.559 \cdot (V_r)^3 - 2.927 \cdot (V_r)^2 + 0.9228 \cdot V_r - 0.9899 = 0$.with the only positive solution $V_r = 0.8897$ from where:

$$V = V_r \cdot V_{pc} = 0.8897 \cdot 116.9 \Rightarrow \mathbf{V = 104.0\ cm^3/mol}$$

Chapter 6
Thermodynamic Models of Condensed Phases

6.1 State of Aggregation

There are two notions – of great general degree, therefore difficult to be defined – namely, the one of *phase* of a macroscopic material system (comprising a very high N number, $N \gg 1$, of microscopic material particles, like molecules) and the one of *state of aggregation* of each phase contained by the system in the given conditions (fixed composition, pressure, and temperature).

The notion of phase is widely used in thermodynamics: definitions introduced in Sects. 2.2 and 2.4 in book on thermodynamics of the same authors (Daneş et al., 2013 [1]) are widely applied for pure substances and poly-component systems, in Chaps. 7 and 13, respectively, of the respective volume.

Conversely, the "state of aggregation" notion is linked to the matter *structure* microscale knowledge, knowledge which is indispensable for the kinetic theory of statistical and molecular physics applications of matter.

The state of aggregation of a finite part from a material system is the most general set of characteristic intensive properties. The most common states of aggregation (which occur at any sufficiently stable substance) are the well-known states of gas, liquid, and solid, differentiated by:

- The existence of an own form (i.e., a reduced sensitivity to deforming factors action) of the *solid*, while the *fluid* states (of aggregation) – gas and liquid – are immediately and easily deformable.
- The nonexistence of an own volume for the *gas* (i.e., an important decrease of the volume when pressure increases and temperature decreases), unlike the *condensed* states (of aggregation) – liquid and solid – whose thermal expansibility and compressibility are so small than in a first approximation it can be considered that their volume is invariable.

Granting the status of *state of aggregation* to other forms of matter existence than the three ones described above (e.g., amorphous substances, glasses, plasma, soils

F. E. Daneş et al., *Molecular Physical Chemistry for Engineering Applications*,
https://doi.org/10.1007/978-3-030-63896-2_6

and gels, smectic or nematic crystals, etc.), although useful in particular situations, is not suitable in the general treatment adopted herein, where it is judicious to treat amorphous substances and glasses as subcooled liquids and plasma – as an ionized gas, the colloidal systems being subsequently treated as a case of special force field systems (here, interfacial interactions) and liquid crystals to be left to special structural studies.

Only the condensed phases are treated in this chapter, namely, only from the equilibrium quantities' point of view, already treated for gases in Chap. 4.

Condensed States of Aggregation

Liquids are more similar thermodynamically to solids than to gases, especially at low temperatures. Thus, near the triple point where the three phases coexist, it is found that liquids and solids have molar volume, isothermal, or adiabatic compressibility, as well as expandability, of sizes close to each other, being hundreds or thousands of times smaller than V, χ_T, χ_S, or α for gases.

Generally, these parameters take intermediate values for liquids, but the Y_L/Y_S ratio between the liquid and solid features exceeds 1 with little, and the Y_G/Y_L ratio is much higher than 1. For example, the relative volume increase at melting is usually only a few percentages, whereas for the evaporation case, the liquid volume can be neglected as compared to the gas (which is valid not only for the triple point but also for nine tenths of the liquid-vapor equilibrium curve).

Also, liquid enthalpy and entropy are much closer to solid values than to the ones of gas, at least at the triple point where the three phases are in equilibrium.

In terms of the thermodynamic features, the liquid is therefore sensitively more similar to a solid than to a gas.

Liquid State Particularities

The liquid resembles a gas in some respects and a solid from other points of view, so the *liquid* state of aggregation presents an intermediate situation between the gaseous and solid (crystalline) states.

Moreover, once the physical state changes – in particular with temperature change – the liquid similarity ratios with the other two states of aggregation also change, but radically: at high pressures as well as at low temperatures (in the vicinity of melting temperature), liquid likeness with the solid is much more accentuated than the liquid-gas similarity, while at high temperatures (critical or supercritical), the liquid is confused with a highly compressed gas.

This intermediate and variable feature of liquid physical quantities results in particularly major difficulties in their explaining and interpreting.

The thermodynamic-statistical and kinetic-molecular calculations are facilitated for gases by molecule movement independence and for solids by the existence of molecule equilibrium positions (or, more generally, of particles in the crystalline lattice nodes), positions that are invariant in time.

Liquids do not benefit however from any of these simplification possibilities, from which the occurrence of a high number of liquid theories (Sect. 6.3.), based on more complex models than for the other two states of aggregation; each of these models may of course only satisfactorily explain certain liquid quantities, or apply to a narrower range of physical states, possibly to a single class of liquids.

Thermodynamic Physical Quantities in Liquids

A fundamental difference between gases on the one hand and the condensed states (liquids together with solids) on the other hand is given by the volume behavior at low pressures: when P tends towards 0, the gas volume tends to infinity. In the kinetic-molecular interpretation, eigenvolume presence or absence is a consequence of a higher or lower intensity of cohesion. Cohesion, intermolecular attraction force manifestation, measured by *internal pressure* (Sect. 6.3), is also present in gases, but its effect is canceled by intense thermal agitation of molecules.

On the other hand, pressure and especially temperature dependence on liquid quantities has a distinct character compared to similar variations of the crystalline solid: once the temperature increases in the (T_{tr}, T_c) range, the V, χ_T, χ_S, or α liquid values also grow – at the beginning slowly, but increasingly faster near T_c – becoming identical at the critical point with the compressed gas quantities. Also, ΔH_v and ΔS_v on the L-G equilibrium curve decrease with T_{tr} towards T_c shifting, at the beginning more slowly, and then increasingly faster, being canceled at the critical point.

The ES of any liquid is identical to its vapor equation, but at pressures much higher than the critical pressure, the Tait ES applies:

$$V = A/(1 + B \cdot P)^C \tag{6.1a}$$

with A, B, and C – positive constants, independent of pressure. The last can be obtained from Mie potential exponents, $u = a/d^m - b/d^n$ from relation (4.21):

$$C = 3/(m + n + 6) \tag{6.1b}$$

Therefore, $C = 1/8$ for the Lennard-Jones potential, where $m = 6$ and $n = 12$.

Heat capacity is also different for the three states of aggregation, but differences are more reduced – C_p (or C_v) value ratio of two states of aggregation rarely exceeds 2.

For liquids, C_p and C_v are almost always higher for solids, and usually they are also higher for gases, so that from the heat capacity point of view, liquids do not present an intermediate feature between solids and gases.

The fact that among all the states of aggregation liquid heat capacity is the highest represents a new indication liquids feature to rapidly change their caloric properties (here: enthalpy, entropy, and internal energy, because at P constant $d\,H = C_p{\cdot}d\,T$ and $dS = C_p{\cdot}d(ln\mathrm{T})$ and at constant volume $d\,U = C_v{\cdot}d\,T$ şi $d\,S = C_p{\cdot}d\,ln\,T$) once that temperature changes.

Liquids also have particularities in terms of heat capacity variation with temperature: while at all gases and solids C_p and C_v increase (slowly) once that temperature increases, the temperature increase for liquids can result in both increases and decreases of the heat capacity, relative to liquid nature and temperature range. Heating decreases of heat capacity are found especially for C_v.

Non-thermodynamic Macroscopic Physical Quantities

Non-thermodynamic equilibrium macroscopic physical quantities also differ according to the state of aggregation. Thus, the refractive index, dielectric constant, and magnetic susceptibility have values very close to 1 for gases, while at liquids and solids, they can take much higher values (which is especially conditioned by the much higher density of the condensed phases). *Surface tension*, the unit of area's energy where the interface between substance and vacuum is formed, is much more important to liquids and solids than to gases.

Transport quantities of the three states of aggregation, which will be discussed in Chapters 7 and 8 (part III of this volume), also allow for a net differentiation of the states of aggregation. The most important differences are in the transport of momentum, which leads to a different way of grouping them than that resulting from comparing the states of aggregation thermodynamic quantities: liquids and gases *flow* – they have the property of being *fluid* – whereas the crystalline solids can not flow, but only can deform.

Any fluid is *isotropic*: no macroscopic quantity varies with direction. The typical crystalline solid is *anisotropic* (excepting solids with cubic lattice): different equilibrium or nonequilibrium physical quantities – expansibility, compressibility, transport coefficients, refractive index, sound speed, etc. – change with direction.

6.2 Molecular Structure of the States of Aggregation

The above states of aggregation macroscopic physical-chemical features are determined by their *structure*. The X-ray diffraction or neutron diffraction across different media provides the most reliable information regarding the particle sorting pattern, the mean distance r between adjacent molecules, and *coordination* number L (defined as the direct neighbors' number for a given molecule).

Diagrams of Interference

Matter structure experimental study by X-ray diffraction, based on propagated beam intensity *I* measurement in relation to distance *R* to the beam center, gave further arguments for the similarity between liquid and solid states while allowing also for their differentiation. The Debye-Scherrer diagram of interference starts from the *radial distribution* function g(r) of the probability g that another molecule is found at a given r distance from the center of a molecule. From g it is made the passage to G (r), the *correlation* function between diffracted radiation intensity, and the intermolecular distance r:

$$G(r) = g/\left(4 \cdot \pi \cdot r^2 \cdot \bar{\mu}\right) \tag{6.2a}$$

where G represents thus the ratio between density $\mu \equiv g/(4 \cdot \pi \cdot r^2)$ of molecules found at distance r from a given molecule and the mean density $\bar{\mu}$ of molecules from the whole sample. The Debye- Scherrer diagram is situated at values G > 0 and $r \geq r_o$ (r_o minimum being a measure of the molecular diameter) and presents an horizontal asymptote at G = 1 (dotted line from figure 6.1), since *lim* $\mu_{r \to \infty} = \bar{\mu}$.

Continuous and Discontinuous Spatial Distributions

Particle distribution within the volume of a crystal is ordered: the particles occupy certain positions, and the number of adjacent neighbors is constant, so that molecule distribution function has a discontinuous dependence on distance.

The quantity Λ from Fig. 6.2a – number of neighbors located at distance r – for the case of discontinuous functions plays the role of continuous radial distribution function g(r) of fluids.

The *G*(r) correlation for liquids, exemplified in Fig. 6.2b, shows intermediate features between gases and solids:

- At small distances (3–4 molecular diameters) there is a regular arrangement of molecules similar to the solid state one.

Fig. 6.1 Debye-Scherer diagram

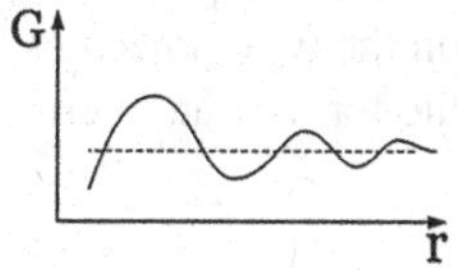

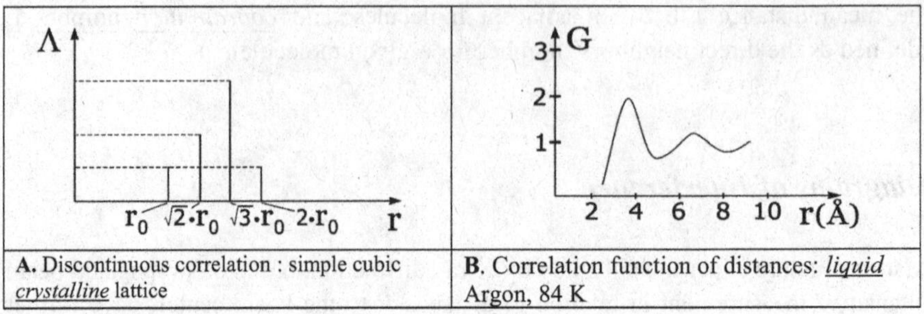

Fig. 6.2 X-ray diffraction correlation with intermolecular distance, for solids and liquids

Table 6.1 Coordination for liquids and solids

Substance	Na	Pb	Zn	Ar
L (liquid)	10	11	10.8	11
L (solid)	12	12	12	12

- At long distances, the probability of distribution tends towards a constant value, similarly to the gas case. However, even at long distances, the molecules are not chaotically distributed as they are in gases, but they are grouped with tens or hundreds into *cybotactic swarms*, where there is a regular arrangement like into a crystal embryo. Under the action of thermal agitation, these swarms break out, and others form, their mean life being 1 ns.

Molecular Order in Liquids

Experimental X-ray study of liquids led to the following conclusions:

- In liquids, molecule order is kept only on a short distance (the nearest order), but the solid state distance order disappears.
- The intermolecular distance at liquids and solids remains unchanged.
- *Coordination number* L – the number of immediate neighbors of a molecule – is with 1 or 2 units lower than for liquids and solids, as can be seen in Table 6.1.

The intermolecular distance is obtained by integration of radial distribution function, as the area corresponding to the first maximum of Fig. 6.3, namely:

$$L = S_1.$$

It is accepted that the mean internal energy u_m of a given molecule depends only on the two molecules' collisions. These latter depend only on distance, thus having the form of an intermolecular potential, u = u(r), as in Sect 4.7 :

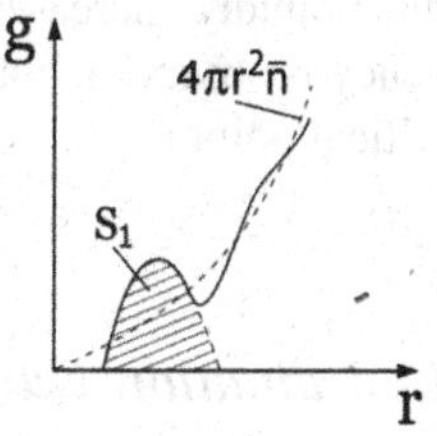

Fig. 6.3 Liquid coordination number

$$u_m = -\int_{r=0}^{\infty} u(r) \cdot g(r) \cdot dr \tag{6.2b}$$

The sign "–" occurs within relation (6.2b) due to the intermolecular potential definition manner, as being the interaction energy, with *changed* sign. For one mole, the internal energy $U^* = -N_A \cdot u_m / 2$ is obtained (factor 2 is introduced in order to avoid that one pair of molecules energy is summed up once per pair member), from where:

$$U^* = -(N_A/2) \cdot \int_{r=0}^{\infty} u(r) \cdot g(r) \cdot dr \tag{6.2c}$$

Quantity U^* represents the difference between internal energy $U(V)$ of the given system and the internal energy $U_{V \to \infty}$ of a system formed from the same particles but having an infinite volume, so that molecules do not interact:

$$U^* = U(V) - U_{V \to \infty} \tag{6.2d}$$

From the correlation function, the (isothermal) compressibility can be also obtained:

$$\chi_T = V \cdot [R \cdot T + 4 \cdot \pi/(k_B \cdot T)] \cdot \int_{r=0}^{\infty} [G(r) - 1] \cdot r^2 \cdot dr \tag{6.3}$$

Coordination Number at Different Temperatures

From radial distribution dependence on temperature, conclusions can be drawn regarding the liquid coordination ratio changes.

For example, in argon ($T_{tr} = 84$ K; $T_f = 86$ K; $T_c = 151$ K), at liquid state displacement from 84 to 146 K on the liquid-vapor co-existence curve, the intermolecular distance r remains unchanged, while L decreases gradually from 11 to 4. Close to the critical point (146 ÷ 151 K), where the molar volume of the fluid rapidly increases with temperature, a positive and rapid temperature dependency is observed, both for r and L, is found, the coordination becoming $L = 6$ at the critical point.

Coordination Number at Liquids and Solids

Information on the liquid structure obtained from X-ray diffraction images is confirmed with approximation by other processes. For example, the value slightly higher than 1 of the L_S/ L_L ratio between coordination numbers of solid and liquid states can be obtained by admitting that the energy required to break a bond is equal for both condensed phases (because the interatomic distance is the same in liquid and solid) and that breaking all bonds leads to formation of a perfect gas. From here:

$$L_S/L_L = \Delta U_{sb}/\Delta U_v \tag{6.4a}$$

are internal energy molar variations, at sublimation (sb) and vaporization (v), respectively. The ΔU quantities are calculated from the similar molar enthalpy variations, λ_{sb} and λ_v:

$$\Delta U_{sb} = \lambda_{sb} - R \cdot T \tag{6.4b}$$

$$\Delta U_v = \lambda_v - R \cdot T \tag{6.4c}$$

And due to the triple point thermodynamic relation, $\lambda_{sb} = \lambda_v + \lambda_t$, where λ_t is the enthalpy of fusion, it results that:

$$L_S/L_L = 1 + \lambda_t/\Delta U_v \tag{6.4d}$$

where $\lambda_t/\Delta U_v$ ratio is small (0.1 ÷ 0.3). This ratio can be obtained using for ΔU_v either the experimental data λ_t or the *Hildebrand* rule, according to which, for any liquid at $T > 150$ K:

$$\Delta U_v\,(J/\text{mol}) = -5900 + 103\,T \tag{6.5}$$

Void Fraction Deduction

Similar conclusions regarding the L_S/ L_L ratio can be deduced from the fact that (at equal interatomic distance) the condensed phase molar volume increases with the increase in the void number (positions remained non-occupied compared to the most compact lattice possible), respectively with the coordination number decrease.

When denoting by θ the lattice void fraction and by considering that these voids have a random distribution, the coordination number becomes $L = L_M \cdot (1 - \theta)$, where L_M is the maximum value of the coordination number. Since $(1 - \theta)$ is also proportional to density, i.e., inversely proportional to the molar volume V, it results (admitting that θ is neglectably small for solids (see Sect. 6.9.)):

$$\theta_L = 1 - V_S/V_L \tag{6.6a}$$

$$L_L = L_S \cdot V_S/V_L \tag{6.6b}$$

When denoting by ε_{Vt} the relative volume increases at fusion, defined by:

$$\varepsilon_{Vt} \equiv V_L/V_S - 1 \tag{6.6c}$$

relations (6.6 a,b) become:

$$\theta_L = \varepsilon_{Vt}/(1 + \varepsilon_{Vt}) \tag{6.6d}$$

$$L_L = L_S/(1 + \varepsilon_{Vt}) \tag{6.6e}$$

For example, for argon ($\varepsilon_{Vt} = 15\%$ and $L_S = 12$) we obtain using relation (6.6e), $L_L = 12/(1 + 0.15) = 10.510.5$. This result is in good accordance with the following values: 11 and 9.9. The first value is obtained from the previous table, using diffractograms. The second value resulted from relation (6.4d) at 84 K and using J/mol expressions, namely: $\Delta U_V = 5470$, and $\lambda_t = 1.180$.

Relations (6.6) are applicable only for compact structures ($L_S \geq 8$); otherwise, it must be admitted that for solid state $\theta_S > 0$, and relation (6.6d) becomes:

$\theta_L/(1 - \theta_S) = \varepsilon_{Vt}/(1 + \varepsilon_{Vt})$

6.3 Liquid Models

ES of a Liquid as an Extremely Compressed Gas

Starting from gaseous state continuity with the liquid state in the critical point area, some of the liquid physical quantities' theoretical substantiation and predictions consider liquid as being a compressed gas. In support of this approach is the fact that real gas ESs from Chap. 4 are also valid for liquids: Van der Waals ES (4.6), $P = R \cdot T/(V - b) - a/V^2$, describes some of the liquid physical quantities (such as density dependence on P and T) satisfactory but only on a qualitative level. For optimal concordance with experimental data, the values of the appropriate α and β constants for gaseous state description must be changed when correlating the liquid physical quantities, especially in the case of temperatures much lower than the critical temperature.

Besides, it is only for gas that corrections for effective volume (b) and effective pressure (a/V^2) are much smaller than volume and pressure. In liquids, these correction values are much higher than the respective corrected quantities' values, so that small constants a and b changes lead to state variables' changes with entire orders of magnitude.

Internal Pressure

Fluids comparison can be made on the basis of the kinetic-molecular notion of *internal pressure*, P_{int}, a quantity which has a parallel dependence on molecular interactions' intensity, being a measure of matter cohesion.

The thermodynamic definition of internal pressure, namely:

$$P_{int} \equiv (\partial U/\partial V)_T \tag{6.7}$$

differs from the usual pressure P definition, corresponding also to internal energy U dependence on volume, but at constant *entropy* and not at constant *temperature*, $P = -(\partial U/\partial V)_S$.

By replacing dS at constant volume with $(C_V/T)\cdot dT$ into the thermodynamic expression $dU = T\cdot dS - P\cdot dV$, it results $dU = C_V\cdot dT + P\cdot(\beta\cdot T - 1)\cdot dV$, from where the following is obtained with relation (6.7):

$$P_{int} = T\cdot P\cdot \beta - P \tag{6.8}$$

where $\beta \equiv (\partial lnP/\partial T)_V$ is temperature and pressure isochore coefficient.

For gases $P_{int} \ll P$ and can be either positive or negative, while for liquids P_{int}, of thousands of atmosphere order, is always positive. The difference is illustrated by the Van der Waals ES: from $P = R\cdot T/(V - b) - a/V^2$ it results that $(\partial P/\partial T)_V = R/(V - b)$, from where $\beta = R/[P\cdot(V - b)]$, and by replacing into relation (6.8), it is found that $P_{int} = R\cdot T/(V - b) - P$ or by replacing $P = f(T, V)$:

$$P_{int} = a/V^2,$$

from where it results that in the usual case, gases, with a molar volume of hundreds of times higher than the one of liquids, will have an internal pressure incomparably lower.

"Gaseous" Type Liquid Models

In the case of "gaseous" type liquid models, the sum-over-states calculation differs from the gas process, since molecule energetic states are no longer independent, but depend on neighboring molecules' position.

Mayer Model for Correlation Functions

In the *correlation function* method, Mayer proposes sum-over-states calculation by summing the 2, 3, 4, ..., N molecule configurations' separate contributions. Share calculation of these *configuration contributions* assumes, on one hand, the molecular potential knowledge according to mutual distances and orientations of constituent particles and, on the other hand, the prior knowledge of all inferior rank configuration shares; therefore, it is generally not possible to establish analytical formulas for sum-over-states calculation, but only to elaborate some numerical procedures, even in the simplest case of monoatomic molecules.

Bogoliubov Model of Molecular Dynamics

The impossibility of analytical formula establishment also characterizes the other method of liquid physical quantities calculation – Bogoliubov *molecular dynamics* method, in which physical quantities' mediation of a particle set from the phase space is replaced by the temporal mediation of given physical quantity values for a single particle.

Unlike the previous method, molecular dynamics method allows not only the calculation of equilibrium quantities but also obtaining transport coefficients' numerical values.

The "gaseous" type methods presented above have yielded satisfactory results in correlating critical quantities, but have not been too effective in predicting caloric, mechanical, or transport quantities of liquids, even the simple ones; for example, on the basis of these theories, there is no concordance with experience for C_V values, equal for monoatomic liquids and for not too low temperatures with the value of $3 \cdot R$ (as for monoatomic *solids* and unlike monoatomic gases, where $C_V = 1.5 \cdot R$).

For liquids with polyatomic molecules, the results are even less satisfactory, probably due to the real intermolecular potential form complexity, which in such cases depends not only on the centers of gravity distance as in Sect. 4.7 but also on molecule orientation, their polarity and shape, as well as electron cloud deformability.

Moreover, with molecule complexity increase, it is found that the radial distribution function from Sect. 6.2. has distinct maxima up to increasing distances so that the configuration number to consider when applying the Mayer method increases.

"Solid" Type Liquid Models

Another group of theories start from the opposite pole, assimilating the liquid with the solid. The main argument in favor of this treatment is the fact that at low distances liquid molecules present the ordered arrangement phenomenon, similar to solids, or it is precisely neighboring molecules that make a more important contribution when summing up intermolecular potentials.

Devonshire Cell Model

In the *Devonshire* cell model, the space occupied by liquid is completely filled with regularly arranged cells, each cell having the volume V/N_A (where V is the molar volume) and containing a single fluid molecule, as in Fig. 6.4. The molecule, a little smaller than the cell, undergoes a translational movement limited to cell dimensions.

But vibration and rotation motions from polyatomic molecules occur in the usual manner known for gases.

Based on cellular theory, satisfactory results can be obtained in enthalpy of vaporization calculation and its dependence on temperature, as well as for the translation component of the heat capacity: because of free cell space small dimensions, the translation over the three directions equals the action of three independent oscillators, which, as in the case of solids, leads to the value: $C_V = 3 \cdot R$ for high temperatures, in accordance with experience.

However, this model provides a slower growth for C_V than in reality, and the calculated values for liquid entropy are significantly lower than the experimental ones, since this model overestimates the liquid degree of order.

The results are even less satisfactory for liquids with particles of shapes highly different than the spherical ones. Cell theory provides for rotational and vibrational sum-over-states values identical to those obtained from gases, although the neighboring molecules' presence leads to breaking of oscillation for the real liquid case and especially to free rotation prevention.

Fig. 6.4 Cell model

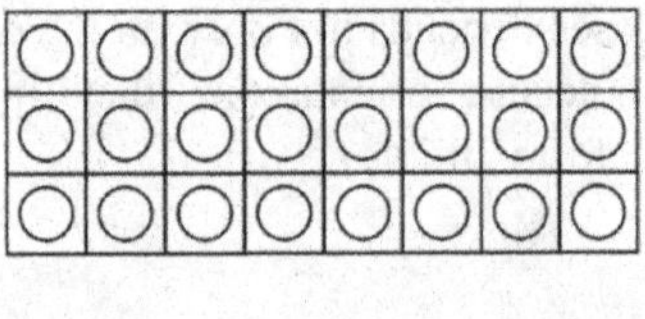

For example, it has been found that the CCl_4 quasi-spherical molecule has an infrared absorption band of a sharper profile than the (similar but more plate) CH_3Cl molecule, which indicates that in the second case the characteristic rotational temperature is significantly higher.

Eyring Free Volume Model

One of the cell model's improved versions is based on the *free volume* theory, where the molecule center of gravity motion within the liquid is limited to an available volume dimensions, of mean value v_f for each molecule, as in Fig. 6.5. The free volume is related to the total cell volume, $v = V/N_A$, by the following relation:

$$(v_f)^{1/3} = c \cdot \left(v^{1/3} - d\right) \tag{6.9}$$

where d is molecule diameter and c – a numeric coefficient depending on the molecule lattice coordination type.

For a simple cubic arrangement, c = 2, the free volume calculation can be made in two different manners:

- Using statistical thermodynamics
- Starting from the sound speed into liquid

Thermodynamic-Statistical Calculation of Free Volume

The calculation starts from the sum-over-states z expression for a molecule:

$$z = z_{rot} \cdot z_{vib} \cdot e^{-\varepsilon_v/(k_B \cdot T)} \cdot v_f \cdot \left(2 \cdot \pi \cdot m \cdot k_B \cdot T/h^2\right)^{3/2} \tag{6.10}$$

where z_{rot} and z_{vib} are rotational and vibrational contributions, m – molecule mass, while ε_V – vaporization internal energy per molecule, linked to the respective molar quantity ΔU_V calculable with relations (6.4c, 6.5), through relation $\Delta U_V = N_A \cdot \varepsilon_V$. Product $v_f \cdot (2 \cdot \pi \cdot m \cdot k_B \cdot T/h^2)^{3/2}$ from relation (6.10) represents translation contribution

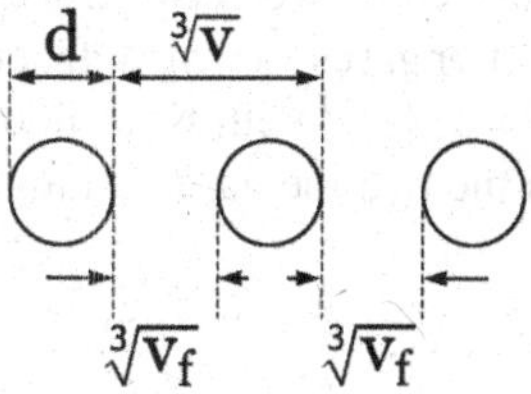

Fig. 6.5 Eyring free volume

for the sum-over-states and can be deducted through the usual procedures from Sect. 2.4.

From relation (1.60) $F - F_c = - R{\cdot}T{\cdot}lnZ$, through the distribution eq. $P = -(\partial F/\partial V)_T$, it is obtained that $P = - R{\cdot}T{\cdot}(\partial lnZ/\partial V)_T$ or since molar quantities V and Z are linked to quantities per molecule v and z through relations $V = N_A{\cdot}v$ and $ln\,Z = N_A{\cdot}ln\,z$:

$$P = -R \cdot T \cdot (\partial\, ln\, z/\partial v)_T \tag{6.11}$$

Relation (6.10) is logarithmized, and then the isotherm is derived by volume (it is assumed that z_{rot} and z_{vib} depend not only on T but also on volume).

The result, $(\partial lnz/\partial v)_T = (\partial lnv_f/\partial v)_T - (\partial \varepsilon_v/\partial v)_T/(k_B{\cdot}T)$, replaced into relation (6.11), leads to $P = (R{\cdot}T/v_f){\cdot}(\partial v_f/\partial v)_T - (R/k_B){\cdot}(\partial \varepsilon_v/\partial v)_T$. Here it is admitted that ε_v is proportional to liquid volume, so that $(\partial \varepsilon_v/\partial v)_T = \varepsilon_v/v$; it results that $P = (R{\cdot}T/v_f){\cdot}(\partial v_f/\partial v)_T - N_A{\cdot}\varepsilon_v/v$ (since $R/k_B = N_A$) or since at not very high pressures $P{\cdot}v \ll N_A{\cdot}\varepsilon_v$:

$$(R \cdot T/v_f) \cdot (\partial v_f/\partial v)_T = N_A \cdot \varepsilon_v/v \tag{6.12}$$

From (6.6a) it results that $v_f = c^3{\cdot}(v^{1/3} - d)^3$, from where by derivation it is found that:

$(\partial v_f/\partial v)_T = 3{\cdot}c^3{\cdot}(v^{1/3} - d)^2/[3{\cdot}(v_f)^{2/3}]$ from where, by substituting from (6.9) $(v^{1/3} - d)$ by $(v_f)^{1/3}/c$, it is found that $(\partial v_f/\partial v)_T = c{\cdot}(v_f/v)^{2/3}$.

By replacing this expression of the $(\partial v_f/\partial v)_T$ derivative into relation (6.12), it is found that $(R{\cdot}T/v_f){\cdot}c{\cdot}(v_f/v)^{2/3} = {\cdot}\varepsilon_v{\cdot}N_A/v$, or $v_f = v{\cdot}(c{\cdot}k_B{\cdot}T/\varepsilon_v)^3$, and by passing to molar quantities through $v = V/N_A$ and $\varepsilon_v = \Delta U_V/N_A$, the following is obtained:

$$v_f = (V/N_A) \cdot (c \cdot R \cdot T/\Delta U_v)^3 \tag{6.13}$$

Calculation of Free Volume from Speed of Sound

Sound speed calculation is based on the hypothesis that oscillations' emission rate within the *free volume*, u', is the same for gas and liquid states, while the *macroscopic* sound speed, denoted by u, is different in the two media. These velocities are linked by their ratio $u/u' = (v/v_f)^{1/3}$, where $v^{1/3}$ and $v_f^{\,1/3}$ are the total distance traveled between two molecules, respective to the free distance.

For gases $v_f = v$, so that $u_L/u_G = (v/v_f)^{1/3}$, where v_f is the liquid free volume and $v = V_L/N_A$, with V_L as liquid molar volume, and u_L and u_G being the emission rates in the two media. It is thus obtained that:

$$v_f = (V_L/N_A) \cdot (u_{G*}/u_L)^3 \tag{6.14a}$$

where u_{G*} designates sound speed in a fictive gaseous medium G* of that substance at the same temperature, $u_{G*} = [(\rho \cdot \chi_S)_{G*}]^{-1/2}$.

But adiabatic compressibility χ_S is related to the isothermal one, χ_T, through the expression $(\chi_S) = (\chi_T \cdot C_v / C_P)$ where C_v and C_P are isochore and isobar molar heat capacities; therefore $u_{G*} = [(\rho \cdot \chi_T \cdot C_v / C_p)_{G*}]^{-1/2}$.

Eyring admits that fictive gas *mechanical* quantities are those of the perfect gas, $\rho_G^* = M \cdot P / (R \cdot T)$ and $(\chi_T)_G = 1/P$ but its *caloric* quantities are those of liquid: $C_{v,G}^* = C_{v,L}$ and $C_{P,G}^* = C_{P,L}$, so that $u_G^* = [R \cdot T \cdot C_{p,L} / (M \cdot C_{v,L})]^{-1/2}$ from where, by replacing u_G^* into equality (6.14a):

$$v_f = (V_L/N_A) \cdot \left[(u_L)^2 \cdot M \cdot C_{v,L} / (R \cdot T \cdot C_{p,L})\right]^{-3/2} \qquad (6.14b)$$

The free volume theory allows not only the ESs for liquid equation establishment but also their viscosity prediction through relation (8.4).

Frenkel Model of Empty Cells

Another theory describing the liquid starting from a solid model is the *empty* cells theory (Frenkel) which is based on the same cell compartment lattice as cell theory, but it only assumes the existence of empty cells "voids," while most cells contain one particle. (Fig. 6.6).

Thus, the liquid is treated as a solid with punctual defects (Frenkel type (see Sect. 6.9)), where, however, defect proportion is 1–2 orders of magnitude larger than in the real solid.

The theory of voids correctly reflects the fluid structure, both in terms of the mean particle distribution (coordination numbers slightly lower than in those for solid) and in terms of particle *dynamics*, which vibrate with their own frequencies of about 10^{11} s^{-1} within a narrow volume (of dimension comparable to a molecule), and approximately one oscillation of one hundred leads to a permanent molecule displacement, at a distance of approximately one molecular diameter, corresponding to a new equilibrium position.

The more elaborated physical fundaments of this model explain the reason of succeeding the best liquid transport quantity assumptions when using the theory of voids.

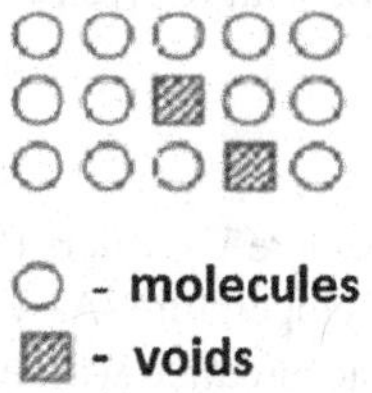

Fig. 6.6 Frenkel model of empty cells

Eyring introduced a complementary sense in the theory of voids, treating structure changes in time not as particle displacements from a busy cell into an unoccupied one, but as chaotic void displacements. The liquid appears as a "reversed gas," within gas the particles moving freely, while in the liquid - they void. This justifies the empirical relation of Cailletet-Mathias, according to which the sum of the densities the equilibrium liquid and vapors is a dependency function of linear temperature and has a slow-growing allure.

Significant Structure Theory: Gas and Solid

From liquid quantity temperature variation point of view, the *significant structure theory* is very successful, which assumes that, from the freedom of movement point of view, the liquid molecules are of two types – "gas" and "solid" type – the first ones being able to perform not only vibrations but also translations or rotations around the center of gravity while those in the second category being able only to perform center of gravity vibrations around the equilibrium position.

The two types of molecules are in dynamic balance, their ratio varying with temperature. The share of "solid" in the totality of the two molecules' structural types, $\alpha \equiv N_S/(N_S + N_G)$, is minimal at the melting temperature, being given by the following expression:

$$\alpha \equiv V_S/V_L \tag{6.15a}$$

In the significant structure theory, the sum-over-states over one liquid mole is:

$$Z = (Z_S)^{\alpha} \cdot (Z_G)^{1-\alpha} \tag{6.15b}$$

where the sum-over-states of "solid" type Z_S and "gaseous" type Z_G correspond to vibration (with characteristic temperature θ_{vib}) and translation, respectively:

$$Z_S = 1/[1 - \exp(-\theta_{vib})/T)];$$
$$Z_G = V \cdot \left(2 \cdot \pi \cdot m \cdot k_B \cdot T/h^2\right)^{3/2}.$$

To these sum-over-states correspond for solid state $C_{v,S} = 3 \cdot R$ (if $T > \theta_{vi}$) while for gas $C_{v,G} = 1.5 \cdot R$, from where, at monoatomic liquid, $C_V = 3 \cdot R \cdot \alpha + 1.5 \cdot R \cdot (1 - \alpha)$ or with α given from relation (6.15a):

$$C_{V,L} = 1.5 \cdot R \cdot (1 + V_{S,t}/V_L) \tag{6.16}$$

where $V_{S,t}$ refers to the solid found at the melting point and V_L and C_V – at any liquid temperature. As V_L increases with T, temperature rising reduces the C_V, which is in accordance with experience.

6.4 Equilibrium Structural Models for Solids

Types of Solids

Due to their variety and their special technical importance, solids form the subject of study of many disciplines: mechanics of rigid bodies, elasticity, strength of materials, physics of the solid state, metallography, materials science, physics of semiconductors, mineralogy, etc. Physical chemistry, however, focuses mainly on the common quantities of all states of aggregation, as has been discussed in Sect. 6.1.

In terms of degree of order, solid models can be:

- *Ideal* models (on solids), with all the *node* – centered molecules – points which are arranged into a perfectly ordered manner within the crystal lattice
- *Real* solids, with an imperfect order

Real solids can in turn be of two types:

(a) *Defective crystals*, in which the distance order is preserved (molecules are centered on nodes, as in the ideal crystal), but there is a certain proportion of "defects" – deviations from atoms regular arrangement

(b) *Noncrystalline solids*, in which the distance order is totally absent. Following the intensity of close order manifestation, noncrystalline solids may be either amorphous or glassy:

- In the *amorphous* solids, the coordination number L is the same as for liquid (see Sect. 6.2.), the difference between amorphous and liquid states consisting mainly of the extremely low particles' mobility within the amorphous substances, implying an incredibly high viscosity (Sect. 6.1.).
- For glassy solids ("glasses"), the coordination number (as well as the thermodynamic quantities) is found to have intermediate values between those of crystalline bodies and amorphous solids.

Crystal Quantity Models

While *thermodynamic* quantities (and the other equilibrium quantities) of solid itself – crystalline – exposed in this chapter are determined in particular by crystalline lattice structure and can be understood on the ideal crystal model basis, their kinetic physical quantities, including those from Chap. 8, depend mainly on solids defects.

The molecular models explain well thermodynamic quantities' dependence on temperature for the *simple* solid case – the one containing identical atoms in the lattice nodes, such are the elements (except for layered lattice metalloids such as P and As or covalent molecules such as O_2 and S_8).

Einstein Vibrations

The thermodynamic-statistical calculation in the *Einstein* model assumes that the N_A atoms from one mole of a simple crystal performs independent vibrational movements in the three spatial directions in the crystalline lattice nodes with the same vibration eigenfrequency, ν_E (the fundamental frequency, corresponding to solid optical spectrum maximum). One mole of solid equates with $3{\cdot}N_A$ harmonic, identical, and independent oscillators, in the Einstein model.

Thermodynamic Functions of Einstein Vibration

One mole of solid equals in the Einstein model with $3{\cdot}N_A$ harmonic, independent, and equal frequency oscillators. The thermodynamic functions (with "rt" index) of such a solid lattice will thus be three times higher than the vibrational contributions to diatomic gas thermodynamic functions (which contains, within one mole, a N_A number of harmonic oscillators (see Sect. 2.6.), namely, by comparing with relations (2.43) for *F, U, S*, and C_v:

$$F_{rt} = 3 \cdot R \cdot T \cdot ln\,(1 - e^{-x_E}) = 3 \cdot R \cdot T \cdot [F_{E,U}(x_E) - F_{E,S}(x_E)] \tag{6.17a}$$

$$U_{rt} = 3 \cdot R \cdot T \cdot x_E/(e^{x_E} - 1) = 3 \cdot R \cdot T \cdot F_{E,U}(x_E) \tag{6.17b}$$

$$S_{rt} = 3 \cdot R \cdot [x_E - ln\,(1 - e^{-x_E})]/(e^{x_E} - 1) = 3 \cdot R \cdot F_{E,S}(x_E) \tag{6.17c}$$

$$C_{v,rt} = 3 \cdot R \cdot x_E^2 \cdot e^{x_E}/(e^{x_E} - 1)^2 = 3 \cdot R \cdot F_{E,C}(x_E) \tag{6.17d}$$

in which $F_{E,C}$, $F_{E,S}$, and $F_{E,U}$ are Einstein functions for heat capacity, entropy, and internal energy, functions defined previously in relations (2.44) through the following notations: E_C, E_S, and E_U. In relations (6.17), F_{rt} ... $C_{v,rt}$ are the crystalline *lattice* contributions to crystal's molar thermodynamic functions F, U, S, and C_v respectively; other contributions are made by the mobile electrons (for metals and alloys case) and by magnetic susceptibility sharp dependencies of temperature, within the temperature range situated near Curie and Neil points of ferromagnetic and ferrimagnetic substances.

Besides, within expressions (6.17), x_E is the nondimensional ratio:

$$x_E = \theta_E/T \tag{6.18a}$$

where θ_E is the *Einstein characteristic temperature*, defined by:

$$\theta_E \equiv h \cdot \nu_E/k_B \tag{6.18b}$$

h and k_B being Planck and Boltzmann universal constants.

For heat capacity, Einstein theory assumes the following:

1. At elevated temperatures (where $\theta_E/T \rightarrow 0$, and $F_{E,C} \rightarrow 1$), $C_{v,rt}$ tends asymptotically to a constant value, namely, 3·*R*:

$$lim_{T\rightarrow\infty} C_V = 3 \cdot R \tag{6.19a}$$

If under these conditions the electronic and magnetic contributions to heat capacity remain negligible, then $C_v = C_{v,rt}$.

According to the equipartition of energy on degree of freedom theorem (Sect. 2.9.), if to an atom it is considered to correspond three vibrational independent motions, each one having two degrees of freedom, a total of six degrees of freedom is obtained, from which a heat capacity $C_v = 6 \cdot (R/2) = 3 \cdot R$.

Dulong-Petit Law

Relation (6.19a) can be only indirectly compared with experimental data, since in practice the measurable calorimetric quantity is C_p and not C_v. For solids, however, at room temperature, the difference $(C_p - C_v)$ is negligible as compared to C_v, where thus $C_p \approx Cv = 3 \cdot 8.3$ or:

$$C_p \approx 25\ J/(mol \cdot K), T = 300\ K \tag{6.19b}$$

which represents the *Dulong-Petit* law, empirically established and well verified by the majority of simple monoatomic solids, at room temperature.

In a series of simple monoatomic solids, the C_p values at room temperature are however much smaller: C-diamond 7, B 11, Be 18, and Al 24. These deviations are qualitatively explained by the relatively high vibration frequency value (and therefore of characteristic temperature) of these substances: at room temperature, the $x_E = \theta_E/T$ ratio is not sufficiently small, and $F_{E,C}$ has values well below 1. But also in these cases, C_p increases with temperature increase, tending towards the value predicted by Dulong and Petit, for example, $C_p = 26.4$ for diamond at 1100°C.

At elevated temperatures (above 1000 K), C_p increases with temperature are observed, especially in the case of metals, which is explained by conductivity *electron* contribution to heat capacity.

Einstein Model at Low Temperatures

When temperature tends towards 0 K (and $x_E \rightarrow \infty$), Einstein model considers that:

$$C_{v,rt} = 0, T = 0.$$

Since, according to phenomenological thermodynamics, $C_p = C_v + \alpha^2 V \cdot T/\chi_T$ and V and χ_T remain positive even at $T = 0$, the following thus results:

$$C_{p,rt} = 0, T = 0 \tag{6.20}$$

Indeed, it is experimentally observed that C_p, as well as C_v, limit is null when temperature tends towards 0.

The $C_v(T)$ dependence provided by Einstein relation (6.17d) is an increasing dependency function of temperature, which generally coincides with experimental data. The form of low temperature $C_v(T)$ dependence is obtained by observing that under these conditions ($T \ll \theta_E$, from where x» 1 and e^x »1) the equality (6.17d) becomes $C_v = 3 \cdot R \cdot (x_E)^2 \cdot \ exp\ (-x_E)$ or by replacing x_E from (6.18a) and by assuming that $C_{v,rt} \cong C_v$: $C_v = 3 \cdot R \cdot (\theta_E/T)^2 \cdot \ exp\ (-\theta_E/T)$, namely, according to the Einstein model, C_v increases exponentially with low temperatures. This conclusion is totally denied by experience; in fact, it is found that under 15 ÷ 30 K, the monoatomic solids exhibit a $C_v(T)$ variation of the following form:

$$C_v = A \cdot T + B \cdot T^3 \tag{6.21}$$

the term of T^3 being negligible under 0.05 ÷ 0.2 K while the term of T becoming negligible above 2 ÷ 4 K. The C(T) dependence at low temperatures presents a major importance, both in a direct way – for calculating the operation of cryogenic installations, and in an indirect one, to obtain solid entropy values at usual temperatures (according to principle III of thermodynamics, for solids $S = \int_0^T (C_p/T) \cdot dT$). For this reason, the Einstein model has been abandoned, seeking a theoretical justification for $C_v(T)$ dependency function (6.21).

It has been established that the term proportional to T is of *electronic* origin (it occurs only for metals, but not for Ar, C, or other simple insulating solids), so that the equality $C_v = A \cdot T + B \cdot T^3$ equals:

$$C_v = C_{v,rt} + C_{v,el} \tag{6.22}$$

$$C_{v,el} = A \cdot T \tag{6.23}$$

$$C_{v,rt} = B \cdot T^3 \tag{6.24}$$

6.5 Debye Vibrations

The lattice heat capacity cubic dependence on temperature explanation is provided by the *Debye* model, which takes into account the interactions between the three vibration modes of a crystal mole: they are no longer equivalent to three *independent* oscillators, but to a system consisting of as many *coupled* oscillators, called *phonons* (since can be determined from the crystal *acoustic* spectrum). Unlike Einstein oscillators, each *phonon* has another frequency.

The frequency of phonons, v, distribution is deduced by Debye using statistical mechanics, which finds that the integral frequency distribution function, G(ν'), the number of phonons with a frequency v lower than a value ν', is proportional to volume $4 \cdot \pi \cdot (\nu')^3 / 3$ of a sphere from the three-dimensional phase of the vibrations:

$$G(\nu') = K \cdot (\nu')^3 \tag{6.5a2}$$

The normalizing constant K is determined from the condition that all $3 \cdot N_A$ phonons frequencies from one mole have the superior limit ν_D, the characteristic Debye frequency, experimentally determined from the acoustic spectrum:

$$\int_{\nu'=0}^{\nu'=\nu_D} dG(\nu') = 3 \tag{6.5b2}$$

By differentiating relation (6.5a2) it results that $d\, G(\nu') = 3 \cdot K \cdot (\nu')^2 \cdot d\nu'$, which after replacement into equality (6.5b2) and integration leads to $K = 1 / \int_{\nu'=0}^{\nu'=\nu_D} (\nu')^2 \cdot d\nu' = 3/(\nu_D)^3$, expression (6.5a2) becoming:

$$G(\nu') = 3 \cdot (\nu'/\nu_D)^3 \tag{6.25}$$

Debye Vibrational Sum-Over-States

Vibrational sum-over-states $Z_{\nu'}$ of a phonon of frequency ν' is obtained from relation (2.39c), $Z_\nu = 1/ [1 - exp(-\theta_{vib}/ T)]$, where θ_{vib} depends of ν' according to Eq. (2.38d), $\theta_{vib} = h \cdot \nu' / k_B$, so that:

$$ln\, Z_{\nu'} = -ln\left\{1 - exp\left[-h \cdot \nu'/(k_B \cdot T)\right]\right\} \tag{6.5c2}$$

The sum-over-states Z for a particle from a node is obtained by averaging according to probability G of phononic frequencies' individual values ν', namely:

$$ln\ Z = \int_{\nu'=0}^{\nu'=\nu_D} (ln\ Z_{\nu'}) \cdot dG_{\nu'}. \tag{6.5d2}$$

But by differentiating relation (6.25), it is obtained that:

$$d\,G_{\nu'} = \left[9/(\nu_D)^3\right] \cdot (\nu')^2 \cdot d\nu'. \tag{6.5e2}$$

By replacing $lnZ_{\nu'}$ and $dG_{\nu'}$ quantities from relations (6.5c2) and (6.5e2) into relation (6.5d2) and then by changing variables (ν', ν_D) through notations (x, x_D):

$$x = h \cdot \nu'/(k_B \cdot T); x_D = h \cdot \nu_D/(k_B \cdot T); \tag{6.5f2}$$

the following relation is obtained:

$$ln\ Z = \left[9/(x_D)^3\right] \cdot \int_{x=0}^{x=x_D} -[ln\,(1 - e^{-x})] \cdot x^2 \cdot dx \tag{6.26}$$

The maximum frequency ν_D is expressible through the material constant θ_D, the Debye characteristic temperature, defined by:

$$\theta_D \equiv h \cdot \nu_D/k_B \tag{6.27a}$$

so that the integration limit x_D from relation (6.26) is:

$$x_D = \theta_D/T \tag{6.27b}$$

Debye and Einstein Phononic Heat Capacities

From *Debye* sum-over-states (6.26) phononic contributions to molar thermodynamic functions (U, S, F, C_v) are obtained using the usual relations (1.58 + 1.61).

It results for $C_{v,rt}$, for example, at the temperatures tending to 0 or ∞, that Debye and Einstein models *limits* are identical ($C_v = 0$ and $C_v = 3 \cdot R$, respectively), but $C_v(T)$ *dependence* is more accurately represented by the Debye model. With Debye sum-over-states introduced through relation (1.61), the molar isochoric heat capacity expression of lattice oscillators becomes:

$$C_{v,rt} = 3 \cdot R \cdot F_{DC}(x_D) \tag{6.28a}$$

where $F_{DC}(x_D)$ is Debye function for (isochoric) heat capacity:

$$F_{DC}(x_D) = \int_{x=0}^{x=x_D} (x_D)^{-3} \cdot x^4 \cdot (1 - e^{-x})^2 \cdot dx \tag{6.28b}$$

At temperatures tending towards 0 or ∞, the Debye and Einstein models *limits* are identical - $C_v = 0$ and $C_v = 3 \cdot R$, respectively, but $C_v(T)$ *dependence* is more accurately represented by the Debye model. By expanding the integrand from definition (6.28b) in MacLaurin series of integer positive powers of x variable, it can be shown after recurrent integration as in Sect. A.3. that F_{DC} dependence is approximated at high temperatures (where x_D «1) through the following series:

$$F_{DC}(x_D) = 1 - (x_D)^2/20 + (x_D)^4/560 - \ldots \tag{6.29}$$

while Einstein model series decomposition under the same conditions leads to:

$F_{EC}(x_D) = 1 - (x_E)^2/12 + (x_E)^4/240 - \ldots$ which predicts a more rapid increase with temperature than the series (6.29), therefore a retarded achievement of high temperatures step $C_{v,rt} = 3 \cdot R$ for the Einstein oscillators than for those considered by Debye.

Debye Thermodynamic Quantities at Low Temperatures

At very low temperatures, $x_D \gg 1$, and (6.28) expressions lead to:

$$C_{v,rt} = 9 \cdot R \cdot I/(x_D)^3 \tag{6.30a}$$

where I is the integral from expression (6.28b), but with changed superior integration limit, i.e., from x_D into ∞, namely:

$$I = \int_{x=0}^{x=\infty} x^4 \cdot e^x \cdot (e^x - 1)^{-2} \cdot dx \tag{6.30b}$$

By expanding $1/(e^x - 1)^2$ into powers series of argument e^{-x}, the following results: $1/(e^x - 1)^2 = e^{-2 \cdot x} + 2 \cdot e^{-3 \cdot x} + 3 \cdot e^{-4 \cdot x} + \ldots = \sum_{k=2}^{\infty} (k-1) \cdot e^{-k \cdot x}$, and by replacing into (6.30b) it results that:

$$I = \sum_{k=1}^{\infty} \left[k^{-3} \cdot \int_{y_k=0}^{\infty} y_k^4 \cdot exp(-y_k) \cdot dy_k \right] \tag{6.30c}$$

where $y_k = k \cdot x$ and k being any positive integer number. All the defined integrals from (6.30c) are identical, therefore having a common value, J, so that:

$$I = J \cdot S \tag{6.30d}$$

where J and S represent, respectively:

$$J \equiv \int_{y=0}^{y=\infty} y^4 \cdot e^{-y} \cdot dy; S = I = \sum\nolimits_{k=1}^{\infty} 1/k^3.$$

Integral J is recurrently calculated by parts, as shown in annex A3, obtaining thus $n = 4$ and $a = -1$, from relation (9.3c), namely:

$F_n(y, a) = (n!/a) \cdot e^{a \cdot y} \cdot \sum_{k=0}^{k=n} \left[(-a)^{-k} \cdot y^{n-k}/(n-k!)\right]$, and J value of

$$J = F_n(4, -1) = -4! \cdot e^{-y} \cdot \left(y^4/4! + y^3/3! + y^2/2! + y/1! + 1\right)\Big|_{y=0}^{y=\infty} = 24.$$

Sum S is equal to $S = \pi^4/90$. By introducing J and S values into relation (6.30d), it results that $I = 4 \cdot \pi^4/15$, which after replacement into (6.30a) leads to:

$$C_{v,rt} = 12 \cdot \pi^4 \cdot R/\left[5\,(x_D)^2\right] \tag{6.31a}$$

By taking into account relation (6.27b), $X_D = \theta_D/T$, it results the proportionality between $C_{v,rt}$ and temperature cubic power, so that dependence (6.24) is obtained at $T \ll \theta_D$ in accordance with experience, $C_{v,rt} = B \cdot T^3$, where B has a dependence only on θ_D:

$$B = 12 \cdot \pi^4 \cdot R/\left[5\,(\theta_D)^2\right] \tag{6.31b}$$

For lattice entropy, internal energy, and Helmholtz free energy at low temperatures, the following expressions are obtained similarly, in accordance with experience, of proportionality with T^3 for S and with T^4 for U and F:

$$S_{rt} = 4 \cdot \pi^4 \cdot R \cdot T^3/\left[5\,(x_D)^2\right] = C_{v,rt}/3 \tag{6.31c}$$

$$U_{rt} = 3 \cdot \pi^4 \cdot R \cdot T^4/\left[5\,(x_D)^2\right] = C_{v,rt} \cdot T/4 \tag{6.31d}$$

$$F_{rt} = \pi^4 \cdot R \cdot T^4/\left[5\,(x_D)^2\right] = U_{rt}/3 \tag{6.31e}$$

Debye Temperature Measurement

Debye characteristic temperature can be experimentally determined through several independent procedures:

1. Thermodynamically through relation (6.31b), by measuring C_p at low temperatures.
2. From wave propagation mean velocity, u_{med}, with Blackman relation, it results:

$$\theta_D = (h/k_B) \cdot u_{med} \cdot [3 \cdot N_A/(4 \cdot \pi \cdot V)]^{1/3} \tag{6.32a}$$

where V is the molar volume and

$$u_{med} = \left\{\left[2/(u_{tr})^3 + 1/(u_{long})^3\right]/3\right\}^{-1/3} \tag{6.32b}$$

where u_{long} and u_{tr} are longitudinal oscillation (sound) propagation velocity and transverse oscillation propagation velocity, respectively.

For anisotropic solids, u_{long} and u_{tr} are in their turn the arithmetic means of values after the three major crystallographic paths. In the absence of experimental determinations, u_{long} and u_{tr} are obtained from density ρ and from solid elastic modulus E and G:

$$u_{long} = \sqrt{E \cdot (4 \cdot E - G)/(\rho \cdot (3 \cdot E - G)} \tag{6.32c}$$

$$u_{tr} = (G/\rho)^{1/2} \tag{6.32d}$$

Young's modulus (of compression) E and the shear modulus G being linked through:

$$G = 0.5 \cdot E/(1 - \mu_P) \tag{6.32e}$$

where Poisson's ratio μ_P is the changed sign ratio – $\varepsilon_{tr}/\varepsilon_{long}$ – of relative transverse dimensions ε_{tr} increases and of longitudinal dimensions ε_{long}, at a solid sample compression. On the other hand, E can be expressed in terms of isotherm compressibility χ_T:

$$E = 3 \cdot (1 - 2 \cdot \mu_P)/\chi_T \tag{6.32f}$$

3. From the melting temperature T_t, using *Lindemann* relation:

$$\theta_D = C_L(T_t/M_n)^{1/2}/(V_{n,s,t})^{1/3} \tag{6.33}$$

Table 6.2 Lindemann's law coefficient

Coordination number, L	4	6	8	12
C_L coefficient from relation (6.33)	26	34	41	43

Table 6.3 Solids' characteristics in Debye temperatures

Body	Li	Na	K	Cs	Be	Mg	Ca	Ba	Al	C *2	Si
θ_D	344	158	91	38	1440	390	230	110	428	2040	640
A *1	1.63	1.38	2.08	3.2	0.17	1.3	2.9	2.7	1.35	/	/
Body	Ge	Sn *3	Pb	Ti	Cr	Fe	Ni	Zn	Au	Hg	Cu
θ_D	230	200	105	420	520	425	410	327	167	72	315
A	/	/	9.42	3.35	1.4	4.95	7.02	0.64	0.73	1.79	/
Body	Ag	Ne	Xe	Stainless steel	NaCl	AgCl	SiC	AlN	ZnS	H_2O	
θ_D	215	75	64	390	281	183	1200	950	440	150	

Notification: *1 A, expressed in $J/(Mol{\cdot}K^2)$, from relation (6.21) *2: Diamond. *3: Sn white

In Lindemann's law, M_n is mean (arithmetic) of node molar mass and Vn, s, and t – solid molar volume (at the melting temperature) divided by the nodes' number from the molecular formula. The coefficient C_L depends on the coordination number L, as seen from Table 6.2.

Lindemann's law expresses the fact that $k_B{\cdot}T_t$ energy needed to break the crystalline lattice bonds – melting – is so much higher than the k_f force constant, and the oscillator $k_BT_t \sim k_f{\cdot}(V/N_A)^{2/3}$ cross section $(V/N_A)^{2/3}$ is higher. In turn, k_f is proportional to a node mean molar mass M_n/N_A, with Debye frequency square $\nu_D = \theta_D k_B/h$ and the reverse of oscillator length $(V/N_A)^{1/3}$ $k_f \sim (M_n/N_A){\cdot}(\theta_D k_B/h)^2/(V/N_A)^{1/3}$.

Debye temperatures for the various solids are shown in Table 6.3.

6.6 Other Contributions to Sum-Over-States

Conductivity Electrons

For conductivity electrons' contribution to the sum-over-states and to thermodynamic quantities, Fermi and Dirac statistics of Sect. 1.2 is used.

Indeed, metal's free electrons are indiscernible particles that respect the Pauli exclusion principle. Their contribution to C_v is:

$$C_{v,el} = 1.5 \cdot z_{el} \cdot R \cdot F_{FC}(x_{el}) \tag{6.34}$$

where z_{el} is the metal's valence electron number, F_{FC} is the Fermi function for the heat capacity, and its argument x_{el} is the ratio:

$$x_{el} = \theta_{el}/T$$

where θ_{el} is metal's characteristic *electronic* temperature. Since $\theta_{el} > 10^4$, at usual temperatures (below 1000 K) the x_{el} ratio is much below 1, and then F_{FC} function expression is simplified to:

$$F_{FC}(x_{el}) = \pi^2/(3 \cdot x_{el})$$

which, after replacement into relation (6.34), allows dependence (6.23) retrieval, $c_{v,el} = A \cdot T$ in accordance with experience, where:

$$A = z_{el} \cdot R \cdot \pi^2/(2 \cdot \theta_{el}) \tag{6.35}$$

At *high* temperatures ($T \gg \theta_{el}$), F_{FC} tends towards 1, so that:

$$C_{v,el} = 1.5 \cdot z_{el} \cdot R, T \to \infty.$$

The characteristic electronic temperature can be obtained from relation:

$$\theta_{el} = h^2 \cdot \left(3 \cdot \pi^2 \cdot N_A \cdot z_{el}/V\right)^{2/3}/(2 \cdot m_{el,ef} \cdot k_B) \tag{6.36}$$

where m_{elef} is the electron's *effective* mass, whose value is not equal to the *real* mass m_{el}, but increases with electron mobility decrease:

$$m_{el,ef} = m_{el}/R_{mob}$$

where R_{mob} is the *mobility* ratio, varying between 0.4 and 3, which can be experimentally obtained from metal behavior in a magnetic field, although θ_{el} deduction from heat capacities is preferred – relations (6.34) and (6.23).

Since $\theta_{el} \gg \theta_D$, $C_{v,el}$ continuously increases with T, although negligible as compared to $C_{v,rt}$ at intermediate temperatures (from 5 ÷ 15 up to 200 ÷ 1000 K), $C_{v,el}$ contribution reigns again at high temperatures. For example, for Platinum at 2000 K, $C_{v,el}$ is 8.5 J /(mol·K), and $C_{v,rt}$ – 25.

Sum-Over-States Magnetic Component

For magnetic metals (Fe, Co, Ni), below Curie temperature occurs an important contribution $C_{v,mg}$, brought by *magnons* (spins ordering states excitation).

For example, for Fe α, ferromagnetic material becoming paramagnetic (Fe β) at 1045 K (Curie point of iron), the C_p value slowly increases between 300 and 1000 K from 26 up to 52 J/(mol·K) while between 1000 and 1045 K rapidly increases from 52 to 80, as shown in Fig. 6.7. For polymorphic transformations $\beta \to \gamma$ and $\gamma \to \delta$, as

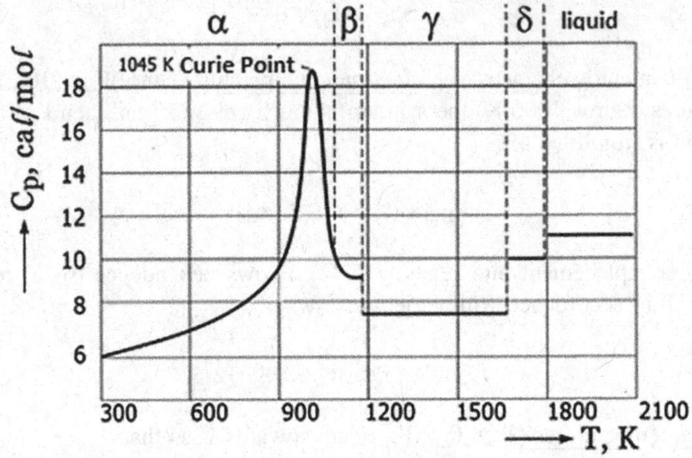

Fig. 6.7 Iron heat capacity dependence on temperature

well as for iron melting δ, there are heat capacity *discontinuous* variations, while for $\alpha \rightarrow \beta$ transformation varies continuously, since this phase transformation is of the second order (see Daneş et al. 2013 [1], § 7.4 and page 136).

Sum-Over-States of Combinations' Crystals

Combinations' thermodynamic quantities can be estimated in a structural manner only if particles' nature from lattice nodes (atoms, molecules, monoatomic ions, polyatomic ions) and chemical bond type, exerted between the nodes, are known.

In any case, C_v and C_p are zero at 0 K and increase with temperature, this increase being rapid at low and slow at high temperatures. Typically, C_v varies proportional to T^3 at low temperatures and becomes constant at fairly high temperatures, while C_p continues to grow slowly even at elevated temperatures, since the difference (C_p – C_v) increases relatively rapid with rising temperature. In electronically conductive compounds (metals, alloys), the electronic component of C_v occurs, which is important at temperatures either very low or very high.

At $T \rightarrow 0$, the ratio of $A \cdot T$ and $B \cdot T^3$ terms from relation (6.21) can be higher than 1, and at very high temperatures, $C_{v,rt}$·is capped, while $C_{v,el}$ continues to increase almost linearly with temperature increase.

If monoatomic particles are found in the lattice nodes, particles which interact through one single type of bonds (alloys, $CaCl_2$ with its ionic lattice, MgO), the Debye model remains valid for one gram atom so that, for 1 mole, the C_p, B, and θ_D values from relations (6.28a, 6.31b, 6.33) are multiplied by the <u>*atomicity*</u> μ – the atom number from the molecular formula:

$$C_{v,rt} = 3 \cdot R \cdot \mu \cdot F_{DC}(x_D)$$

$$\theta_D = C_L \cdot (T_t \cdot \mu/M)^{1/2} \cdot (V_t)^{1/3}; \quad B = 2.4\,\pi^4 \cdot R \cdot \mu/(\theta_D)^3$$

If within nodes there are monoatomic particles interacting with several types of bonds (graphite, Van der Waals and metallic bonds; metallic sulfo-arsenides with S-As, S-Me, As-Me bonds), the substance is characterized by so many characteristic temperatures as the number of existing bonds – two in graphite, three in sulfo-arsenides, etc.

In the case of polyatomic nodes (K_2SO_4 with K^+ and SO_4^{2-} nodes; organic substances), an "internal" component of vibrational thermodynamic functions supplementary occurs, related to the same node atom vibration.

Approximately, C_p at room temperature is obtained by summing several atomic increments k_i (generally different from elements C_p values) as the ones from Table 3.2 of the book of (Daneş et al., 2013 [1]):

$$C_{p,} = \Sigma_i (\mu_i \cdot k_i) \tag{6.37}$$

where μ_i is the atom number of I element in the molecular formula.

This additivity law – *Newman-Kopp* law – is also valid for liquids, with k_i increments somehow modified.

6.7 Lattice Energy from Molecular Interactions

Lattice energy is a measure of interaction intensity of particles found in the lattice nodes. By definition, the *lattice energy* E, of a crystal, is the internal energy variation, from the dissociation reaction at 0 K of a crystal mole within the constitutive particles, which are ultimately located at an infinite distance of one another (thus, in gaseous state of aggregation):

$$\text{Crystal} \rightarrow \text{Constitutive particles (gas)} \tag{6.38}$$

"Constitutive particles" are understood here as the groups located in the lattice nodes, groups whose nature depends on crystal's chemical bond type:

- For ionic crystals, the constituents are mono- and/or polyatomic ions:

$$NaCl \rightarrow Na^+(G) + Cl^-(G); K_2SO_4 \rightarrow 2\,K^+(G) + (SO_4)^{-2}(G);$$
$$TiO_2 \rightarrow Ti^{4+}(G) + 4\,O^-(G)$$

- For covalent crystals – the atoms:

$$SiC \rightarrow Si\,(G) + C\,(G); C\,(\text{diamond}) \rightarrow C\,(G),$$

- For molecular crystals – the respective molecules itself:

$$C_6H_6\,(S) \rightarrow C_6H_6\,(G); H_2O\,(S) \rightarrow H_2O\,(G),$$

- For metallic crystals, two versions of the lattice energies can be defined, E'_{rt} and $\hat{E}_{rt}''$, depending on the consideration made regarding node atoms, whether they are neutral or positive ions; for example, for calcium (solid), the lattice energy is as follows:

$$E'_{rt} \ldots Ca\,(S) \rightarrow Ca\,(G); E_{rt}'' \ldots Ca\,(S) \rightarrow Ca^{2+}(G) + 2\,e^-.$$

Sometimes both ion and molecules occur prior to dissociation, for example, in the case of crystalohydrates:

$$CuSO_4 \cdot 5\,H_2O\,(S) \rightarrow Cu^{2+}(G) + (SO_4)^{-2}\,(G) + 5\,H_2O\,(G).$$

Lattice Energy Determination Methods

The lattice energy can be obtained:

(a) From the u (r) expression of intermolecular potential u dependence on distance r between lattice nodes particles, a method which is presented below.
(b) Thermodynamically, from calorimetric determinable quantities, using Hess's law, as in Sect. 6.8. Often the reaction thermodynamic quantities, required for this calculation, are not known at 0 K and 0 atm (the last condition corresponds to an infinite distance between the gaseous particles resulting from dissociation) but only under standard thermodynamic conditions, 25 ° C and 1 atm. The difference is, however, relatively small compared to thermodynamic data accuracy, so that the exact definition:

$$E_{rt} \equiv \Delta^r U(0\,K, 0\,atm)$$

can be replaced by the approximate definition:

$$E_{rt} \cong \Delta^r H\,(298.15\,K, 1\,atm) \tag{6.39}$$

where both $\Delta^r U$ and $\Delta^r H$ refer to reaction (6.38).

The comparison of E_{rt} values obtained from routes (a) and (b) allows obtaining information about lattice connections' nature (covalent bonds degree of polarity or the metal electrons degree of delocalization), obtaining u(r) function's parameters, or calculating – using the relation (6.43b) – the electron affinity, which is difficult to measure directly.

Lattice Energy from the Born-Landé Potential

The intermolecular *Born-Landé* potential, usable for *ionic lattices*, is:

$$u(r) = -k_C \cdot z' \cdot z'' \cdot (e_o)^2/r - C/r^n \tag{6.7a2}$$

where e_o is the elementary electrical charge, z' and z" are valences of the two ions which interact, r is the distance between these two ions in the crystal, $k_C \cong 8.988$ kg/(m·A^2) is Coulomb's constant, and C (> 0) and n ($\geq$ 3) are two material constants independent of distance. The repulsive term C/ r^n is of a quantum nature, and the term $z' \cdot z'' \cdot (e_o)^2/r$ corresponds to Coulomb (electrostatic) interaction of the two ions (which have the charges $z' \cdot e_o$ and $z'' e_o$) which may be positive (attraction) for opposite sign ions or negative if the ions have the same sign of their charge.

In the particular case of a simple cubic lattice with equal charge ions (z' = – z"= z), the internal interaction energy for one mole will be:

$$U = -0.5 \cdot N_A \cdot \sum_{i=1}^{N_A - 1} u(r_i) \tag{6.7b2}$$

where $u(r_i)$ is the interaction potential between the given ion and another ion located at distance r_i. When taking into account the lattice arrangement, $r_i^2 = R^2 \cdot (X^2 + Y^2 + Z^2)$, where R is the distance between two immediate neighbors and (X, Y, Z) are three integer numbers among which at least one is different from zero. Relation (6.7a2) becomes therefore, for any ion:

$$u(r_i) = k_C \cdot (z \cdot e_o)^2 \cdot (-1)^{X+Y+Z} / \left(R \cdot \sqrt{X^2 + Y^2 + Z^2}\right) - C / \left(R \cdot \sqrt{X^2 + Y^2 + Z^2}\right)^n$$

the electrostatic term being positive for the odd sum X + Y + Z (opposite sign ions) and negative for the even value of the sum X + Y + Z (identical sign ions). By replacing this expression $u(r_i)$ into relation (6.7b2), the following expression is obtained:

$$U(R) = -(N_A/2) \cdot \left[S_1 \cdot (-1)^{X+Y+Z} k_C \cdot (z \cdot e_o)^2/R - S_n \cdot C/R^{n*}\right] \tag{6.7c2}$$

where n* > 1 and

$$S_n = \sum X \sum Y \sum Z (X^2 + Y^2 + Z^2)^{-n/2} \tag{6.40a}$$

The S_n summation from relation (6.40a) is effectuated for all integer values X, Y, and Z from $-\infty$ up to + ∞, provided that X, Y, and Z are not simultaneously zero.

The lattice energy corresponds to the minimum U(R) value, with changed sign. By denoting with d the distance R equilibrium value between two adjacent neighbors:

$$E_{rt} = -U_{R=d} \tag{6.40b}$$

The condition of minimum equals with $(d\,U/\,d\,R)_{R\,=\,d} = 0$ or when considering (c) $[-S_1 \cdot k_C \cdot (z \cdot e_o)^2/\,R^2 + n \cdot S_n \cdot C/\,R^n]_{R\,=\,d} = 0$, from where $C = [S_1 \cdot k_C \cdot (z \cdot e_o)^2/\,(n \cdot S_n \cdot)] \cdot d^{\,n\,-\,1}$.

By replacing C into expression (6.7c2), the following expression results:

$$U(R) = -\left[N_A \cdot S_1 \cdot k_C \cdot (z \cdot e_o)^2/(2 \cdot R)\right] \cdot [1 - (d/R)^n/n] \tag{6.7d2}$$

with the condition of minimum (6.40b) $E_{rt} = N_A \cdot k_C \cdot S_1 \cdot (z \cdot e_o)^2 \cdot (1\text{–}1/n)/(2 \cdot d)$ or

$$E_{rt} = N_A \cdot k_C \cdot A_M \cdot (z \cdot e_o)^2 \cdot (1 - 1/n)/d \tag{6.7e2}$$

where A_M is the *Madelung* constant defined, with S_1 from relation (7.40a), as

$$A_M = S_1/2 \tag{6.40c}$$

Lattice Geometry and Madelung Constant

The reasoning for the cubic lattice may be extended to other lattice geometries, but Madelung constant value will be different. Among the usual equal valence ion lattices, the following values can be quoted: 1.638 for sphalerite (ZnS β), 1.641 for würtzite (ZnS α or sphalerite), 1.748 for halite (NaCl), and 1.762 for CsCl α. It can be observed that A_M increases with the coordination number (for the four listed lattices, L is 4, 4, 6, and 8).

Relation (e) expresses the lattice energy relative to the ion arrangement geometry (A_M), to the intermolecular potential repulsive part exponent n, and to ion valence. For crystals with two types of unequal valence ions, z' and – z'', relation (6.7e2) is replaced by:

$$E_{rt} = N_A \cdot k_C \cdot (e_o)^2 \cdot \left|z' \cdot z''\right| \cdot A_M \cdot (1 - 1/n)/d \tag{6.41a}$$

where A_M has the following general expression:

$$A_M = \left|z' - z''\right| \cdot \mu_I \cdot f_A \tag{6.41b}$$

where μ_I is the ion number from the molecular formula, and f_A is a *geometric accommodation* factor, slightly subunitary ($f_A = 0.69 \div 0.88$), which increases with the coordination number and depends on the lattice geometry.

By comparing the lattice energy values from relation (6.41a) with the thermochemically obtained ones, as in Sect. 6.8, information concerning exponent n can be obtained (this can be directly obtained from compressibility). For n we obtain values increasing with principal quantum numbers mean n_p of the respective ion atoms. They are ranging from n = 5 - for the ions with $n_p = 1$ (H^-, Li^+, Be^{2+}) up to the value n = 12 - for ions with $n_p = 6$ (the radon shell).

Lattice Energy from Mie Potential

While dissociating into atoms (or neutral molecules) of covalent, metallic, or molecular lattices, the lattice energy is calculated in a similar way, but starting from Mie potential from relation (4.21), $u(r) = A/r^m - B/r^n$, $1 < m < n$:

$$E_{rt} = C \cdot (1 - m/n)/d^m \tag{6.42}$$

where d is the equilibrium distance between lattice nodes and C depends on constant A of the attractive potential $A \cdot r^m$ and of the lattice geometry:

$$C = N_A \cdot A \cdot S_m/2$$

where S_m results in intermolecular distances summing up to infinity upon three directions, divided by d and raised to -m power.

6.8 Hess's Law Lattice Energies

In ionic crystals, the lattice energy can be obtained from Hess's thermodynamic law in volume I (Daneş et al., 2013 [1], in § 4.3), through the *Born-Haber* cycle, exemplified in the case of aluminum trioxide Al_2O_3 from Fig. 6.8

Born-Haber Cycle Steps

It is possible to reach from the initial state – Al and O elements in standard state Al (S) and O_2(g) – to the final state (gaseous ions Al^{3+} and O^{2-}) on two different paths, I and II, each consisting of two processes in series A and B.

On path I, the intermediate step is constituted by the Al_2O_3(S) crystal; the IA process, namely, 2 Al(S) + 1.5 O_2(g) → Al_2O_3(S), has therefore the thermal effect

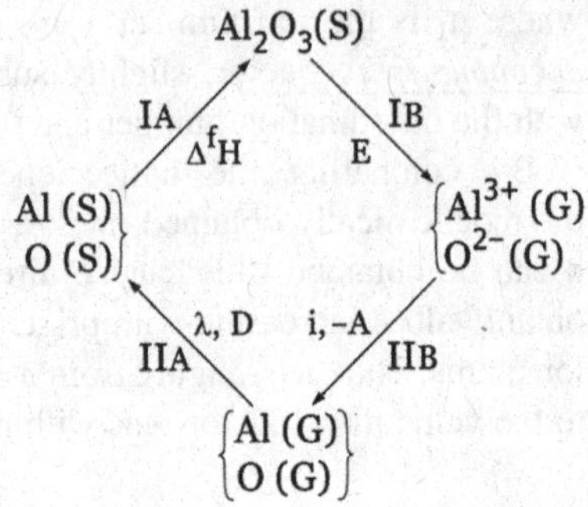

Fig. 6.8 Hess's law lattice energy

equal to the aluminum oxide formation enthalpy, $\Delta^fH(Al_2O_3, S)$, while the IB process, namely, $Al_2O_3(S) \rightarrow 2\ Al^{3+}(g) + 3\ O^{2-}(g)$, will have as thermal effect the lattice energy E_{rt} itself.

On path II, the transformation intermediate step consists of nodes particles, neutralized and passed into gaseous state, i.e., in the given case 2 Al(g) + 3 O(g).

- Process IIA therefore will be $2\ Al(S) + 1.5\ O_2(g) \rightarrow 2\ Al(g) + 3\ O(g)$. Obviously, this process is obtained by summing up two concomitant processes:
  - IIA_1) Al (S) → Al (g), the thermal effect being represented by aluminum latent heat of sublimation, $\lambda_{sb}(Al)$
  - IIA_2) $O_2(g) \rightarrow 2\ O(g)$, the thermal effect being represented by diatomic oxygen molecule dissociation reaction enthalpy into the free atoms; this quantity is denoted by $D(O_2)$ – ("dissociation" enthalpy)
- Process IIB – gaseous atoms Al and O transformation into ions – is also the result of these concomitant processes summing up:
  - (IIB_1) $Al(g) \rightarrow 3\ e^- + Al^{3+}(g)$
  - (IIB_2) $O(g) + 2\ e^- \rightarrow O^{2-}(g)$

Aluminum ionization (IIB_1) is carried out in steps; the n-th step being characterized by its *ionization potential* to whom corresponds the n-th energy of ionization, $E_{I,n} = N_A \cdot e_o \cdot i_n$ (with e_o – elementary charge):

- $Al(g) \rightarrow Al^+(g) + e^-$, with the molar energy variation $N_A \cdot e_o \cdot i_1$
- $Al^+(g) \rightarrow Al^{2+}(g) + e^-$, with the molar energy variation $N_A \cdot e_o \cdot i_2$
- $Al^{2+}(g) \rightarrow Al^{3+}(g) + e^-$, with the molar energy variation $N_A \cdot e_o \cdot i_3$

The X neutral particle n-th ionization potential i_n is the tension difference (expressed in V), required for pulling off the n-th electron of the neutral particle:

$$X^{n-1} \rightarrow X^{n-2} + e^- \tag{6.43a}$$

As an energetic effect of oxygen ionization, affinity changed sign value for two electrons of the oxygen atom corresponds. The affinity *for n electrons* of the neutral particle Y is the changed sign thermal effect of the process:

$$Y + n \cdot e^- \rightarrow Y^{-n} \ldots \text{energy variation} - A_n \quad (6.43b)$$

In the case of process II B, the affinity for two electrons of the oxygen atom O occurs.

Born-Haber Lattice Energy for Aluminum Oxide

In summary, the overall process thermal effect is obtained as follows:

- On path I: $\Delta^r H_I = \Delta^r H_{IA} + \Delta^r H_{IB}$

$$\Delta^r H_I = \Delta^f H(Al_2O_3, S) + E_{rt}(Al_2O_3) \quad (6.7f2)$$

- On path II: $\Delta^r H_{II} = \Delta^r H_{IIA} + \Delta^r H_{IIB}$ (6.7g2)

where, considering that two ions Al^{3+} and three ions O^{2+} occur within the crystal:

$$\Delta^r H_{IIA} = 2 \cdot \Delta^r H_{IIA1} + 1.5 \cdot \Delta^r H_{IIA2} \text{ or : } \Delta^r H_{IIA} = 2 \cdot \lambda_{sb}(Al) + 1.5 \cdot D(O_2) \quad (6.7h2)$$

and similarly: $\Delta^r H_{IIB} = 2 \cdot \Delta^r H_{IIB1} + 3 \cdot \Delta^r H_{IIB2}$ or:

$$\Delta^r H_{IIB} = 2 \cdot e_o[i_1(Al) + i_2(Al) + i_3(Al)] - 3 \cdot A_2(O) \quad (6.7i2)$$

By replacing into relation (6.7g2) the thermal effects $\Delta^r H_{IIA}$ and $\Delta^r H_{IIB}$ from expressions (6.7h2) and (6.7i2), respectively, the following expression is obtained:

$$\Delta^r H_{II} = 2 \cdot [\lambda_{sb}(Al) + 2 \cdot e_o \cdot [i_1(Al) + i_2(Al) + i_3(Al)] + 1.5 \cdot D(O_2) + 3 \cdot A_2(O)] \quad (6.7j2)$$

According to Hess's law, the thermal effects on the two paths are equal:

$$\Delta^r H_I = \Delta^r H_{II}.$$

From where, using the expressions for $\Delta^r H_I$ and $\Delta^r H_{II}$ from (6.7f2) and (6.7j2), respectively:

$$E_{rt}(Al_2O_3) = -P_{cr} + 2 \cdot P_{Al} + 3 \cdot P_O \quad (6.44)$$

where P_{cr} groups the thermodynamic quantities that refer to the given crystal, P_{Al} are the ones involving aluminum element, and P_O characterizes the other component element, oxygen:

$$P_{cr} \equiv \Delta^r H(Al_2O_3, S); P_{Al} \equiv \lambda_{sb} + e_o \cdot (i_1 + i_2 + i); P_O \equiv 0.5 \cdot D(O_2) - A_2(O)$$

Relation (6.44) is the Born-Haber expression of lattice energy for Al_2O_3.

Born-Haber Cycle for Other Ionic Crystals

For other ionic crystals, appropriate changes occur. For example, for NaCl, having in the molecular formula one cation and one anion – both monovalent – the lattice energy becomes:

$$E_{rt}(NaCl) = -\Delta^f H(NaCl, S) + [\lambda_{sb}(Na) + I_1(Na)] + [0.5 \cdot D(Cl_2) - A_1(Cl)];$$

and in bromides, where the anion is liquid under the standard state, the bromine latent heat of vaporization occurs in addition, $E_{rt}(KBr) = - \Delta^f H(KBr, S) + [\lambda_{sb}(k) + I_1(K)] + [\lambda_v(Br_2) + 0.5 \cdot D(Br_2, G) - A_1(Br)]$.

The Born-Haber cycle is used in multiple manners:

(a) If an is known – which is in principle measurable, as well as the other quantities from member II of the relation (6.44) – lattice energy numerical values are obtained, which allows the evaluation of ionic crystal chemical bond strength.
(b) If a E_{rt} value calculated from the intermolecular potential (as in Sect. 6.7.) is introduced into member I of Born-Haber relation, one of member II quantities can be calculated. This procedure is often used to evaluate electron affinity, since its experimental determination is particularly inaccurate.

Atomic, Molecular or Metallic Lattice Crystals

For molecular or atomic crystals, the lattice energy is identical to sublimation heat:

$$E_{rt} = \lambda_{sb} \tag{6.45a}$$

provided that gaseous particles resulted from the crystal-vapor transformation are identical to crystal nodes particles. The condition is fulfilled in most cases. An exception is constituted by phosphorus, whose sublimation gives vapors the molecular formular P_4, so that in this case:

$E_{rt}(P) = \lambda_{sb}(P) + 1/4 \cdot D(P_4, G)$where $D(P_4, G)$ is dissociation thermal effect $P_4(G) \rightarrow 4\,P(G)$.

For metallic crystals case, the lattice energy E'*rt* of neutral atoms formation is calculated similarly to relation (6.45a):

$$E'rt = \lambda_{sb}, \tag{6.45b}$$

while the lattice energy of solid metal formation from gaseous cations and electrons has the following expression (where z is metal's valence electrons number):

$$E'_{rt} = \lambda_{sb} + e_o \cdot \sum\nolimits_{n=1}^{z} I_n \tag{6.45c}$$

6.9 Real Crystal Lattice Defects

Punctiform Defect Generation

From the point of view of their **generation**, the zero-dimensional defects in simple crystals are classified mainly in two types: Schottky defect and Frenkel defect.

Schottky Defect

It is a simple empty (unoccupied) node and is generated by successive outward displacement of atoms from different nodes, starting from crystal surface, as in Fig. 6.9.

Frenkel Defect

It is a "vacant node + occupied internode" pair, generated by migration of an atom leaving an occupied node and settling in within the <u>*internode*</u> – the interstitial space between the neighboring nodes.

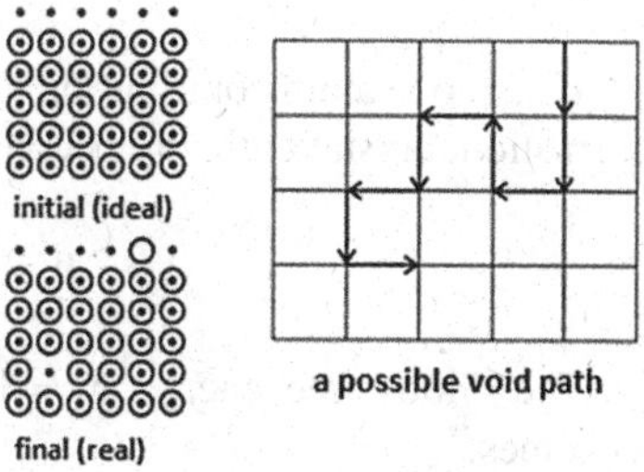

Fig. 6.9 Schottky defect

While Schottky defects reduce the crystal density, the Frenkel type ones do not influence it.

Both defect type occurrence produces not only an increase of crystal energy but also of its entropy.

The share of different defects depends on temperature, as it is shown furtherly in the analysis following defect generation into simple lattice crystals.

Thermodynamics of Schottky Defect Formation

Let N_1 be the occupied node number and N_2 – the unoccupied node number. The thermodynamic probability is given in this case by formula (1.7a):

$$W = (N_1 + N_2)!/[(N_1!) \cdot (N_2!)],$$

the particles being indistinguishable, and entropy – by Boltzmann relation (1.20), $s = k_B \cdot ln\ W$:

$$s = k_B \cdot \{ ln\,[(N_1 + N_2)!] - ln\,(N_1!) - ln\,(N_2!)\}$$

or, using the third Stirling approximation (1.6c):

$$s = k_B \cdot \{[(N_1 + N_2) \cdot \ln (N_1 + N_2) - (N_1 + N_2)] - [N_1 \cdot \ln N_2 - N_1]$$
$$-[N_2 \cdot \ln N_2 - N_1]\}$$

After term regrouping, the following is obtained:

$$s = k_B \cdot [(N_1 + N_2) \cdot ln\ (N_1 + N_2) - N_1 \cdot ln\ N_1 - N_2 \cdot ln\ N_2].$$

By denoting with x the empty node versus atom (unoccupied nodes) weight, i.e., $N_2 = N_1 \cdot x$, the last expression becomes:

$$s = N_1 \cdot k_B \cdot [(1 + x) \cdot ln\ (1 + x) - x \cdot ln\ x] \tag{6.46a}$$

If ε' is one atom contribution to enthalpy and ε_g – the energy necessary for a void formation, crystal enthalpy becomes $h = N_1 \cdot \varepsilon' + N_2 \cdot \varepsilon_g$ or:

$$h = N_1 \cdot \varepsilon' + N_1 \cdot x \cdot \varepsilon_g \tag{6.46b}$$

The Gibbs free energy $g = h - s \cdot T$, with s and h from relations (6.46a, b) becomes:

$$g = N_1 \cdot \varepsilon' + N_1 \cdot x \cdot \varepsilon_g - N_1 \cdot k_B \cdot T \cdot [(1+x) \cdot ln(1+x) - x \cdot ln\, x] \quad (6.46c)$$

At equilibrium, g is minimum, from where $dg/dx = 0$, or by applying derivative into (6.46c)

$\varepsilon_g = k_B \cdot T \cdot [\ln(1 + x) + 1 - \ln x - 1]$, from where $1 + 1/x = exp\, [\varepsilon_g/(k_B \cdot T)]$ or, since $x \ll 1$ therefore $1 + x \cong 1$:

$$x = exp\left(-\varepsilon_g/k_B \cdot T\right) \quad (6.47)$$

Void share is null at 0 K and increases exponentially, reaching its maximum at the melting temperature.

Thermodynamics of Frenkel Defect Formation

It is denoted by N_1, N_2, N_3, and N_4 the numbers of occupied nodes, nonoccupied nodes, occupied internodes, and unoccupied internodes. The thermodynamic probability W is the product $W_1 \cdot W_2$ of thermodynamic probabilities:

$W_1 = (N_1 + N_2)\,!\,/[(N_1!) \cdot (N_2!)]$ and $W_2 = (N_3 + N_4)\,!\,/[(N_3!) \cdot (N_4!)]$of microstate W_1 of *node* occupancy and macrostate W_2 of *internode* occupancy. Since $s = k_B \cdot lnW$ it results that:

$$s = k_B \cdot \{ ln\,[(N_1 + N_2)! \cdot (N_3 + N_4)! \cdot] - ln\,[(N_1!) \cdot (N_2)! \cdot (N_3)! \cdot (N_4)! \cdot]\}$$

By approximating $ln\,(A!) \cong A \cdot ln\, A - A$, it results after simplification:

$$s = k_B \cdot \{ ln\left[(N_1 + N_2)^{N_1+N_2} \cdot (N_3 + N_4)^{N_3+N_4}\right] - \sum\nolimits_{i=1}^{4} N_i \cdot ln\, N_i \quad (6.48a)$$

Let N be the atom number, z – internode and node number ratio (ratio depending only of lattice geometry), and x – defect and atom number ratio so that, by definition:

$$z \equiv (N_3 + N_4)/(N_1 + N_2) \text{ and } x \equiv N_2/N.$$

From balances $N_1 + N_2 = N$ (nodes number equal to atoms number) and $N_2 = N_3$ (unoccupied node number equal to occupied internode number), it is possible to express N_1, N_2, N_3, and N_4, according to x and z ratios:

$$N_1 = N \cdot (1 - x); N_2 = N \cdot x; N_3 = N \cdot x; N_4 = N \cdot (z - x).$$

With these expressions for (N_1, N_2, N_3, N_4), relation (6.48a) becomes:

$$s = N \cdot k_B \cdot [z \cdot ln\, z - (1 - x) \cdot ln\,(1 - x) - 2 \cdot x \cdot ln\, x - (2 - x) \cdot ln\,(2 - x)] \quad (6.48b)$$

By denoting with (ε, ε_g, ε_{io}) the energies corresponding respectively to an atom, unoccupied node, and occupied internode, crystal enthalpy is $h = N{\cdot}\varepsilon + N_2{\cdot}\varepsilon_g + N_3{\cdot}\varepsilon_{io}$ or, after replacing $N_2 = N_3 = x$ as above:

$$h = N \cdot \left[\varepsilon + x \cdot \left(\varepsilon_g + \varepsilon_{io}\right)\right] \tag{6.48c}$$

From relations (6.48b,c) crystal Gibbs free energy can be obtained, $g = h - s{\cdot}T$:

$$g = N \cdot \{\varepsilon + \left(\varepsilon_g + \varepsilon_{io}\right) \cdot x - k_B \cdot T \cdot [z \cdot \ln z - (1 - x) \cdot \ln(1 - x)$$
$$-2 \cdot x \cdot \ln x - (z - x) \cdot \ln(z - x)]\}$$

from where, using the equilibrium condition $d\,g/d\,x = 0$, $0 = \varepsilon_g + \varepsilon_{io} - N{\cdot}k_B{\cdot}T{\cdot}[-\mathit{ln}(1 - x) + 1 - 2{\cdot}\mathit{ln}\,x - 2 - \mathit{ln}(2 - x) + 1]$ or:

$$x^2/[(1 - x) \cdot (z - x)] = \mathit{exp}\left[-\left(\varepsilon_g + \varepsilon_{io}\right)/(k_B \cdot T)\right] \tag{6.48d}$$

Defect Ratio Dependence on Temperature

Since z is of unity order, and $x \ll 1$, denominator $(1 - x){\cdot}(z - x)$ can be approximately replaced by z, and relation (6.48d) is approximated thus by:

$$x = \sqrt{z} \cdot \mathit{exp}\left[-\left(\varepsilon_g + \varepsilon_{io}\right)/(2 \cdot k_B \cdot)\right] \tag{6.48e}$$

As in the case of Schottky defects, the Frenkel defect share is null at 0 K and increases exponentially with temperature. The defect activation energies are for Schottky (E_S) defects and Frenkel defects (E_F), respectively:

$$E_S = E_g \tag{6.49a}$$

$$E_F = 0.5 \cdot \left(E_{io} + E_g\right)/2 \tag{6.49b}$$

where E_g and E_{io} are the *molar* enthalpies of formation of unoccupied nodes and occupied internodes, respectively, and $E_g = N_A{\cdot}\varepsilon_g$ and $E_{io} = N_A{\cdot}\varepsilon_{io}$, where N_A is Avogadro's number.

6.10 Seven Worked Examples

Ex. 6.1

Calculate approximately the X ratio between liquid (P_{IL}) and vapor (P_{IG}) benzene internal pressures of benzene, under equilibrium conditions of the two phases, at P = 12.9 kPa and T = 299 K.

Liquid benzene density at this temperature is $\rho_L = 872\ kg/m^3$, and molecular mass is for benzene: M = 0.0781 kg/mol.

Both phases respect Redlich-Kwong ES.

Solution

$$V_L = M/\rho_L = 0.0781/872 = 8.96 \cdot 10^{-5}\ m^3/kg$$

is liquid molar volume.

On the other hand, from R-K ES (4.8a):

$$P = R \cdot T/(V - b) - a/\left[V \cdot (V + b) \cdot \sqrt{T}\right] \quad (6.9a2)$$

The following is obtained through isochore derivation:

$$(\partial P/\partial T)_V = R/(V - b) + a/\left[2 \cdot V \cdot (V + b) \cdot T^{3/2}\right] \quad (6.9b2)$$

Relation (6.8) becomes, when replacing β by its definition, $\beta \equiv (\partial P/\partial T)_V/P$: $P_I = T \cdot (\partial P/\partial T)_V - P$, from where, with P and $(\partial P/\partial T)_V$ from (6.9a2) (6.9b2), respectively:

$$P_I = 3 \cdot a/\left[2 \cdot V \cdot (V + b) \cdot T^{3/2}\right] \quad (6.9c2)$$

Customizing relation (6.9c2) for liquid (P_{IL}, V_L) and vapors (P_{IG}, V_G), the following expression results after division for $X \equiv P_{IL}/P_{IG}$ ratio:

$$X = V_G \cdot (V_G + b)/[V_L \cdot (V_L + b)] \quad (6.9d2)$$

For the approximate calculation of X ratio, it is considered that vapors have a perfect gas behavior (thus $V_G \cong R \cdot T/P$) and that $P_{IL} \gg P$, from where it results that $R \cdot T/(V_L - b) \gg P$, so that $b \cong V_L$.

By replacing V_G and b into (d), the following expression results:

$$X = (R \cdot T/P) \cdot (R \cdot T/P + V_L)/[V_L \cdot (V_L + V_L)] \text{ from where,}$$

since $V_L \ll R \cdot T/P$: $X = [R \cdot T/(P \cdot V_L)]^2/2$ or:

$$X = \left[8.31 \cdot 299/\left(12900 \cdot 8.96 \cdot 10^{-5}\right)\right]^2/2a \Rightarrow \mathbf{X = 2.31 \cdot 10^6}$$

Ex. 6.2
Liquid acetone (molecular mass M = 0.0461 kg/mol) presents, at T = 293 K, the following features:

- Isobaric molar heat capacity $C_p = 135.1$ J/ (mol·K)
- Density $\rho = 788$ kg/ m^3, expansibility $\alpha = 0.00143$ K^{-1}
- Sound speed $u_L = 1190$ m/ s
- Coordination number L = 10.5
- Isothermal compressibility $\chi_T = 1.24 \cdot 10^{-9}$ m^2/ N

It is required to calculate Y ratio between acetone free and total volumes:

1) On thermodynamic-statistical path
2) From sound speed

Solution
In the statement, Y is defined as:

$$Y = N_A \cdot v_f/V \tag{6.9a3}$$

where V is the liquid phase molar volume. A molecule's free volume v_f can be found using the two methods presented in Sect. 6.3.:

1. By replacing dependency (6.13) into relation (6.9a3), the free volume dependency obtained on a thermodynamical-statistical path, $v_f = (V/\ N_A) \cdot (c \cdot R \cdot T/\ \Delta U_v)^3$, a first expression is found, denoted by Y_A:

$$Y_A = (c \cdot R \cdot T/\Delta U_v)^3 \tag{6.9b3}$$

where c is evaluated from L by $c = L^{1/3} = 10.5^{1/3}$ and ΔU_v – from relation (6.5), $\Delta U_v = -5860 + 103 \cdot T = -5860 + 103 \cdot 293 = 24200$ J/mol. By replacing into (b) it results that $Y_1 = 10.5 \cdot (8.31 \cdot 293/\ 24200)^3 \rightarrow \mathbf{Y_1 = 100 \cdot 10^{-4}}$.

2. Sound speed-based calculation, from (6.9b3) and from relation (6.14a), $v_f = (V/\ N_A) \cdot [R \cdot T \cdot C_P/(M \cdot C_V)]^{-1.5}/\ (u_L)^3$, gives the second value of X:

$$Y_B = \left[(u_L)^2 \cdot M \cdot C_V/(R \cdot T \cdot C_P)\right]^{-3/2} \tag{6.9c3}$$

The isochoric heat capacity C_v is obtained with the thermodynamic relation (3.30b) from the book of (Daneş et al., 2013 [1]): $C_V = C_P - V \cdot T \cdot \alpha^2/\chi_T$ or, with $V = M/\ \rho$, $C_V = C_P - M \cdot T \cdot \alpha^2/\ (\rho \cdot \chi_T) = 135.1 - 0.0581 \cdot 293 \cdot 0.00143^2/$

$(788 \cdot 1.24 \cdot 10^{-9}) = 99.5$ J/ (mole·K). By replacing into (c) the values T, C_p, M, C_v, and u_L it results that:

$$Y_2 = \left[1190^2 \cdot 0.0581 \cdot 99.5/(8.31 \cdot 293 \cdot 135.1)\right]^{-3/2} a \Rightarrow \mathbf{Y_2 = 81 \cdot 10^{-4}}$$

with 19% lower than value Y_1 obtained by the first method.

Ex. 6.3

Indium is under ambient conditions ($T_1 \cong 293$ K, $P_1 \cong 1$ bar) a solid metal, with density $\rho_1 = 7667$ kg/ m^3 and volume relaxation coefficient $\alpha_o = 69 \cdot 10^{-6}$ K^{-1}, melting at $T_t = 460$ K. At temperature $T_2 = 1300$ K liquid density is $\rho_2 = 5955$ kg/mol.

It is required to evaluate liquid molar isobaric heat capacity C_{p2} at this later temperature.

Solution

Being an evaluation, it can be admitted that solid expansibility is independent of temperature, on the range between ambient conditions and melting. Solid density $\rho_{S,t}$ at the melting temperature is therefore:

$$\rho_{S,t} = \rho_1 / \exp\left[\alpha_0 \cdot (T_t - T_l)\right] = 7667/\exp\left[69 \cdot 10^{-6} \cdot (460 - 293)\right] \text{ or}$$

$$\rho_{S,t} = 7595 \text{ kg/m}^3$$

Relation (6.16) $C_{V,L} = 1.5 \cdot R \cdot (1 + V_{S,t}/V_L)$ of significant structure theory gas + solid into the liquid monoatomic substance becomes, when considering that density is connected to molar volume V through relation $V \cdot \rho = M$ and molar mass M is constant under statement's conditions:

$$C_{V,L} = 1.5 \cdot R \cdot \left(1 + \rho_t/\rho_{S,t}\right),$$

and using the statement notations, the isochoric molar heat capacity $C_{V,2}$ at 1300 K is obtained:

$$\begin{aligned} C_{V,2} &= 1.5 \cdot R \cdot \left(1 + \rho_2/\rho_{S,t}\right) \\ &= 1.5 \cdot 8.31 \cdot (1 + 5955/7595) \\ &= 22.2 \text{ J/(mol} \cdot \text{K)}. \end{aligned}$$

For condensed phase case, isochoric and isobaric heat capacity slightly differ; thus:

$$C_{p,2} \cong C_{V,2} => \mathbf{C_{p,2} \cong 22\ J/(mol \cdot K)}$$

Ex. 6.4

It is required to calculate molar isobaric heat capacity, C_p, for aluminum at 20°C, knowing the following physical quantities:

- Compression and shear modulus expressed in N/ m², $E = 7{\cdot}10^{10}$, and $G = 2.7{\cdot}10^{10}$
- Thermal relaxation (volumic) coefficient, $\alpha = 76{\cdot}10^{-6}\ K^{-1}$
- Molar volume, $V = 10^{-5}\ m^3/mol$

Solution

From Table 6.3: $A = 0.00135{\cdot}J/(mol{\cdot}K^2)$ and $\theta_D = 428$ K.

From relation (6.23) it is obtained, with $T = T_o + t = 273 + 20 = 293$ K:

$$C_{v,el} = A \cdot T = 0.00135 \cdot 293 = 0.40\ J/(mol \cdot K);$$

From relation (6.27b) $x_D = \theta_D/T = 428/293 = 1.46$.

From relation (6.29), $F_{DC} = 1 - (x_D)^2/20 + (x_D)^4/560$, it results that $F_{DC} = 1 - 1.46^2/20 + 1.46^4/560 = 0.901$, and from relation (6.28a):

$$C_{v,rt} = 3 \cdot R \cdot F_{DC} = 3 \cdot 8.31 \cdot 0.901 = 22.5\ J/(mol \cdot K).$$

From relation (6.22): $C_v = C_{v,el} + C_{v,rt} = 0.40 + 22.5 = 22.9$ J/ (mol·K).

On the other hand, from relations (6.32e) and (6.32f) it results that:

- $\mu_P = 0.5{\cdot}E/G - 1 = 0.5{\cdot}7/2.7 - 1 = 0.296$
- $\chi_T = 3{\cdot}(1-2{\cdot}\mu_P)/E = 3(1-2{\cdot}0.296)/(7{\cdot}10^{10}) = 1.75{\cdot}10^{-11}\ m^2/N$

Heat capacity difference can be now obtained: $C_p - C_v = \alpha^2{\cdot}V{\cdot}T/\chi_T = (76{\cdot}10^{-6})^2{\cdot}10^{-5}{\cdot}293/(1.75{\cdot}10^{-11}) = 0.97$ J/ (mol·K). It results that:

$$C_p = C_v + (C_p - C_v) = 22.9 + 0.97 => \mathbf{C_p = 23.9\ J/(mol \cdot K)}$$

Calorimetric experimental value is practically the same: 24 J/ (mol·K).

Ex. 6.5

It is required to calculate cesium chloride polymorphic transformation entropy, the transformation temperature being $T_{tr} = 725$ K. The form α (established at low temperatures) molar volume extrapolated at 0 K is of 41.7·cm³/mol, while internuclear distance d between the closest ions Cs^+ and Cl^- is identical in both crystallin forms, as well as heat capacity.

Elemental cell of form α is a centered cube (e.g., with one Cl^- ion in the middle and one Cs^+ ion in each of the eight corners), and form β crystallizes in NaCl lattice.

Coefficients n of the quantum part from Born-Landé potential are 9.5 and 11.5 for electronic cloud ions of argon (Cl^-) and xenon (Cs^+), respectively.

Solution

Madelung constant A_M is 1.762 and 1.748, respectively, for α and β forms, as mentioned in Sect. 6.7.

Exponent n of CsCl is obtained through harmony in mean of values for Cl^- and Cs^+: $n = 2/(1/n_{Ar} + 1/n_{Xe}) = 2/(1/9.5 + 1/11.5) = 10.41$.

Elemental cell large diagonal D' of CsCl α is 2·d, so that the cell's side is $a = D'/\sqrt{3} = 2 \cdot d/\sqrt{3}$. Cell's volume $v_c = a^3$; therefore $v_c = \left(2 \cdot d/\sqrt{3}\right)^3$ from where the molar volume $V = N_A \cdot V_c = N_A \cdot \left(2 \cdot d/\sqrt{3}\right)^3$.

From the last relation, it is found that $d = \sqrt{3} \cdot (V/N_A)^{1/3}/2$; therefore $d = \sqrt{3} \cdot \left[41.7 \cdot 10^{-6}/\left(6.02 \cdot 10^{23}\right)\right]^{1/3} = 2.47 \cdot 10^{-10}$ m, from where can be obtained through relation (6.41a), $E_{rt} = N_A \cdot k_C \cdot (e_o)^2 \cdot |z' \cdot z''| \cdot A_M \cdot (1-1/n)/d$, the lattice energies of polymorphic changes α and β. Excepting Madelung constant, the right member factors are identical for the two lattices, so that the difference between the two lattice energies is, with valences z' = 1 and z'' = −1:

$$E_\alpha - E_\beta = N_A \cdot k_C \cdot (e_o)^2 \cdot |z' \cdot z''| \cdot (A_{M,\alpha} - A_{M,\beta}) \cdot (1 - 1/n)/d = 6.02 \cdot 10^{23} \cdot \ldots$$

$8.99 \cdot 10^{-9} \cdot (1.602 \cdot 10^{-19})^2 \cdot |1| \cdot (1.762 - 1.748) \cdot (1 - 1/10.41)/(2.47 \cdot 10^{-10}) = 7135$ or $E_\alpha - E_\beta \cong 7.1$ kJ/mol.

The difference $E_\alpha - E_\beta$ represents the latent *internal* energy of polymorphic transformation, ΔU_{tr}, to 0 K.

If $C''_v = C''_v$, or $\Delta C_{v,rt} = 0$, then ΔU_{tr} does not depend on temperature, taking the same value of 7.1 kJ/mol at T_{tr}. Equilibrium thermal effect of α → β transformation is therefore $\lambda_{tr} = \Delta U_{tr} + P \cdot \Delta V_{tr}$. It can be considered that term $P \cdot \Delta V_{tr}$ is negligible when compared to λ_{tr}, so that $\lambda_{tr} \cong \Delta U_{tr} = 7.1$ kJ/mol. According to relation (7.1b) from the book Daneş et al., 2013 [1]:

$$\Delta S_{tr} = \lambda_{tr}/T_{tr} = 7100/725 \Rightarrow \mathbf{\Delta S_{tr} = 9.8\ J/(mol \cdot K)}.$$

The experimental value ΔS_{tr} is 8.8 J/(mol·K).

Ex. 6.6

When heating a polycrystalline aluminum sample from $T_1 = 293$ up to $T_2 = 930$ K, crystal relative elongation is A = 1.902%. Within the same temperature range, relative increase of aluminum lattice constants is A' = 1.880%. It is required to calculate:

1) Activation molar energy E_S of Schottky defects

2) The x_1 and x_2 proportions of the Schottky defects, at T_1 and T_2 temperatures, respectively.

Solution

1. Let V'_1 and V'_2 be an ideal crystal volume at the two temperatures and L'_1 and L'_2 – elementary cell side under the respective conditions.

   From geometric considerations it results that:
   $V'_2/V'_1 = (L'_2/L'_1)^3$ or, since $L'_2 = L'_1 \cdot (1 + A')$:

$$V'_2/V'_1 = (1 + A')^3 \tag{6.9a4}$$

When denoting by V_1 and L_1 sample volume and length at T_1, and with V_2 and L_2 the corresponding quantities at T_2, the following quantities are obtained similarly: $V_2/V_1 = (L_2/L_1)^3$ or

$$V_2/V_1 = (1 + A)^3 \tag{6.9b4}$$

since $L_2 = L_1 \cdot (1 + A)$.

For N atoms and N·x voids, the real crystal volume is proportional to (N + N·x), while the second one with N, so that their ratio is (1 + x): $V_2 = V'_2 \cdot (1 + x_2)$; $V_1 = V'_1 \cdot (1 + x_1)$, from where

$V_2/V_1 = (V'_2/V'_1) \cdot (1 + x_2)/(1 + x_1)$ from where, by replacing V_2/V_1 and V'_2/V'_1 from (b) and (a), respectively:

$$(1 + A)/(1 + A')^3 = (1 + x_2)/(1 + x_1) \tag{6.9c4}$$

Since A, A', x_2, and x_1 are all much lower than 1, relation (6.9c4) can be approximated by

$1 + 3 \cdot A - 3 \cdot A' = 1 + x_2 - x_1$ from where, since $x_2 \gg x_1$:

$$x_2 = 3 \cdot (A - A') = 3 \cdot (1.902 - 1.880)/100 \Rightarrow \mathbf{x_2 = 0.00066}$$

2. When applying the logarithm to relation (6.47), it results that $\varepsilon_g = -k_B \cdot T_2 \cdot \ln x_2$ or, passing to the molar value E_g of empty nodes forming $E_g = -R \cdot T_2 \cdot \ln x_2$

   But $E_g = E_S$ according to relation (6.49a), so that:

$$E_S = -R \cdot T_2 \cdot \ln x_2 \tag{6.9d4}$$

therefore, expressed in KJ: $E_S = 0.001 \cdot 8.31 \cdot 930 \cdot \ln 0.00066 \rightarrow$ **E_S = 56.6 kJ/ mol**

3. Similarly to relation (6.9d4), at temperature T_1 it results that:

$E_S = -R \cdot T_1 \cdot \ln x_1$ from where
$x_1 = exp\ (-E_s/\ (R \cdot T_1)] = exp\ (-56600/(8.31 \cdot 293)] \rightarrow \mathbf{x_1 = 8.0 \cdot 10^{-11}}$
It is observed that at room temperature, voids proportion x_1 is negligible, while near the aluminum melting temperature (which is of 931 K), a void returns to $10^4/6.6 \cong 1500$ atoms.

Ex. 6.7
From the structure of crystalline cupric selenium, a lattice energy of 1891 kJ/mol was calculated using the Born-Landé potential. The heat of $Cu_2Se(S)$ of −61 kJ/mol is known, while the sublimation heat of Cu and Se is 340 and 111 kJ/mol. The first ionization potential of copper is 7.72 V.
It is required to determine the affinity of selenium for two electrons.

Solution
The first ionization energy of Cu (transforming the metal into cuprous cation) is as in Sect. 6.8, $E_{I,1}(Cu) = N_A \cdot e_o \cdot i_1(Cu)$ or, passing from J to kJ:

$$E_{I,1}(Cu) = 6.02 \cdot 10^{23} \cdot 1.602 \cdot 10^{-19} \cdot 7.72/1000 = 745\ \mathrm{kJ/mol}.$$

The crystalline lattice being ionic, the Born-Haber cycle is used. Since both Cu and Se are solid, while their vapors are monoatomic:

$$\Delta^f H(Cu_2Se, S) + E_{rt}(Cu_2Se) = 2 \cdot \lambda_{sb}(Cu) + \lambda_{sb}(Se) + 2 \cdot E_{I,1}(Cu) - A_2(Se)$$

from where metalloid affinity for two electrons:
$A_2(Se) = 2 \cdot 340 + 111 + 2 \cdot 746 + 61 - 1891 \Rightarrow \mathbf{A_2(Se) = 453\ kJ/mol.}$

Part III
Transport Phenomena and Their Mechanism

Disequilibrium and Evolution

The thermodynamic models discussed in Part II of this book, referred to *equilibrium state* system characteristics, defined by maintaining constant with time the macroscopic state parameters, without an external force intervention.

Part III, consisting of Chaps. 7 and 8 and concluding the herein presented book, is devoted to *non-equilibrium states*, which can only be analyzed by the proposed methods of *kinetics*, which are to be discussed in § 7.1.

This Part III discusses in particular the *transport* phenomena of a conservative physical quantity from a given area of a system towards other area. Transport takes place in a *homogeneous* media (a phase part without space discontinuities) but *non-uniform* (with quantities varying from one point to another).

Transport phenomena are inevitable when the studied system is not under equilibrium conditions, so that there are spatial distribution gradients of system matter quantities. Transport consists of this distribution non-uniformity reduction through spatial differences gradual reduction process:

- of translational velocity (through momentum transport, due to the viscous flow phenomenon, produced by internal friction), but also
- of temperature (transport of energy by heat transmission), as well as
- of composition (mass transport through the diffusion process).

Transfer

Otherwise, there are also other six important paths for the physical system uniformization, which are grouped under the general name of *transfer*:

- convective heat transfer,
- radiative transfer,

- sound transfer,
- interfacial transfer,
- electrochemical transfer and
- photochemical transfer.

In all transfer cases, media non-uniformity reduction is not realized through a *dispersive* mechanism based on neighboring particles collision (usually, molecules), unlike the transports case.

System parameters equalization rate is given in the case of transfers by complex, empirical or difficultly predictable that are studied in non-chemical technical disciplines (the first three type of transfer) or in specific physical chemistry chapters of force fields, as for the last three types of transfer.

Transport

Part III is composed of two chapters:

- Chapter 7 presents the general results of transports' physico-chemical study, detailing the phenomenological (macroscopic) transport laws, as well as the molecular mechanism – allowing material constants prediction based on these laws – in the simple case of gaseous media.
- Chapter 8 extends the statistical mechanics results to more complex cases of transport in condensed phases – solids and liquids.

Chapter 7
General Laws of Transport in Gases

7.1 Physical Kinetics

Equilibrium and Disequilibrium—Kinetics and Thermodynamics

Kinetics is temporal macroscopic variations science of quantities that locally or globally characterize a certain physical system. Among these dimensions, system and its parts position or orientation are not included, these ones being the object of study of a classical mechanics branch referred to as kinematics.

A technological process kinetics study can determine, for example, what is the necessary period of time to achieve the desired final state of a transformation taking place within a technical system. Indeed, the central object of kinetics is not the system *state*, but its *change* (without external exchanges contribution).

It is understood that *thermodynamics* is not appropriate for such a study, at least as long as it remains within the classical frame—equilibrium thermodynamics—of this discipline. Logically, the equilibrium aspects are *separately* studied from non-equilibrium ones, with a notable exception: *electrochemistry*—the study of the electric field and current effects on physical systems local macroscopic quantities—which gathers system state characterization thermodynamic methods with the kinetic ones, of processes study.

Nuclear, Chemical, and Physical Kinetics

In kinetics, it is necessary to distinguish between *chemical* kinetics and *physical* kinetics, depending on whether there exists or not some proportional change of

F. E. Daneş et al., *Molecular Physical Chemistry for Engineering Applications*,
https://doi.org/10.1007/978-3-030-63896-2_7

molecules types existing within the system, in other words a system chemical composition change. This eventual change, studied in the book devoted to chemical kinetics of Daneş and Ungureanu, 2009 [3]—is referred to as *reaction*, or *chemical* reaction more precisely, in order to distinguish it from the *nuclear* reaction, where not only molecules but atomic nuclei themselves change. Plasma phenomena occur under somewhat similar laws with nuclear kinetics ones. However, their study is frequently incorporated to physical kinetics, according to Landau & Lifshitz 1976–1989 book of 10 volumes. In the strictly adopted in this book sense, *physical kinetics* is limited to temporal change phenomena study of global or local physical quantities, without any change of molecules or atoms nature. These processes are regrouped as *transfer phenomena*.

Transfer and Stationarity

The notion of *transfer* designates the extensive conservative physical quantities displacement—namely, a certain substance mass, the amount of motion (momentum) and energy—within a given media. Transfer phenomena variety from the technical field is widely depicted by Lightfoot, Bird, and Stewart, 2014.

In principle, for any physical quantity subjected to a conservation law, this displacement can be highlighted—the transport from one side to another of the system. For example, besides the above-mentioned quantities, both moment of rotation (which only occurs in the case of solids) and electric charge (whose transport is electrochemistry's object of study) can be conserved—and thus transported or transferred.

Transfer phenomena can occur into a system only in the case of concentration, velocity or temperature gradients phases, or if discontinuities of any intensive physical quantity exist—like in the case of pressure, temperature or chemical potential of one of components at any interface.

Transfer takes place only in systems that are not at equilibrium. But transfer laws are simpler to study for systems that are found in a *stationary* state.

Staţionarity is defined as a *time* invariability. The equilibrium state is both stationary and uniform, *uniformity* means invariance in *space* invariability.

At steady non-equilibrium state, maintaining the parameters constants in time is not spontaneous: stationary state achievement in this case by modifying a system's physical quantity (density, velocity, temperature) in a certain system's part, under several external forces influence. Except for rest periods or regime entry/exit ones, technological processes usually take place under stationary conditions.

Transfer: Location and Mechanism

The different types of transfer differ not only in the transported extensive physical quantity nature—mass, momentum, energy, electric charge, or usually several of them simultaneously—but mainly by other two elements:

- locating the process within the physical system, and
- the change mechanism for conservative quantities spatial distribution.

Transfer Location
Four transfer mechanisms can be distinguished, as following:

- either within a homogeneous *volume*—namely into a system phase
- either two system phases separating *interface*, or two physical system sub-systems, or separating the system from the surrounding media.

Transfer Mechanism
Four transfer mechanisms can be distinguished, as following:

1. *Dispersive* mechanism, where system quantities homogeneity propagates to microscale, through direct contact between the system infinitesimal parts, either in a homogeneous media or at interfaces. *Dispersion* typically occurs by particle collision—usually molecules, rarely microscopic or colloidal particles.
2. *Convective* transfer mechanism, which is possible in flowing media (homogenous or heterogeneous). At *convection*, system quantities homogeneity is due to matter currents, generated either by external forces action (convection itself or *forced*) or by system-specific gradients (especially gravitational, and rarely thermic or of composition) to free convection phenomenon, also called *advection*.
3. *Radiative* mechanism—transfer by *radiation*—where energy is transported by the electromagnetic field between the two distant bodies separated by a media not totally absorbing the field energy.
4. *Sonic* mechanism of *sound* transfer, more rarely encountered, where the oscillating field transferring the energy is of mechanical nature.

Transport and Transfer
A transfer is a *transport* if two conditions are simultaneously met:

- the location is made within a phase volume (and not at an interface)
- the transport mechanism is the dispersive one.

Otherwise, the phenomenon is called a *transfer*. Physical chemistry studies the transfer phenomena in three cases, regrouped in a book of the same authors (under preparation),

a) mass transfer at the contact interface separating two phases, studied by the

superficial physical chemistry of superficial layers and important in nanotechnologies
b) electric charge transfer, studied by *electrochemistry*, which also includes as object of study the mass transfer due to electrical potential and chemical reactions differences under these conditions
c) mass transfer and chemical transformations due to system interaction with an electromagnetic field, in *photochemistry*.

The rate of equalization for the system parameters is given for other transfers by complex relations, difficult to predict and usually studied in non-chemical technical disciplines, such as:

- Energetics, for interface transfer between system and its surroundings
- Thermotechnics, for convective or radiative energy transfer
- Technical Fluids Mechanics, for the convection of momentum or mass

7.2 Transport Phenomenology

- If within a homogeneous system a concentration difference along one certain direction (concentration gradient) exists, then a mass transport from the higher to the smallest concentration takes place and consequently there is a composition equalization, this phenomenon being referred to as *diffusion*.
- If a temperature gradient exists, a kinetic energy transport occurs—*thermal conduction*.
- If a gradient velocity exists, which is perpendicular to the flow direction, then a momentum transport occurs—*viscous flow*.

Transport phenomena occur in all states of aggregation, except the viscous flow, absent for crystalline solids. But the transport quantities are completely different for the three states of aggregation.

Viscous Flow

This phenomenon occurs in any fluid under motion due to *internal* friction—friction between fluid layers moving at different velocities.

Considering two parallel solid plates, located at a small distance one from the other and oriented perpendicularly to the Oz axis.

If one of the plates is fixed and the other—mobile, moving with velocity v in the Ox direction, then the fluid layers will move at different velocities in this direction, depending on the distance from the mobile plate.

Fig. 7.1 Viscous flow

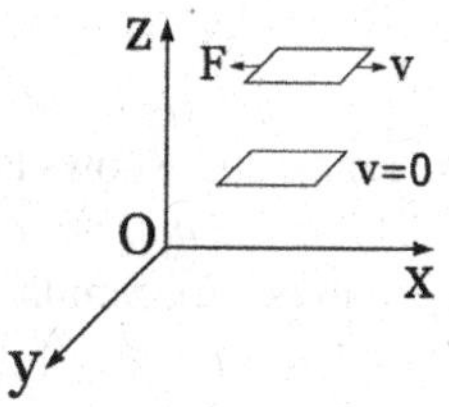

It is denoted by dv/dz the velocity gradient in Oz direction; this gradient is proportional to the momentum gradient $d\,(m\cdot v)/dz$ (Fig. 7.1).

According to *Newton's law*, f—friction force in direction z per unit area—is proportional to velocity gradient:

$$f = -\eta \cdot dv/dz \tag{7.1}$$

where η is the *dynamic* viscosity, or *actual* viscosity, with dimensions kg/(m·s), and v—fluid motion velocity, on a direction perpendicular to direction z. The minus sign from relation (7.1) indicates the opposite sense of the friction force in relation to particle velocity gradient.

Newton's law is always respected in gases. A *non-newtonian* behavior can occur in suspensions and in certain liquids, where viscosity depends only on velocity gradient (see Sect. 8.5).

The *kinematic* viscosity ν, expressed in m^2/s, is the following ratio:

$$\nu = \eta/\rho \tag{7.2}$$

where ρ designates the volumic mass (density). Similar to dynamic viscosity, kinematic viscosity is a *state function*, which depends on pressure, temperature, chemical composition, and phase of the liquid.

Heat Conduction

This phenomenon occurs when there is a temperature gradient and consists of a molecules uncoordinated energy transport, referred to as *heat*.

Heat flow *density* $j_q \equiv s^{-1}\cdot\delta q/\delta z$, where q is heat, t—time and s—area perpendicular on direction z, is the heat quantity transported in direction z on the unit of time per unit area.

j_q is proportional, according to *Fourier's* law (1882), with temperature gradient in the same direction:

$$j_q = -\lambda \cdot dT/dz \tag{7.3}$$

the minus sign shows heat transmission from the hot zone to the cold one. In relation (7.3) λ is the *thermal conductivity* and it is expressed in J/(m·s·K), which is equates to kg·m/(s^3·K). Similar to the other transport coefficients, λ depends of state parameters, $\lambda = \lambda\left(P, T, \vec{X}\right)$.

Thermal conductivity is much lower for gases than for condensed states of aggregation. However, considering heat transmission by convection, for low viscosity fluids, the best industrial insulators are high porosity solids, such as glass wool (in this case air is the actual insulator, the wool fibers reducing the possibility of mixing currents occurrence). For high quality thermal insulation, the optimum solution is represented by vacuum (e.g., in Dewar vessels).

Fourier equation (7.3) integration is difficult since dT/dz, λ and heat capacity C (entering the left member of equation, through heat) depend all of the temperature T.

Stationary Heat Transport and the Isolated System Transport

The two simple cases of *heat transport* are the stationary transport and the non-stationary one into isolated adiabatic system.

Stationary transport is characterized by temperature field invariance in time: $dT/dz = 0$. In this case, the heat transferred during time τ through a section of a certain area s perpendicular to direction z is:

$$q = \lambda \cdot s \cdot \tau \cdot (dT/dz) \tag{7.4}$$

since $q = j_q \cdot s$, and $dq/dt = q/\tau$.

The transport *into an isolated adiabatic system* is non-stationary, so that temperature T depends on both space and time, according to *Fourier-Kirchhoff* equation:

$$c \cdot \rho \cdot \frac{\partial T}{\partial t} = \frac{\partial}{\partial x} \cdot \left(\lambda \cdot \frac{\partial T}{\partial x}\right) + \frac{\partial}{\partial y} \cdot \left(\lambda \cdot \frac{\partial T}{\partial x}\right) + \frac{\partial}{\partial z} \cdot \left(\lambda \cdot \frac{\partial T}{\partial z}\right) \tag{7.5a}$$

obtained by combining Fourier's law with an energy balance. If there are heat "sources" or "wells" (corresponding, e.g., to some exothermic, respectively, endothermic chemical reactions), a positive (source) or negative (well) flow is added to member II of equation (7.5a). For unidirectionally heat transport, the Fourier-Kirchhoff equation becomes:

$$c \cdot \rho \cdot (\partial T/\partial t) = [\lambda/(c \cdot \rho)] \cdot \left(\partial^2 T/\partial z^2\right)/\partial z \tag{7.5b}$$

where c is the specific heat, and if λ does not depend on T the following is obtained:

Fig. 7.2 Diffusional flow direction

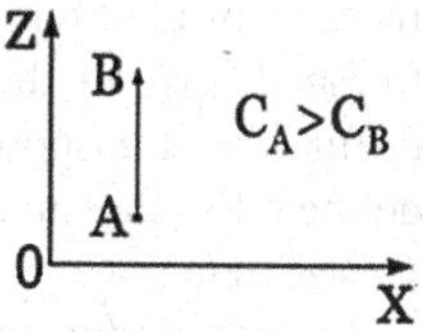

$$c \cdot \rho \cdot (\partial T/\partial t) = [\lambda/(c \cdot \rho)] \cdot \left(\partial^2 T/\partial z^2\right)]/\partial z \qquad (7.5c)$$

Mass Transport

Diffusion is molecules displacement in relation to each other. If molecules belong to the same species, the phenomenon is called *self-diffusion*. Due to molecules motions, two different substances in contact intertwine, they "diffuse" to one another, and a polycomponent system composition is equalized, this phenomenon being referred to as *interdiffusion*.

Transport by diffusion has *diffusional flow density* as measure, defined as $j_d \equiv -s^{-1} \cdot (dn/dt)$ and representing the substance quantity (expressed in moles) transported in the unit of time per unit area.

Interdiffusion's driving force (its potential) is the *concentration gradient dC/dz*, where C is expressed in mol/m^3 is the diffused substance concentration.

According to the first *Fick's law* (1855), j_d and dC/dz are proportional and—as seen from Fig. 7.2—of inverse sign:

$$j_d = -D \cdot dC/dz \qquad (7.6)$$

where D, expressed in m^2/s, is referred to as *diffusivity* or *diffusion coefficient.*

In Eq. (7.6), it is possible to pass from substance quantity expressed in moles to molecules number, by expressing j_d in $m^{-2} \cdot s^{-1}$ as the molecules number crossing the unitary surface in the unit of time and by replacing concentration C with *molecular number density*—the molecules number from the unit of volume.

Diffusivity

In Fick's first law, D does not depend on concentration gradient, but it is a state quantity, $D = D\left(P, T, \vec{X}\right)$ and it depends very much on the state of aggregation, decreasing with a few orders of magnitude when passing from gases to liquids or from them to solids. In the case of crystalline solids, diffusion is generally anisotropic, diffusivity depending on direction (as well as thermal conductivity λ) so that diffusional transport is characterized by *three* principal diffusion coefficients, defined for the main crystallographic axes (only two for the higher symmetry crystallographic systems and one for crystals from the cubic system).

For the unidimensional stationary diffusion, relation (7.6) becomes:

$$n/(s \cdot t) = D \cdot (dC/dz)$$

where n is the substance quantity transported through surface s (perpendicular on direction of transport z) in time t.

The Second Fick's Law Closed system material balance is:

$$\partial c/\partial t = \partial(D \cdot \partial c/\partial x)/\partial x + \partial(D \cdot \partial c/\partial y)/\partial y + \partial(D \cdot \partial c/\partial z)/\partial z$$

or for diffusion over one direction:

$$\partial c/\partial t = \partial[D \cdot (\partial c/\partial z)]/\partial z \tag{7.7a}$$

The reaction velocity (substance intake velocity) is added and, respectively, subtracted from the second member reaction rate for substance "sources" and "wells" (corresponding to reactions or a mass change with the exterior). When diffusivity is independent of composition, *Fick's second law* is obtained:

$$\partial c/\partial t = D \cdot \left(\partial^2 c/\partial z^2\right) \tag{7.7b}$$

Interdiffusion

The mass transport relations outlined here are valid in their simple form only for self-diffusion (e.g., for isotopic composition homogenizing of a pure substance). *Interdiffusion* is more practically important. In homogeneous media with several components, distinguished by i (i = 1, 2, ..., C, where C is the number of components) each component will have an *actual diffusion coefficient* "within mixture" or "in solution", D'_i, and first Fick's law becomes the following system:

$$\forall_i \in [1, 2, \ldots C]: \; j_{d,i} = -D'_i \cdot (dC_i/dz) \tag{7.7c}$$

In a binary system, therefore, two types of equations exist (7.7c). At perfect binary ideal gas mixture, the two concentrations sum is constant, $c_1 + c_2 = (n_1 + n_2)/v = R \cdot T/P$, so that, differentiating after direction z:

$$d(C_1 + C_2)/dz = 0$$

Through any fixed section of the gaseous system, the total flow density must be null (if more molecules pass from left to right than in the opposite direction, the total gas amount increases in the right compartment, which would lead to a mechanical disequilibrium—lowering the left pressure and increasing it in the right part of the separation surface): $j_{d1} + j_{d2} = 0$ or, by replacing with indices 1 and 2, respectively, into relation (7.7c) and by summing:

$$D'_1 \cdot dc_1/dz + D'_2 \cdot dc_2/dz = 0$$

where after replacing $d\, c_2/d\, z$ from relation ($d(C_1 + C_2)/dz = 0$), the common value $D_{1,2}$ is obtained, which is referred to as *interdiffusion coefficient* of gases 1 and 2 in their mixture:

$$(C_1 + C_2) = \text{const} \rightarrow \{(D'_1 = D_{12}) \cap (D'_2 = D_{12})\} \tag{7.7d}$$

Transport Phenomena Similarity

By comparing the viscous flow, heat, and diffusion conduction it can be observed that in any transport phenomenon an equation of the following form is valid:

$$j_g = -K_g \cdot grad\Psi_g \tag{7.8}$$

Transport potential, degree Ψ_g, is specific to quantity g. Ψ is:

- for viscous flow, *momentum* (and gradient direction is perpendicular to the flow direction).
- for diffusion, substance 1 *concentration* (expressed kg/m^3, mol/m^3 or molecule/m^3, as g is the mass, number of moles or molecule).
- for heat transfer, Ψ is *temperature.*

Transport phenomenological coefficient K_g is η, D or λ for momentum, mass, and heat transport, respectively.

The sense of law (7.8) consists in describing the proportionality between flow density and potential coupled with it: their ratio $K_g = j_g/grad\ \Psi_g$ is independent of variables value j_g and $grad\ \Psi_g$.

Transport flow density, j_g, is the *flow density* of an extensive conservative quantity (i.e., proportional with matter quantity from the system) g_{ext} representing the quantity of g crossing, in the unit of time, per unit area, of the reference plan:

$$j_g \equiv -s^{-1} \cdot dg/dt \tag{7.9}$$

The g quantity passing through the surface area s in time t is:

- momentum—for viscous flow (force F is momentum derivative in relation with time and the specific force is the ratio f = F/s)
- mass of a given substance—at diffusion (the mass is therefore proportional with molecules number, or with substance quantity) and
- heat—for thermal conduction phenomenon.

General Law of Transport

Newton, Fourier şi Fick (I) laws are particular cases of density flow and potential proportionality (7.9). In irreversible processes thermodynamics it is shown that it corresponds to entropy growth rate and the equilibrium deviation proportionality for the analyzed system.

The existence of a common transport law is the basis of several physical quantities study through simulation: direct flux measurement or a physical quantity variation in time and space can be replaced by measuring the flow or variation of a different nature physical quantity, for example, instead of composition distribution (often more difficult to be determined), it is possible to measure velocity or temperature distribution (macroscopic).

The most common procedure is simulating different transport phenomena through electric charge transport phenomenon. This since for electricity case it is an easy task to maintain constant the potential (V/cm) or the flow (A/cm) values, and also the measure of different quantities.

Transport phenomena phenomenological treatment cannot establish the *numerical* values of phenomenological coefficients η, D, and λ, nor their variation dependence on state parameters. Kinetic-molecular theory may, however, express the transport coefficients according to molecular quantities, pressure and temperature, as it will be seen in the following paragraphs.

In the case of moderate pressure gases, the calculation accuracy is satisfactory (5%), but for solids and liquids case, the errors can exceed 25%.

Fig. 7.3 Molecular transport geometry

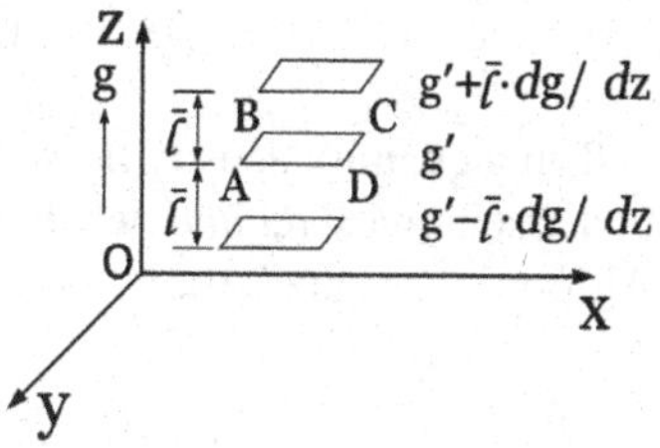

7.3 Transport in Perfect Gases

Free Path and Transport in Gases

In the kinetic-molecular theory, the μ, λ, and D coefficients molecular significance in the gas transport phenomena phenomenological equations can be found. Due to similarity, gas transport phenomena can be treated unitarily, with subsequent general formula customizations.

For a certain gas in which extensive physical quantity g is transported, whose values increase in Oz direction, so that a positive gradient of the intensive corresponding quantity G is maintained along the Oz axis, $dG/dz > 0$.

By considering as in Fig. 7.3 three thin gas layers, parallel to each other and to xOy plane found at a distance equal to the mean free path $\overline{\ell}$, so that it can be admitted that molecules leaving a layer are not deviated in the space between layers, but they reach the next layer; here they collide, yielding every g quantity that they transport.

Each molecule from the middle plan (ABCD surface) is characterized by G' value of the g quantity per molecule. Due to the increase of the G quantity in the Oz direction, each molecule in the upper layer (located at a distance l) will be characterized by the quantity $G' + 1 \cdot (dG'/dz)$, and each molecule in the lower layer – by the quantity: $G' - 1 \cdot (dG'/dz)$. The transported quantity's balance when a molecule passes through the ABCD layer, while moving from the upper layer to the lower one, is:

$$\Delta G' = [G' - l \cdot (dG'/dz)] - G' + l \cdot (dG'/dz) = -2 \cdot l \cdot (dG'/dz) \qquad (7.10a)$$

All directions are equiprobable, so that one third of the total molecules number can be displaced in Oz direction, while half of them move in the positive direction. The area layer s will be crossed by a number of $s \cdot \mu/6$ molecules, of medium velocity $\overline{v}$ and molecular density μ (expressed in molecules/m^3). Through the layer will pass a number of N_m molecules in time Δt, given by the following relation:

$$N_m = s \cdot \mu \cdot \overline{v} \cdot \Delta t/6 \qquad (7.10b)$$

The transported quantity from quantity g is equal to $\Delta g = \Delta G' \cdot N_m$ – the product between one molecule transported quantity ΔG' and the total molecules number crossing surface N_m from relations (7.10a) and (7.10b), respectively, as follows:

$$\Delta g = -(\mu \cdot \overline{v} \cdot \overline{\ell} \cdot s \cdot \Delta t/3) \cdot (dG'/dz) \tag{7.10c}$$

But g quantity density flow, defined by $j_g = -s^{-1} \cdot dg/dt$—relation (7.9)—becomes under stationary regime, where $(dg/dt) = \Delta g/\Delta t$: $j_g = -\Delta g/(s \cdot \Delta t)$ or, after replacing Δg from relation (7.10c):

$$j_g = -(1/3) \cdot \mu \cdot \overline{v} \cdot \overline{\ell} \cdot (dG'/dz) \tag{7.11a}$$

which represents the *general relation* between density flow and potential for *transport phenomena in ideal gases*.

General Equation of Transport in Perfect Gases

The previous deduction has been simplified; in reality, molecules move in any direction, with velocities and free paths differing from one molecule to another, and $(\mu, \overline{v}, \overline{\ell})$ may differ from one layer to another. For example, it is possible to replace the number of N_m crossing molecules from relation (7.10b) with the following relation: $N_m = Z_s \cdot s \cdot \Delta t$, according to Sect. 3.5, where it was been shown that Z_s, that is, the number of gas molecules collisions per unit surface in the time unit is $Z_s = \mu \cdot \overline{v}/4$; it is found, therefore, that $N_m = s \cdot (\mu/4) \cdot \overline{v} \cdot \Delta t$, leading to a numerical coefficient of 1/2, which is 50% higher than the previous one, from relation (7.11a), the correct form being therefore, generally, the following:

$$j_g = -K \cdot \mu \cdot \overline{v} \cdot \overline{\ell} \cdot (dG'/dz) \tag{7.11b}$$

where K is the dimensionless <u>*transport factor*</u>, specific to the transported quantity – momentum, energy, mass with the respective indices $g \in [\eta, \lambda, D]$:

$$j_g = -K_g \cdot \Phi \cdot (dG'/dz) \tag{7.11c}$$

Φ from relation (7.11d), expressed in $m^{-1} \cdot s^{-1}$, designates the following product:

$$\Phi \equiv \mu \cdot \overline{v} \cdot \overline{\ell} \tag{7.11d}$$

where the following expressions are introduced from relations (3.16c), (3.12b), and (3.33c), respectively:

$\mu = N_A \cdot P/(R \cdot T)$, $\overline{v} = \sqrt{8 \cdot k_B \cdot T/(\pi \cdot m')}$, and $\overline{\ell} = R \cdot T/(\sqrt{2} \cdot \pi \cdot N_A \cdot P \cdot \delta^2)$.

It results in: $\Phi = [2 \cdot (k_B \cdot T)^{1/2}/[\delta^2 \cdot (m' \cdot \pi^3)^{1/2}]$ or, passing from microscopic constants to the macroscopic ones through $k_B = R/N_A$ and $m' = M/N_A$:

$$\Phi = 2 \cdot (R \cdot T/M)^{1/2} / \left(\delta^2 \cdot \pi^{3/2}\right) \tag{7.11e}$$

Intermediate variable Φ elimination between relations (7.11c) and (7.11e) leads to:

$$j_g = -2 \cdot K_g \cdot \left[R \cdot T/(\pi^3 \cdot M)\right]^{1/2} \cdot \delta^{-2} \cdot (dG'/dz) \quad (7.12)$$

Equation (7.12) is similar to the general phenomenological equation (7.8) of transport phenomena so that, by comparing the terms of these two relations—customized for viscous flow, thermal conduction, and diffusion—the expressions of phenomenological transport coefficients (η, λ, D) as pressure, temperature, and molecular functions can be deducted.

Viscosity of Gases

The transported quantity in this case is momentum $m \cdot v$, where v is the (macroscopic) velocity of displacement in a direction perpendicular to gradient z direction; thus $g = m \cdot v$, and $G' = m' \cdot v$, where m′ is molecule mass $m' = M/N_A$, from where $G' = (M/N_A) \cdot v$. Relation (7.12) is in this case:

$$j_{m \cdot v} = -2 \cdot K_\eta \cdot \sqrt{M \cdot R \cdot T/\pi^3} \cdot \left[N_A \cdot \delta^2\right]^{-1} \cdot dv/dz \quad (7.13a)$$

Numerical coefficient exact calculation (Cooling & Chapman) leads to:

$$K_\eta = 5 \cdot \pi/32 \cong 0.4908 \quad (7.13b)$$

and density flow definition (7.9) becomes $j_{m,v} = s^{-1} \cdot d(m \cdot v)/dt$. But force F is the time-derived momentum, and making the ratio of force to surface gives the specific force f per unit area:

$d(m \cdot v)/dt = F$; $F/s = f$; thus $j_{m \cdot v} = f$. Expression (7.13a), therefore, becomes $f = (M \cdot R \cdot T/\pi)^{1/2} \cdot [3.2 \cdot N_A \cdot \delta^2]^{-1} \cdot (dv/dz)$.

By comparing to relation (7.1), $f = -\eta \cdot dv/dz$, it is found that:

$$\eta = (M \cdot R \cdot T/\pi)^{1/2} / (3.2 \cdot N_A \cdot \delta^2) \quad (7.13c)$$

The molecular diameter is obtained from this relation through viscosity measurements.

Gas Viscosity Dependence on Different Factors

Gas nature has a poor influence on viscosity: it is found, therefore, the following approximation:

$$\delta \sim M^{1/5} \quad (7.14)$$

so that $\eta \sim M^{0.1}$.

Intuitively, when pressure increases, leading to intermolecular distances reduction, the internal friction for gas flow should also increase. However, pressure does not influence at all viscosity. Confirmation by experience of this surprising conclusion has been a strong argument in favor of accepting the kinetic-molecular theory.

Temperature is positively correlated to viscosity. Experimentally, viscosity increase is becoming less pronounced as temperature rises. Thus, at low temperatures $\eta \sim T^{0.5}$ according to relation (7.14), while at sufficiently elevated temperatures it is observed that $\eta \sim T^{0.15}$.

Thermal Conductivity in Gases

Molecule internal energy U' is the molecular quantity equivalent to heat q. But the volume between the three layers of Fig. 7.3 is invariable, and from phenomenological thermodynamics it is known that at constant volume the heat is equal to extensive internal energy u. Relation (7.11c), $j_g = -K_g \cdot \Phi \cdot (dG'/dz)$, therefore becomes, by customizing K_g in K_λ:

$$j_q = -K_\lambda \cdot \Phi \cdot (dU'/dz) \tag{7.15a}$$

Derivative dU' from expression (7.15a) can be replaced by $dU' = (C_v/N_A) \cdot dT$ where C_v, the isochoric molar heat capacity, does not depend on layer altitude z but only on temperature T. Thus, $j_q = -(K_\lambda \cdot C_V \cdot \Phi/N_A) \cdot (dT/dz)$ from where, by comparing to relation (7.3), $j_q = -\lambda \cdot dT/dz$, it results that:

$$\lambda = K_\lambda \cdot C_v \cdot (\Phi/N_A) \tag{7.15b}$$

K_λ takes different values for the different heat capacity components, as shown by Eucken. Relation (7.15b) must be therefore replaced with:

$$\lambda = (K_{\lambda,tr} \cdot C_{v,tr} + K_{\lambda,int} \cdot C_{v,int}) \cdot \Phi/N_A \tag{7.15c}$$

where index *tr* designates the translational motion, while index *int* corresponds to the rest of the sum-over-states—the "internal" motions—rotational, vibrational and electronic ones. $C_{v,tr} = 1.5 \cdot R$ from relation (2.28c), therefore: $C_{v,int} = C_v - 1.5 \cdot R$.

But $C_v = C_p - R$ according to Robert Meyer relation from the perfect gas thermodynamics, and the Eucken's coefficients are: $K_{\lambda,tr} = 25 \cdot \pi/64$, $K_{\lambda,int}$ Φ $5 \cdot \pi/32$, relation (7.15c) thus becoming: $\lambda = (C_p + 1.25 \cdot R) \cdot \pi \cdot \Phi/(6.4 \cdot N_A)$ or, with $\Phi = 2 \cdot (R \cdot T/M)^{1/2}/(\delta^2 \cdot \pi^{3/2})$ from Eq. (7.11e):

$$\lambda = (R \cdot T/M)^{1/2} \cdot (C_p + 1.25 \cdot R)/\left(3.2 \cdot \pi^{1/2} \cdot N_A \cdot \delta^2\right) \tag{7.16}$$

From relation (7.16) for perfect gas case, it is concluded that:

- λ does not depend on pressure (since C_p does not depend on pressure).
- λ increases with temperature increase. Conductivity coefficient increase with temperature is more rapid than for viscosity, since heat capacity increases with temperature.
- λ decreases with molecular mass increase: since $\delta \sim M^{1/5}$ according to relation (7.14), it can be concluded that $\lambda \sim M^{-0.9}$.

Diffusion

The density flow from relation (7.9), $j_g \equiv -(dg/dt)/s$ becomes, for the i species component transferred mass m_i: $j_{ni} = -(dm_i/dt)/s$ from where, when passing from mass to transferred molecules number N_i:

$$J_{N,i} = -(dN_i/dt)/s \tag{7.3a2}$$

In relation (7.11c), $j_g = -K_g \cdot \Phi \cdot dG'/dz$, j_g is replaced by $J_{N,i}$ from equality (a), coefficient K_g being customized at K_D for diffusion, while quantity G′ found within the gradient represents the type I molecules number N_i, that is, molecules that contribute to the motion, reported to the total molecules number, N: $dN_i/dt)/s = -K_D \cdot \Phi \cdot [d(N_i/N)/dz)]$ or, after replacing N_i by $N_A \cdot n_i$ and N by $N_A \cdot n$, where n_i is I substance quantity (in moles), and n – the matter total quantity: $s^{-1} \cdot dn_i/dt = -(K_D \cdot \phi/N_A) \cdot [d(n_i/n)/dz)]$.

By replacing in the right member of the expression, $v \cdot C_i$ (where $v = n \cdot R \cdot T/P$ is the total volume) instead of n_i leads to: $s^{-1} \cdot dn_i/dt = [K_D \cdot \Phi \cdot R \cdot T/(P \cdot N_A)] \cdot dc_i/dz$.

When comparing with the first Fick's law (7.6), $s^{-1} \cdot dn_i/dt = -D \cdot dC/dz$, it is found that:

$D = K_D \cdot \overline{v} \cdot \overline{\ell}$, where the following expressions are introduced from relations (3.12b) and (3.33c), respectively, that is, $\overline{v} = \sqrt{8 \cdot k_B \cdot T/(\pi \cdot m')}$ and $\overline{\ell} = R \cdot T/\left(\sqrt{2} \cdot \pi \cdot N_A \cdot P \cdot \delta^2\right)$.

The mechanic-statistical calculation determines that $K_D = 3 \cdot \pi/16$, so that the following is obtained:

$$D = 3 \cdot \left[R^3 \cdot T^3/(\pi \cdot M)\right]^{1/2} / \left(8 \cdot N_A \cdot \delta^2 \cdot P\right) \tag{7.17}$$

The order of magnitude of gas diffusivities is of 10^{-5} m²/s. Thus, for a typical gas, with $M = 0.03$ kg/mol, $\delta = 3.5 \cdot 10^{-10}$ m, under usual conditions ($T = 300$ K, $P \cong 10^5$ N/m²) from relation (7.17) resulting that: $D = 1.95 \cdot 10^{-5}$ m²/s.

It is noticed that gas diffusivity is inversely proportional to pressure, increases rapidly with temperature and decreases when molar mass increases: with $\delta_0 \sim M^{1/5}$ from relation (7.14), it results that $D \sim M^{0.9}$.

Molecular Diameter Dependence on Temperature

The *apparent* molecular diameter δ, qualified also as *kinetic* diameter, depends on temperature, according to Sutherland's (3.32a), $\delta = \delta_0 \cdot (1 + C_S/T)^{1/2}$, where the limit-diameter (at infinite temperature) δ_0 and Sutherland temperature C_S are basically independent of T.

Relations (7.11e, 7.12, 7.13c, 7.16, 7.17), therefore, become:

$$\Phi = 2 \cdot \left(R \cdot T^3/M\right)^{1/2} / \left[\pi^{3/2} \cdot (\delta_0)^2 \cdot (C_S + T)\right] \tag{7.18a}$$

$$j_g = -2 \cdot K_g \cdot \sqrt{R \cdot T^3/(M \cdot \pi^3)} \cdot \left[(\delta_0)^2 \cdot (C_s + T)\right]^{-1} \cdot dG'/dz \tag{7.18b}$$

$$\eta = \left(M \cdot R \cdot T^3\right)^{1/2} / \left[3.2 \cdot \pi^{1/2} \cdot N_A \cdot (\delta_0)^2 \cdot (C_S + T)\right] \tag{7.18c}$$

$$\lambda = \left(R \cdot T^3/M\right)^{1/2} \cdot (C_p + 1.25 \cdot R) / \left[3.2 \cdot \pi^{1/2} \cdot N_A \cdot (\delta_0)^2 \cdot (C_S + T)\right] \tag{7.18d}$$

$$D = 3 \cdot \left[R^3 \cdot T^5/(\pi \cdot M)\right]^{1/2} / \left[8 \cdot N_A \cdot (\delta_0)^2 \cdot P \cdot (C_S + T)\right] \tag{7.18e}$$

The Sutherland procedure is a successful correction, yet empirical, to temperature dependence of transport coefficients resulting from the simpler mechanical-statistical model of the "vertical wall" type intermolecular potential from Fig. 4.10a.

More precise values of transport coefficients could be obtained from the attractive-repulsive molecular interactions models, as those presented in Fig. 4. 10b, models in which the intermolecular potential, the two molecules interaction potential, u (energy, with changed sign) depends on d—the distance between molecules centers.

Collision Integrals Calculation

The calculations have been completely effectuated only for Lennard-Jones intermolecular potential, $u(d) = 4 \cdot \varepsilon \cdot [(\sigma/d)^6 - (\sigma/d)^{12}]$ from relation (4.22b), where ε (maximum molecular energy of attraction) and σ (distance between molecules centers, at this maximum) are the two microscopic features, depending only on the considered substance nature, and not also on the physical conditions such as pressure and temperature. The result is the replacement:

$$\delta = \sigma \cdot \Omega_g(\tau) \tag{7.19a}$$

of *apperent* molecular diameter δ with parameter σ, considered the "true" molecular diameter, in relations (7.12, 7.13c, 7.16, 7.17), which, therefore, become the Curtiss-Hirschfelder relations

Table 7.1 Collisions integrals calculation

Formula		Relation
$\Omega_\eta(\tau) = \Omega_\lambda(\tau) = A/\tau^B + C/e^{D\cdot\tau} + E/e^{F\cdot\tau}$		(7.20b)
Where	A = 1.16145; B = 0.14174; C = 0.52487; D = 0.77320; E = 2.16178; F = 2.43787	
$\Omega_D(\tau) = A'/\tau^{B'} + C'/e^{D'\cdot\tau} + E'/e^{F'\cdot\tau} + G'/e^{H'\cdot\tau}$		(7.20c)
Where	A′ = 1.06036; B′ = 0.15610; C′ = 0.19300; D′ = 0.47635; E′ = 1.03587; F′ = 1.52996; G′ = 1.76474; H′ = 3.89411	

$$j_g = -2 \cdot K_g \cdot \left[R \cdot T/(\pi^3 \cdot M)\right]^{1/2} \cdot \left[\sigma^2 \cdot \Omega_g(\tau)\right]^{-1} \cdot (dG'/dz) \tag{7.19b}$$

$$\eta = (M \cdot R \cdot T/\pi)^{1/2}/\left[3.2 \cdot N_A \cdot \sigma^2 \cdot \Omega_\eta(\tau)\right] \tag{7.19c}$$

$$\lambda = [R \cdot T/(\pi \cdot M)]^{1/2} \cdot (C_p + 1.25 \cdot R)/\left[3.2 \cdot N_A \cdot \sigma^2 \cdot \Omega_\lambda(\tau)\right] \tag{7.19d}$$

$$D = 3 \cdot \left[R^3 \cdot T^3/(\pi \cdot M)\right]^{1/2}/\left[8 \cdot N_A \cdot P \cdot \sigma^2 \cdot \Omega_D(\tau)\right] \tag{7.19e}$$

Ω_g, with g ϵ (η, λ, D), are the *Curtiss* collision integrals, dimensionless, depending only on the *Lennard-Jones reduced* temperature, τ, defined as:

$$\tau \equiv k_B \cdot T/\varepsilon \tag{7.20a}$$

dimensionless, proportional to actual temperature. Dependencies $\Omega_g(\tau)$ are slightly decreasing – they decrease approximately five times from 2.7 down to 0.5 when τ increases for more than a thousand times from 0.3 up to 400. The numerical approximation with 0.02% accuracy is possible with formulas from Table 7.1.

Dimensionless Transport Criteria

These criteria correlate the transport coefficients (η, λ, D) for a given media and under the same physical conditions, with molar mass M, density ρ and isobaric heat capacity C_p. The most commonly used criteria are Prandtl [Pr] and Schmidt [Sc] criteria, defined by:

$$[Pr] \equiv \eta \cdot C_p/(\lambda \cdot M) \tag{7.21a}$$

$$[Sc] \equiv \eta/(\rho \cdot D) \tag{7.21b}$$

where M, ρ, and C_p are mass, density, and isobaric heat capacity, respectively.

Table 7.2 Thermal conductivity calculation from viscosity at 25 °C

Gas	He	Ne	Kr	Xe	H_2	O_2
$\lambda/(\eta \cdot C_v)$	2.45	2.50	2.54	2.58	2.00	1.92
$1+9 \cdot R/(4 \cdot C_v)$	2.50	2.50	2.50	2.50	1.92	1.89
Gas	CO	Aer	CO_2	CH_4	C_3H_8	
$\lambda/(\eta \cdot C_v)$	1.81	1.96	1.64	1.74	1.66	
$1+9 \cdot R/(4 \cdot C_v)$	1.90	1.90	1.67	1.72	1.31	

Prandtl Criterion

From the perfect gas kinetic molecular theory, it is obtained for [Pr], defined by relation (7.21a), from expressions (7.13c) and (7.16) giving η and λ, the following expression:

$$[\mathrm{Pr}] \equiv C_p/(C_p + 1.25 \cdot R) \quad (7.22a)$$

or, by replacing the isobaric heat capacity C_p with $C_v = C_p - R$:

$$\lambda/(\eta \cdot C_v) = 1 + 9 \cdot R/(4 \cdot C_v) \quad (7.22b)$$

Relations (7.22) allow the avoidance of difficult and inaccurate determinations of thermal conductivity in gases. This is made by their replacement with specific viscosity and heat data that is measured with an accuracy higher than 5% (see Table 7.2). However, the more complex molecules are excepted: e.g. for C_3H_8 the accuracy is of 20%.

Schmidt Criterion

The diffusion coefficients, experimentally determined by mass flow measurement (in the self-diffusion case, by recording a radioactive isotope activity), but usually D can be obtained from more easily accessible viscosity measurements, using the Schmidt criterion from relation (7.21b) $[\mathrm{Sc}] \equiv \eta/(\rho \cdot D)$, whose value is $5/6 \cong 0.83$ for any gas, at any (P, T) values, as it results from the comparison of relations (7.13c) and (7.17).

In Table 7.3, the experimental values of viscosity, thermal conductivity, and diffusivity for a series of gases are presented, under what it is appropriate to designate as *normal thermomechanical* conditions (term which is often abbreviated by "*normal conditions*"): 0 °C and 1 atm, together with calculated density values under these conditions from the perfect gas molar volume.

From Table 7.3 the following can be observed:

- D varies with gaseous nature less than with density, but more than with λ and especially than with η. By excluding the two quantum gases (H_2, He) with very

Table 7.3 Transport coefficients of gases under "normal"conditions

Gas	He	Xe	H_2	O_2	Cl_2	HCl	H_2O*	CO_2	CH_3OH	C_3H_8
$10^8 \cdot \eta$	1885	2107	835	1911	1227	1313	861	1375	865	746
$10^4 \cdot \lambda$	1435	50	1753	243	80	129	153	142	151	141
$10^7 \cdot D$	1386	47.6	1604	192	58.1	121	276	106	109	57.5
$10^3 \cdot \rho$	178.6	5858	89.9	1428	3164	1627	804	1964	1430	1967
$10^3 \cdot [Sc]$	762	756	579	697	668	667	387	659	555	659

low molecular mass, the ratio between the maximum and minimum value is of (2.5:3:5:6) for η, λ, D, respectively, ρ (or M).

- With the exception of polar gases (H_2O, CH_3OH, HCl) or of quantum ones, the [Sc] criterion value remains sensitively constant, namely—with mean deviations of 5%—[Sc] = 0.71, a value 15% lower than the theoretical one.

7.4 Transport in Mixtures of Perfect Gases

Wilke's relations express the mixture viscosity η or thermal conductivity λ, depending on pure components viscosities η_i and thermal conductivities λ_i at the same temperature and pressure (i = 1, 2, . . ., C):

$$\eta = \sum_{i=1}^{C} \left(X_i \cdot \eta_i / \sum_{j=1}^{C} X_j \cdot r_{i,j} \right); \tag{7.23a}$$

$$\lambda = \sum_{i=1}^{C} \left(X_i \cdot \lambda_i / \sum_{j=1}^{C} X_j \cdot r_{i,j} \right) \tag{7.23b}$$

where X_i is component I molar fraction and $r_{i,j}$ is a dimensionless dependency on components I and j viscosities and molar masses:

$$r_{ij} = \left[1 + (\eta_i/\eta_j)^{1/2} / (M_i/M_j)^{1/4} \right]^2 / \left[8 \cdot (1 + M_i/M_j) \right]^{1/2} \tag{7.23c}$$

If indices i and j are identical, it results, as can be observed, that $r_{i,i} = 1$.

In the case of mass transport in gas mixtures, the situation is more complex, since generally it is not advantageous to express D_{ij} diffusion coefficients from a mixture—even binary one—depending on the self-diffusion coefficients D_i, which are difficult to measure.

Interdiffusion in Binary Gas Mixtures

As seen from the relation (7.7d), the D_{ij} and D_{ji} interdiffusion coefficients are equal. If diffusional flow calculation considers that only different molecular collisions show a diffusion effect (identical type molecules collisions do not contribute to concentrations equalization), for the interdiffusion coefficient in a binary mixture, the following expression is found:

$$D_{12} = K_D \cdot (\overline{v}_{r12})^2 \cdot x_2/z_{1,2} = K_D \cdot (\overline{v}_{r21})^2 \cdot x_1/z_{2,1} \tag{7.24a}$$

where $\overline{v}_{rij}$ is type i molecules mean relative velocity toward type j molecules, and $z_{i,j}$—type i molecule collisions number with type j molecules in the unit of time. When replacing in relation (7.24a): $\overline{v}_{r12} = \overline{v}_{r21} = \sqrt{(\overline{v}_1)^2 + (\overline{v}_2)^2}$ and $z_{i,j} = \pi \cdot (\delta_{ij})^2 \cdot \mu_j \cdot \sqrt{(\overline{v}_1)^2 + (\overline{v}_2)^2}$, it results that:

$$D_{12} = K_D \cdot x_1 \cdot \sqrt{(\overline{v}_1)^2 + (\overline{v}_2)^2} / \left[\pi \cdot \mu_1 \cdot (\delta_{12})^2\right] \tag{7.24b}$$

But $\overline{v}_i = \sqrt{8 \cdot R \cdot T/(\pi \cdot M_i)}$, $\mu_1 = X_1 \cdot \mu$, and $\mu = N_A \cdot P/(R \cdot T)$, from where:

$$D_{12} = K_D \cdot \sqrt{(2 \cdot R \cdot T/\pi)^3 \cdot (1/M_1 + 1/M_2)} / \left[N_A \cdot P \cdot (\delta_{12})^2\right] \tag{7.24c}$$

and after replacing δ_{12} by apparent kinetic diameters arithmetic mean and with $K_D = 3 \cdot \pi/16$ as in Sect. 7.3, it is found that:

$$D_{12} = 3 \cdot \sqrt{(R \cdot T)^3 \cdot (1/M_1 + 1/M_2)/(2 \cdot \pi)} / \left[N_A \cdot P \cdot (\delta_1 + \delta_2)^2\right] \tag{7.25a}$$

with δ_i from dependency (5.57b) on liquid molar volume V_{Lf} at boiling temperature $\delta = (0.955 \cdot V_{Lf}/N_A)^{1/3}$ the *Gilliland* correlation results:

$$D_{12} = 1.234 \cdot \sqrt{(R \cdot T)^3 \cdot (1/M_1 + 1/M_2)} / \left[P \cdot \sqrt[3]{N_A} \cdot \left(\sqrt[3]{V_{Lf1}} + \sqrt[3]{V_{Lf2}}\right)^2\right] \tag{7.25b}$$

and with Sutherland dependency (3.33b) $\delta = \delta_0 \cdot (1 + C_s/T)$ of the apparent diameter on temperature, *Arnold* relation is obtained, which describes somehow better interdiffusion coefficient dependence on temperature:

$$D_{12} \cong 1.234 \cdot \left[T^5 \cdot (1/M_1 + 1/M_2)\right]^{1/2} / A_D \tag{7.25c}$$

$$A_D \equiv P \cdot \sqrt[3]{N_A} \cdot \left(\sqrt[3]{V_{Lf1}} + \sqrt[3]{V_{Lf2}}\right)^2 \cdot (T + C_S) \quad (7.25d)$$

Sutherland constant for mixture, denoted by C_{S12}, is pure components constants geometric mean, C_{S1} and C_{S2}, obtained from relation (3.32b), $C_{S1} = 1.47 \cdot T_{f,i}$, from the boiling temperatures:

$$C_{S12} = (C_{S1} \cdot C_{S2})^{1/2} \quad (7.25e)$$

Interdiffusion and Self-diffusion

Relations (7.25) are suitable for self-diffusion coefficients calculation, if $D_{12} = D$, $\delta_1 = \delta_2 = \delta$, $M_1 = M_2 = M$, $C_{S1} = C_{S2} = C_S$, $V_{f1} = V_{f2} = V_f$:

$$D = 3 \cdot \sqrt{(R \cdot T)^3} / \left(4 \cdot N_A \cdot P \cdot \delta^2 \cdot \sqrt{\pi \cdot M}\right) \quad (7.26a)$$

$$D = 0.436 \cdot \sqrt{(R \cdot T)^3 / M} / \left[P \cdot \sqrt[3]{N_A} \cdot (V_{Lf})^2\right] \quad (7.26b)$$

$$D = 0.436 \cdot \sqrt{T^5 / M} / \left[P \cdot \sqrt[3]{N_A} \cdot (V_{Lf})^{2/3} \cdot (T + C_S)\right] \quad (7.26c)$$

Conversely, the interdiffusion coefficient can be expressed in relation to self-diffusion coefficients D_i at the same pressure and temperature, if σ_i and V_{Lf} from relations (7.26a) and (7.26b) are replaced in relations (7.25a) and (7.25b):

$$D_{12} = \sqrt{8 \cdot (1/M_1 + 1/M_2)} / \left[1/\left(\sqrt{D_1} \cdot \sqrt[4]{M_1}\right) + 1/\left(\sqrt{D_2} \cdot \sqrt[4]{M_2}\right)\right]^2 \quad (7.26d)$$

Relation (7.26d) is used when self-diffusion data are available.

Diffusion in Mixtures with More than Two Components

While in binary mixtures mass flows can be completely characterized by a single diffusion coefficient, D_{12}, in mixtures with C components ($C > 2$), according to irreversible processes thermodynamics there exists a number of $C \cdot (C - 1)/2$ independent diffusion coefficients at $i \neq j$.

Diffusion streams can be approximated by a Wilke method, which introduces component i diffusivities *against the mixture*, D_i', defined by:

$$j'_{D,i} = -D'_i \cdot dc_i / dz \quad (7.27)$$

where mass transport density flow does not refer to an *immobile* reference plan, as in (7.6) $j_{D,i} = -D_i \cdot dc_i/dz$, but to a reference plan *joint with* the other components mixture *center of gravity*.

Diffusivities D'_i coincide at the binary mixture with the usual ones, $D'_1 = D'_2 = D_{12}$ as in relation (7.7d), but at mixtures with $C > 2$:

$$D'_i = (1 - X_i)/\left[\sum_{j=1}^{j=C} (X_j/D_{ij})\right], \forall j \notin i \tag{7.28a}$$

For example, for component i diffusion toward the other components of a ternary mixture, relation (7.27b) becomes:

$$D'_1 = (X_2 + X_3)/(X_2/D_{12} + X_3/D_{13}) \tag{7.28b}$$

For coefficients D'_1 calculation using relations (7.28), D_{ij} coefficients are used, which are obtained from relations (7.25b) and (7.25c) or (7.26b) and (7.26c).

7.5 Pressure Effect on Transport in Gases

The above-deducted transport laws validity depends on two conditions:

1. The negligible ratio of molecule diameter and L_f—the *physical* length, equal to the mean distance between the centers of two nearby molecules:

$$L_f >> \sigma \tag{7.29a}$$

Length L_f is assimilable to a cube side in which there is one single molecule, in average scale. Since the space occupied by one gas mole is R·T/P, and the number of molecules inside it is N_A, volume $(L_f)^3$ of the respective cube will be $(R \cdot T/P)/N_A$; or, since $R = N_A \cdot k_B$:

$$(L_f)^3 = k_B \cdot T/P \tag{7.29b}$$

At high pressures where condition (7.29a) is not fulfilled, molecules are subjected much of the time to intermolecular interaction forces: perfect gases law, Maxwell–Boltzmann distributions of molecular velocities and energies, sum-over-states expression for translational motion (2.25), etc. are no longer valid, the ideal gas becoming a real gas—furtherly treated in Sect. 7.8.

2. The mean free path from relation (3.33c) between intermolecular collisions, $\bar{\ell} = R \cdot T/(\sqrt{2} \cdot \pi \cdot N_A \cdot P \cdot \sigma^2)$, to be negligible in relation to the mean path $\bar{\ell}_K$ from Sect. 3.8 between two collisions with the wall; this path, denoted here by L_g

as a *geometrical* length, is $L_g = 2 \cdot d_{car}/3$, where d_{car} is the harmonic mean of this volume dimensions in the three main directions:

$$L_g \equiv 2/(1/d_1 + 1/d_2 + 1/d_3) \tag{7.29c}$$

These dimensions are being denoted by $d_1 \leq d_2 \leq d_3$. For example,

- for two infinite plates at distance a: $L_g = 2/(1/a + 2/\infty)$: $L_g = 2 \cdot a$
- for an infinite cylinder of diameter, a: $L_g = 2/(2/a + 1/\infty)$: $L_g = a$
- for gas passage between two compartments separated by a hole of dimensions (a_1, a_2): $L_g = 2/(1/a_1 + 1/a_2)$

Relations for transport in perfect gases from (Sect. 7.3) are valid only if $L_g >> \bar{\ell}$. Since $\bar{\ell} = k_B \cdot T/\left(\sqrt{2} \cdot \pi \cdot P \cdot \sigma^2\right)$, and when according to relation (7.29b) the following notation is introduced: $P = k_B \cdot T/(L_f)^3$, the condition becomes:

$$L_g \cdot \sigma^2/(L_f)^3 >> 1/\left(2^{1/2} \cdot \pi\right) \tag{7.29d}$$

When condition (7.29d) is not fulfilled, the gas is in the Knudsen regime described in Sect. 3.8, where the relations in Sect. 7.3 can no longer be applied for expressing η, λ, D in relation to gas features.

Conditions (7.29a) and (7.29d) are numerically concretized by replacing the approximate expression A *much smaller than* B or "A $\ll$ B" through inequality"$A \leq K \cdot B$", where K is a dimensionless variable parameter, $K > 1$:

$$K_s \leq L_f/\sigma \tag{7.30a}$$

and $K_q/(2^{1/2} \cdot \pi) \leq L_g \cdot \sigma^2/(L_f)^3$ or, for an explicit L_g/σ ratio:

$$(L_f/\sigma)^3 \leq \left(2^{1/2} \cdot \pi/K_q\right) \cdot (L_g/\sigma) \tag{7.30b}$$

These equalities are linearized by change of variables:

$$X \equiv lg(L_g/\sigma); \tag{7.31a}$$

$$Y \equiv lg(L_f/\sigma) \tag{7.31b}$$

Transport Regimes Applicability

The validity of transport coefficients expressions from this paragraph is synthesized in Fig. 7.4, where s′ and q′ lines delimit, for all physical conditions, a number of four fields, conventionally designated by: A, moderate pressures field; B, low pressures field; C, high pressures field and D, micropores gas field.

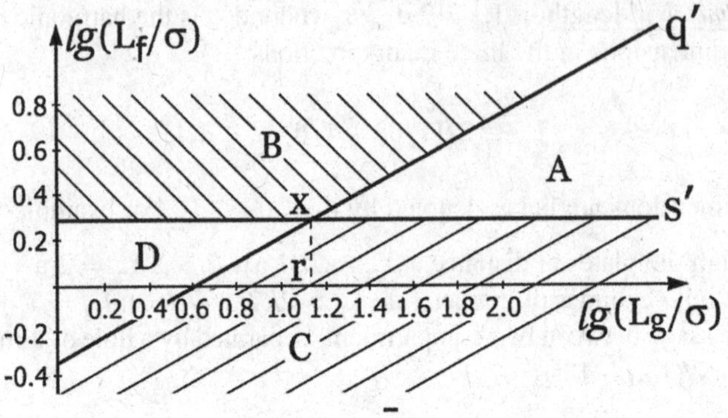

Fig. 7.4 Transport models practice

Values $K_s = \sqrt[6]{40}$ and $K_q = 10$ have been adopted for inequalities (7.30), leading to the following equations of s′ and q′ lines from Fig. 7.4:

$$Y = (1/6) \cdot lg(40); Y = (1/3) \cdot \left[X + lg\left(2^{1/2} \cdot \pi/10\right)\right] \text{ or} \tag{7.31c}$$

$$Y \cong 0.267 \text{ (ss')}; Y \cong X/3 - 0.1174\text{(qq')} \tag{7.31d}$$

with intersection at x point, of coordinates (X = 1.15, Y = 0.27).

Only in the A field *all* laws are valid—both equilibrium and transport law—deductible so far for perfect and ideal gas. In the other fields, these laws are infringed, partially or totally, for reasons that may vary from one field to another. Therefore,

- for C field, statistical thermodynamics perfect gas laws are no longer valid. The models from Chap. 4 will no longer be valid, some changes being necessary—see Sect. 7.7.—which could reflect the real gas case of intermediate state of aggregation situated between the liquid and the perfect gas.
- for B field, the ideal gas thermodynamic—statistical laws remain valid, so that equilibrium phenomena from field B do not differ from those from moderate pressures A field. Instead, transport coefficients expressions as dependencies on T, P, and σ change (as will be shown below), as a result of replacing mean free path expression $\overline{\ell} = 1/\left(\sqrt{2} \cdot \pi \cdot \sigma^2 \cdot \mu\right)$ with the relation: $\overline{\ell} \cong L_g$.

for D field, both types of previous deviations occur.

Pressure Effect on Transport Regime

For a given physical system (L_g and σ are constant), under isothermal conditions, the gradual pressure increase is equivalent to a downward displacement ($L_f \sim P^{-1/3}$) of the representative point from Fig. 7.4, on the vertical line corresponding to the L_g/σ ratio. If this vertical line is located to the right side of the x-point, pressure increase involves a gradual transition from B to A field, then to C field, from where the three fields names ("low," "moderate," and "high" pressures).

By solving the system of the two inequalities (7.30) in which the pressure in relation (7.29b) $(L_f)^3 = k_B \cdot T/P$ was made explicit, the minimum and maximum pressure limits (P_{min}, P_{max}) are obtained, in the "moderate pressures" A field, where both ideal gas models—the thermodynamic one (Chap. 3) and the transport one (Sect. 7.3)—are valid:

$$P_{min} = k_B \cdot T \cdot K_q / \left[(L_g) \cdot \sigma^2 \cdot \pi \cdot \sqrt{2} \right] \leq P \leq P_{max} = k_B \cdot T / \cdot (\sigma \cdot K_s)^3 \quad (7.32)$$

If the limit-values do not meet the $P_{min} \leq P_{max}$ condition, the double inequality (7.32) cannot be fulfilled, the vertical line $X = const.$ is located in the left part of point x from Fig. 7.4, and when pressure drops, the gas passes from fields B to D and then to C, so that the deducted ideal gas transport laws are not practicable at any pressure. The vertical line $X = X_x$ of point x is found from relations (7.32) by equaling P_{min} with P_{max}. The limit value R_{lim} of R ratio is found, which is the ratio between the molecular diameter σ and the "geometrical" limit dimension $(L_g)_{lim}$ of the space containing the gas, defined by the relation (7.29c): $(\sigma/L_g)_{lim} = \pi \cdot \sqrt{2} / \left[K_q \cdot (K_s)^3 \right]$. For example, with the multiplicative parameters $K_s = K_q = 10$, for gases with diameters exceeding 0.5% of the pore size, it is found that gas is found in Knudsen field at any pressure. Otherwise, the A-range limits, calculated from relations (7.32) are, at 27 °C and at typical value of 0.35 nm of gas molecule diameter: $0.76/(L_g/\mu m) \leq [P/bar] \leq 0.97$.

7.6 Knudsen Transport Field

The *Knudsen* transport field includes not only the low-pressure field C from Fig. 7.4, but also the small-scale systems D field in which (as in tree shafts capillaries) the pressure is much higher.

Transport coefficients from C field, obtained as in Sect. 7.3, are:

$$\eta = K_\eta \cdot \rho \cdot \overline{v} \cdot \overline{\ell} \quad (7.33a)$$

$$\lambda = \rho \cdot \overline{v} \cdot \overline{\ell} \cdot (K_{\lambda 1} \cdot C_{v1} + K_{\lambda 2} \cdot C_{v2})/M \quad (7.33b)$$

$$D = K_D \cdot \overline{v} \cdot \overline{\ell} \quad (7.33c)$$

where K_η, $K_{\lambda 1}$, $K_{\lambda 2}$, K_D are numerical coefficients of the order of unity, M and ρ are the molar mass and density, $\overline{v}$ the average arithmetic speed, $\overline{\ell}$ the mean free path, and

C_{v1} and C_{v2} the internal and translational heat capacity components at constant volume. The following quantities

$$\begin{aligned} \rho &= M \cdot P/(R \cdot T), \\ \overline{v} &= \sqrt{8 \cdot R \cdot T/(\pi \cdot M)}, \\ C_{v1} &= C_p - 5 \cdot R/2 \text{ and} \\ C_{v2} &= 3 \cdot R/2 \end{aligned}$$

have, in the Knudsen field, the same expressions as the ones at moderate pressures.

However, the mean free path is different, being equal to the available space geometric feature L_g, calculated with relation (7.29c) from dimensions harmonic mean. Moreover, also the numerical coefficients K_η, K_λ, K_D slightly change.

Knudsen Viscosity Field

Relation (7.33a) becomes, in the Knudsen field: $\eta = K_\eta \cdot \rho \cdot \overline{v} \cdot L_g$ or, by expressing $(\rho, \overline{v})$ in relation to (M, P, T):

$$\eta = K_\eta \cdot L_g \cdot P \cdot \sqrt{8/(\pi \cdot R \cdot M \cdot T)} \qquad (7.34a)$$

From relation (7.34a) a dependency $\eta \sim P^1 \cdot (T \cdot M)^{-1/2}$ is observed, which is totally different from the one of moderate pressures field, where $\eta \sim P^0 \cdot (T \cdot M \cdot \sigma)^{1/2}$. For example, at 300 K the "normal" viscosities of H_2 and O_2 from Table 7.3 are found in the 1:2.3 ratio, while in the very low pressures field the two viscosities ratio is of 4:1 (inversely proportional to the quadratic mean of respective molecular masses ratio 2:32), so that ratio value increases 9 times when passing from the usual field to the Knudsen one.

During this displacement the numerical coefficient K_η also changes, which will depend on wall material nature, molecule nature and vessel geometry.

For example, at the flow between two parallel plates located at distance a, where $1/L_g = (1/a + 2/\infty)/2$ and $L_g = 2a$, the following is found:

$$K_\eta = \cdot 3 \cdot \pi/16 \cong 1.47 \qquad (7.34b)$$

which is 70% higher than the value $K_\eta = 5 \cdot \pi/32$, of "normal" viscous flow from Sect. 7.3.

When flowing through cylindrical pores of diameter a, to which $1/L_g = (2/a + 1/\infty)/2$ or $L_g = a$, it is found similarly that:

$$K_\eta = 3 \cdot \pi/64 \cong 0.0368 \qquad (7.34c)$$

Value K_η is 4 times lower than for the case of flowing between two parallel plates, showing its dependence on the available space form.

Law of Cosines

Relations (7.34) are valid only if the molecule adheres so strongly to the wall that it gives to it all its tangential momentum. Then, molecules leaving from the wall are distributed in random directions, which represents the law of cosines.

The probability density of molecules directions from a solid angle element $d\omega$ that makes the (plan) angle θ with wall surface normal line is proportional to: $\cos\theta$: $\text{prob}\omega[\theta \in (\theta_1, \theta_1 + d\theta)] \sim (\cos\theta_1)\cdot d\omega$, which corresponds to a distribution probability density after angle (since $d\omega = 2\cdot\pi\cdot\sin\alpha\cdot d\,\alpha$) proportional to the product $(\sin\alpha\cdot\cos\alpha)$:

$$\text{prob}\,[\theta \in (\theta_1, \theta_1 + d\theta)] \sim (\cos\theta_1)\cdot(\sin\theta_1)\cdot d\theta$$

or, by introducing the normalization factor $C = 1/\int_0^{\pi/2} \cos\theta \cdot \sin\theta \cdot d\theta = 2$:

$$\text{prob}\,[\theta \in (\theta_1, \theta_1 + d\theta)] = 2\cdot(\cos\theta_1)\cdot(\sin\theta_1)\cdot d\omega$$

By integrating θ_1 between values 0 and θ', range of values, it results the probability in Eq. 7.35a. It describes the probability to have a value inferior to θ' for the angle formed between the molecule direction and the normal perpendicular line to the wall.

$$\text{prob}(\theta < \theta') = (\sin\theta')^2 \tag{7.35a}$$

Mechanical Accommodation Coefficients

The law of the cosines is actually a limit law, respected only if molecule–wall interactions are particularly strong. Usually, molecule direction after collision depends somewhat on the direction it had before collision, so the probability of small angles normal line is greater than the one given by relation (7.35a). Consequently, the real free path is reduced as compared with L_g, a phenomenon that it is shown by multiplication of K_η, K_λ or K_D factors with a subunitary factor in the right member of relations (7.33)—the so-called mechanical accommodation coefficient, α_K:

$$K_G = K_{G,33}\cdot\alpha_K \tag{7.35b}$$

Table 7.4 Mechanical accommodation coefficients (low, 20 °C)

Gas	H_2	He	CH_4	N_2	Ar	CO_2	Kr	Hg
M, g/mol	2	4	16	28	40	44	84	201
α_K, %	66	70	77	84	89	91	100	100

where $0 < \alpha_K \leq 1$ and $K_{G,33}$ is the coefficient given by relations (7.33), which assumed $\alpha_K = 1$. Therefore, for flowing through cylindrical pores case, where $L_g = 3 \cdot a/2$ with $K_{G,33}$ from relation (7.34c), the following is found from relations (7.33a) and (7.35b):

$$\eta = 3 \cdot \pi \cdot \rho \cdot \overline{v} \cdot a \cdot \alpha_K / 256 \tag{7.35c}$$

α_K is measured from rotational oscillations damping in the rarefied gas of a suspended quartz wire. A value $\alpha_K \cong 1$ is obtained for metallic or rough surfaces and relatively low coefficients for smooth plates of ionic or atomic material. α_K increases with gas molecular mass, tending toward 1 for sufficiently large molecules (see Table 7.4).

Based on pressure P dependence on viscosity η in the Knudsen regime, Garthsen's "friction gauge" allows to measure pressures of $10^{-10}/10^{-7}$ atm by recording the frequency ν_{tors} of torque oscillation for a quartz wire from which a mica plate is suspended. This frequency is inversely proportional to the Knudsen viscosity coefficient—which in turn is proportional to the pressure according to dependence (7.34b)—so that $\nu_{tors} = K_{Gar}/P$, where constant K_{Gar} depends on the gaseous nature.

Thermal Conductivity and Diffusion at Low Pressures

The kinetico-molecular calculation of heat transfer between two surfaces at different temperatures, $T_1 > T_2$, for the Knudsen regime proves that: $K_{\lambda 1} = 1/8$; $K_{\lambda 2} = -1/2$. Since $C_{v1} = C_p - 5 \cdot R/2$ and $C_{v2} = 3 \cdot R/2$, by replacing $\overline{\ell}$ with L_g in relation (7.33b), $\lambda = \rho \cdot \overline{v} \cdot \overline{\ell} \cdot (K_{\lambda 1} \cdot C_{v1} + K_{\lambda 2} \cdot C_{v2})/M$, it results that:

$$\lambda = \rho \cdot \overline{v} \cdot L_g \cdot (C_p - R/2)/(8 \cdot M)$$

and when replacing $\rho = M \cdot P/(R \cdot T)$, $\overline{v} = \sqrt{8 \cdot R \cdot T/(\pi \cdot M)}$ and taking into account of the mechanical accommodation subunitary coefficient α_K, it results that:

$$\lambda = P \cdot (C_p - R/2) \cdot L_g / \cdot \left[(8 \cdot \pi \cdot M \cdot R \cdot T)^{1/2} \cdot (2/\alpha_K - 1)\right] \tag{7.36a}$$

The thermal flow j_q between two surfaces of temperatures T_1 and T_2 does not depend on the distance between plates; when replacing into relation (7.3),

$j_q = -\lambda \cdot dT/dx$ gradient $dT/dx = (T_1 - T_2)/L_g$ and λ with expression (7.36a), it is found that:

$$J_q = P \cdot (C_p - R/2) \cdot (T_2 - T_1)/ \cdot \left[(8 \cdot \pi \cdot M \cdot R \cdot T_m)^{1/2} \cdot (2/\alpha_K - 1)\right] \quad (7.36b)$$

Diffusion

Relation (7.33c), $D = K_D \cdot \overline{\ell} \cdot \overline{v}$, where $\overline{\ell} = L_g$ and $K_D = 1/3$, becomes:

$$D = L_g \cdot [8 \cdot R \cdot T/(9 \cdot \pi \cdot M)]^{1/2} \quad (7.37a)$$

According to relation (7.37a), diffusion process rate does not depend on pressure in the Knudsen field (while at moderate pressures $D \sim 1/P$).

But D increases with temperature; $D \sim T^{1/2}$ for Knudsen diffusion, showing a slower dependence on temperature than at usual pressures, where $D \sim T^{3/2}$ according to relation (7.17) $D = 3 \cdot [R^3 \cdot T^3/(\pi \cdot M)]^{1/2}/(8 \cdot N_A \cdot \delta^2 \cdot P)$.

Diffusion flow density does not depend on the distance between the two surfaces, in the Knudsen field; when replacing into relation (7.6), $j_D = -D \cdot dc/dx$, the ratio $dc/dx = \Delta c/L_g$ and when introducing D from (7.37a), it results that:

$$j_D = [8 \cdot R \cdot T/(\pi \cdot M)]^{1/2} \cdot \Delta c/3 \quad (7.37b)$$

where ($\Delta c = c_1 - c_2$) is concentrations difference, while at moderate pressures the mass transport flow is inversely proportional to distance between planes for which concentration takes the values c_1 and c_2.

Effusion

The Knudsen diffusion coefficient is inversely proportional to molecular mass quadratic mean ($D \sim M^{-1/2}$), which allows gaseous substances (e.g., isotopes) separation by ultrafine pores diffusion process, referred to as "effusion" and based on different gas diffusion rates of different molecular mass gases, like H_2, HD, D_2, or $^{235}UF_6$, and $^{238}U\ F_6$.

In an effusion installation, A and B gases' mixture ($M_A > M_B$), marked by white circles and black points respectively, in Fig. 7.5., circulated with constant velocity through compartment 1, which is separated through a membrane (with ultrafine pores) from compartment 2, where gas is found at the same temperature, but to a much lower pressure.

The flows through membranes are given by relation (7.37b), where $j_D = dn/dt$ (t = time, n = moles number) and Δc represents concentration c in compartment 1:

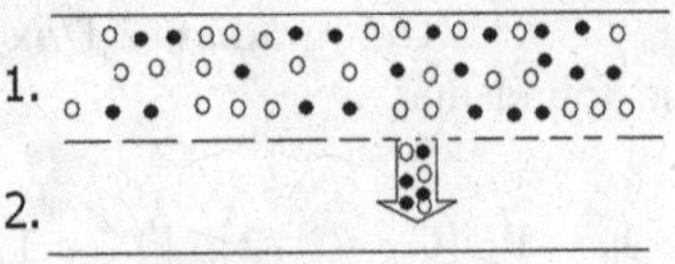

Fig. 7.5 Effusion

$$\forall i \in \{1,2\}: \quad dn_i/dt = (2 \cdot s \cdot c_i/3) \cdot [2 \cdot R\,T/(\pi \cdot M_i)]^{1/2} \text{ from where :}$$

$$dn_A/dn_B = (c_A/c_B)/(M_A/M_B)^{1/2} \tag{7.38}$$

Flows are proportional to concentrations during effusion; therefore, they are proportional to the time; hence c_A/c_B ratio increases in compartment 1 and decreases in compartment 2.

7.7 Pure Real Gases at Moderate Pressures

Knudsen and "Normal" Simultaneous Transport

When the mean free paths – "normal" bounded by intermolecular and Knusen collisions caused by molecular/wall collisions – have comparable values, transport coefficients (η, λ, D) calculation must take into account both mechanisms. Since (η, λ, D) are inversely proportionate to the two collisions types frequencies and these latter occur independently, it follows that the transport coefficients reverses, through the two mechanisms, are additive, namely,

$$\forall y \in \{\eta, \lambda, D\}: 1/y = 1/y_n + 1/y_K \tag{7.39}$$

where y_n is the "normal" value of transport coefficients from relations (7.13) ... (7.19), while y_K—the Knudsen value, given, for example, by equations (7.34) for η, eventually considering the accommodation, or by relations (7.36) and (7.37) for λ and diffusivity, respectively.

Corresponding States for Transport Phenomena

"Moderate" pressures are defined in Sect. 7.5. as pressures with not so low values. This signifies, on one side, that wall collisions and intermolecular collisions frequencies are comparable, and on the other side, that the perfect gas model is totally erroneous in their case as not so high in pressure value. In order to obtain (η, λ, D) at moderate pressures, the PVT dependencies of real gases from Sect. 4.1 can be used for molecular density μ calculation. However, this implies algebraic or even exponential equations solving, so that a more expeditious process is preferred in practice.

It is the corresponding states one, that allows not only the rapid calculation of thermodynamic features from Sect. 5.6, but also transport coefficients estimation from critical coordinates with satisfactory accuracy. Meanwhile, however, in Sect. 7.3, neither η, λ nor D·P depend on pressure, but only on temperature.

The simple principle (5.41b) of corresponding states becomes thus:

$$\eta_{red} = F_\eta(T_{red}); \lambda_{red} = F_\lambda(T_{red}); (D \cdot P)_{red} = F_{D \cdot P}(T_{red})$$

where F_η, F_λ, and $F_{D \cdot P}$ are dependency functions (independent of composition) of reduced temperature, denoted by T_{red}. Although the corresponding states principle in Sect. 5.6 implies in general *two* independent variables, the transport quantities will depend *only one*—the reduced temperature.

The Reference State

The critical temperature T_c is chosen as reference temperature, and T_{red} becomes, thus the ratio $T_r \equiv T/T_c$ from definition (5.42). But at the critical point, the perfect gas laws are far from being valid, since the critical pressure is too elevated. Therefore, in order to define a "*reference* transport coefficient" (η_c*, λ_c*, D_c*), the gas state at $T = T_c$ is chosen and at pressure tending toward zero:

$$\forall y \in \{\eta, \lambda, D \cdot P\} : Y_c* \equiv Y\,(P \rightarrow 0, T = T_c) \tag{7.40a}$$

The reduction relations, therefore, becoming in this case:

$$\eta = \eta_c * \cdot F\eta(T_r); \lambda = \lambda_c * \cdot F_\lambda(T_r); D \cdot P = (D \cdot P)_c * \cdot F_{D \cdot P}(T_r) \tag{7.40b}$$

Analytical or graphic functions F_η (F_λ, $F_{D \cdot P}$) can be used, either universal or valid for similar substances groups (according to the *extended* principle of corresponding states from Sect. 5.7). Therefore:

I. When both T_c and η_c* (or λ_c*) are known, the universal reduced temperature functions F_λ and F_η from Fig. 7.6 allow to obtain directly η or λ at any temperature.

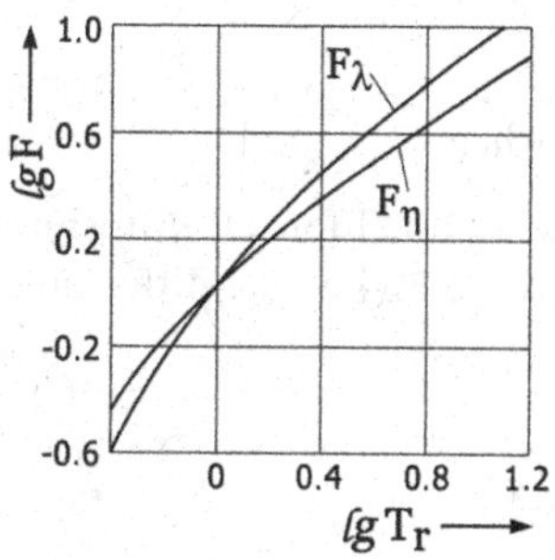

Fig. 7.6 Transport coefficients reduction, at moderate pressures

The empirical procedure of Golubev can be also used for η, where F_η is analytically given, namely,

$$T_r \leq 1 \rightarrow F_\eta = (T_r)^{0.965}; T_r \geq 1 \rightarrow F_\eta = T_r^{\,0.71+0.29/T_r} \quad (7.41)$$

II. When T_c is known but η_c* (or λ_c* or G_c*) are not known, the unknown quantity is eliminated: Y_2 (value at T_2) can be obtained from Y_1 value at temperature T_1 and from universal function F_Y by applying relation (7.40b) at temperatures (T_1, T_2): $Y_1 = Y_c{*}\cdot F_Y(T_{r,1}) \cap Y_2 = Y_c{*}\cdot F_Y(T_{r,2})$, from where:

$$\forall y \in \{\eta, \lambda, D.P\} : Y_2 = Y_1 \cdot F_Y(T_{r,2})/F_Y(T_{r,1}) \quad (7.42)$$

III. When no experimental value of transport coefficients is known but P_c and T_c are known, the η_c* or D_c* values (as well as λc*) from relations (7.44a, b) of Trautz, or Slatery, respectively, can be evaluated on the basis of kinetic-molecular theory with good results for nonpolar molecules:

$$\eta_c{*} = 0.631 \cdot \left[M^3 \cdot (P_c)^4\right]^{1/6} / \left[(N_A)^2 \cdot R \cdot T_c\right]^{1/6} \quad (7.43a)$$

$$(D \cdot P)_c{*} = 0.826 \cdot \left[(R \cdot T_c)^5 \cdot (P_c)^4 \cdot\right]^{1/6} / \left[(N_A)^2 \cdot M^3\right]^{1/6} \quad (7.43b)$$

Slatery's relation is useful not only for self-diffusion but also for 1 and 2 gases interdiffusion. In this case, P_c, T_c represent geometric means and M the harmonic mean of pure components values:

$$T_c = (T_{c,1} \cdot T_{c,2})^{1/2}; P_c = (P_{c,1} \cdot P_{c,2})^{1/2}; M = 2/(1/M_1 + 1/M_2) \quad (7.44)$$

For the "water vapors + nonpolar gas" interdiffusion, the coefficient of 0.826 of Slatery's relation becomes 1.094.

Functions F_η and F_G are for viscosity—Golubev function (7.41), while for diffusion—Slatery's power-function:

$$F_G = (T_r)^b \quad (7.45)$$

where

b = 1.823 for self-diffusion and interdiffusion at nonpolar gases and
b = 2.334 at interdiffusion in water vapors mixture with a nonpolar gas.

Corresponding States Intermolecular Potential

Curtiss and Hirschfelder relations (7.19c), (7.19d) and (7.19e), which starting from Lennard Jones (L.J.) intermolecular potential, show more accurately the transport coefficients dependence on temperature and also have an equivalent in the corresponding states theory: expressing the maximum Lennard Jones potential ε through critical temperature, in accordance with dependence (4.22c), $\varepsilon = 0.751 \cdot T_c$—Curtiss and Hirschfelder (C. & H.) relations can be brought to the form (7.40b), where:

$$F_\lambda(T_r) = \left[(C_p)^\circ / (C_{p,c})*\right] \cdot (T_r)^{1/2} \cdot \varphi_\lambda(T_r/0.751)$$

$$F_\eta(T_r) = (T_r)^{1/2} \cdot \varphi_\eta(T_r/0.751), F_G(T_r) = (T_r)^{1/2} \cdot \varphi_G(T_r/0.751),$$

and where $C_{pc}{}^*$ is the heat capacity at $P \to 0$ and $T = T_c$, while $(C_p)^\circ$—heat capacity at pressure zero and temperature T.

Values η_c*, λ_c*, and G_c* depend on M, on (C_{pc})* and—through constant σ from relation (4.22e), $\sigma = 0.841 \cdot (V_c/N_A)^{1/3}$—of the critical volume V_c.

Transport in Mixtures of Real Gases

Material constants combination procedure is widely used for real gases mixtures, as in Wilke relations (7.28).

The viscosity is given by the semi-empirical formula of Buddenberg and Wilke, whose mean error is 2%, but who requires corrections for polar or too elongated form molecules, as for the quantum gases:

$$\eta = \sum_{i=1}^{i=C} \frac{X_i \cdot \eta_i}{\sum_{j=1}^{j=C} X_j \cdot \left(1 + \sqrt{\eta_i/\eta_j} \cdot \sqrt[4]{M_i/M_j}\right)^2 / \sqrt{8 \cdot (M_i/M_j)}} \tag{7.46a}$$

Constituents and mixture viscosities are considered at the same temperature and at low pressures. For a binary mixture, the relation becomes:

$$\eta = X_1 \cdot \eta_1/(X_1 + X_2 \cdot \Psi_{1,2}/\eta_2) + X_2 \cdot \eta_2/(X_2 + X_1 \cdot \Psi_{12}/\eta_1) \tag{7.46b}$$

where

$$\Psi_{1,2} \equiv \left[(M_1)^{1/4} \cdot (\eta_1)^{1/2} + (M_2)^{1/4} \cdot (\eta_2)^{1/2}\right]^2 / [8 \cdot (M_1 + M_2)]^{1/2} \tag{7.46c}$$

7.8 Transport in Pure Gases at High Pressure

As pressure and real gas deviation magnitude from the ideal behavior increase, there is also a more pronounced scattering of experimental transport coefficients values as compared with the predicted values from the perfect gas kinetic theory, presented in Sect. 7.3. Therefore, there will be furtherly presented the main thermodynamic characteristics resulting from this behavior.

Isothermal Pressure Dependence While at moderate pressures η and λ are independent of pressure, at high pressures both quantities increase when pressure increases. This increase becomes more pronounced at high pressures and is more rapid for viscosity than for thermal conductivity; in both cases transport coefficients increase with pressure faster than density ρ increases with pressure. Therefore, η/ρ and λ/ρ, ratios, which are proportional to 1/P in the moderate pressure range, pass through a *minimum* value when pressure increases (see Fig. 7.7), after which they become again increasing pressure functions.

The existence of this minimum is justified in Enskog model through a detailed molecular scale analysis of momentum and energy transport.

These quantities are transferred between two neighboring plans located at distance $\overline{\ell}$ in Fig. 7.3 not only by the molecules whose *center* from a plan reaches the neighboring plan—as to the simplified analysis that led to relation (7.10a)—but also by molecules whose edge reaches the neighboring plan molecules edge.

Consequently, transport coefficients are in fact proportional not to the mean free path $\overline{\ell}$, but with $(\overline{\ell} - f \cdot \delta)$, where f is a numerical coefficient close to value 1, and δ – the molecular diameter.

Enskog, therefore, obtains for viscosity dependence on density (which in turn c depends on pressure) the following expression:

$$\eta = \eta_0 \cdot \left[1 + 0.175 \cdot \beta \cdot \rho + 0.865 \cdot (\beta \cdot \rho)^2\right] \tag{7.47a}$$

where η_0 is viscosity value at the same temperature but at moderate pressures, ρ—density and β is the correction (în m^3/kg) for the own volume of molecules, calculable using the following relation:

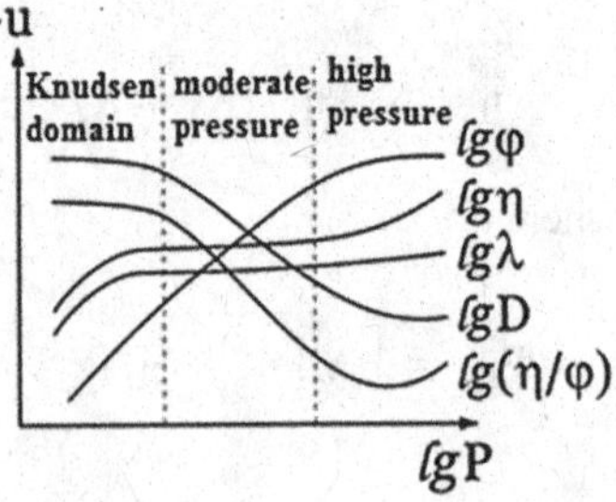

Fig. 7.7 Pressure effect on transport phenomena

$$\beta = \left[(R \cdot T)^3/M\right]^{1/4} \cdot \left[N_A \cdot (\eta_0)^3\right]^{-1/2} \tag{7.47b}$$

where M is the molar mass and T—temperature. The (actual) *dynamic* viscosity η increases with density in the Enskog model—therefore, with pressure.

Kinematic Viscosity Minimum Value

The kinematic viscosity given by the following definition: $\nu = \eta/\rho$, (7.2), presents a minimum for its pressure dependence by using these notations for all constants independent of density $A = \eta_0$, $B = 0.175 \cdot \beta \cdot \eta_0$, and $C = 0.865 \cdot \beta^2 \cdot \eta_0$ it results that:

$$\nu = A/\rho + B + C \cdot \rho \tag{7.8a2}$$

and by deriving relation (a) relative to ρ it is found that: $d\nu/d\rho = -A/\rho^2 + C$ and $d\nu^2/d\rho^2 = A/\rho^3$, from where: $d\nu/d\rho = 0 \rightarrow \rho = (A/C)^{0.5}$, so that:

$$\rightarrow \rho_{\nu=\min} = 1.075/B \tag{7.47c}$$

The kinematic viscosity minimum value, therefore, is: $B + 2 \cdot (A \cdot C)^{1/2}$ or

$$(\nu)_{\min} = 2.04 \cdot \beta\, \eta_0 \tag{7.47d}$$

At pressures of hundreds of atmospheres, Enskog model (7.47) is no longer practicable. The liquids momentum transport model from Sect. 8.2 is applied relatively well to the highly compressed gas.

The pressure dependence of diffusivity also changes in the high-pressure domain: the sharp drop ($D \sim 1/P$) from the moderate pressure range is followed at high pressures by a much less pronounced decrease. At extremely high pressures (of thousands of atmospheres), diffusivity—decreasing—variation with pressure increase becomes again more pronounced and is satisfactorily quantitatively described by diffusion in liquids relations from Sect. 8.7.

Transport Coefficients Dependence on Temperature

The isobaric temperature dependence of the transport coefficients for real gases is also complex and differs from that of ideal gases.

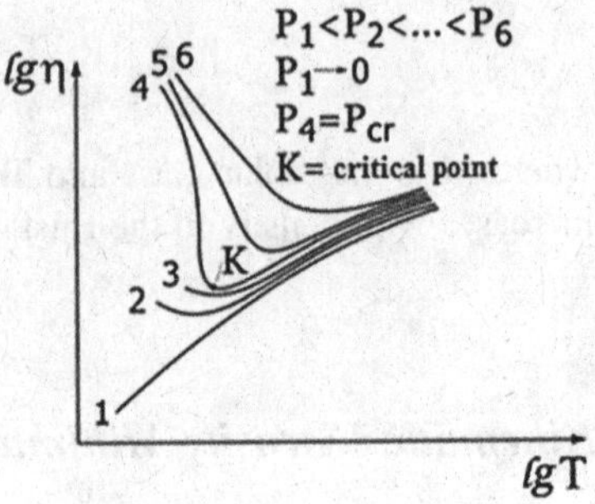

Fig. 7.8 Viscosity dependence on temperature at constant pressure

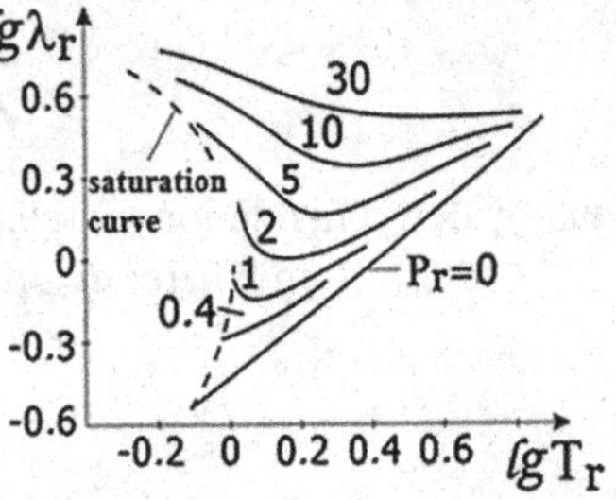

Fig. 7.9 Reduced conductivity dependence on reduced state parameters

At moderate pressures, η is an increasing dependency function of T, but at P >10 atm, the η(P) function presents a minimum value, as in Fig. 7.8. As pressure increases, the minimum is displaced toward more elevated T values.

Minimum's depth increases also with pressure at subcritical pressure values, while at $P > P_c$ the minimum becomes less pronounced when pressure increases.

If critical point quantities η_c, λ_c, and D_c are known, it is recommended to use the plots of the *reduced* transport coefficients, defined by ratios to the critical point:

$$\eta_r = \eta/\eta_c;\ \lambda_r = \lambda/\lambda_c;\ (D \cdot P)_r = D \cdot P/(D \cdot P)_c \tag{7.48a}$$

The reduced coefficients from (7.48a) are graphically represented, as dependency functions of temperature and reduced pressure, with universal plots, like $\eta_r = F_\eta(T_r, P_r)$ from Fig. 7.9.

$$\eta_r = F_\eta(T_r, P_r);\ \lambda_r = F_\eta(T_r, P_r);\ G_r = F_G(T_r, P_r) \tag{7.48b}$$

but η_c and λ_c are evaluated from η_c*, λ_c* values at T_c and moderate pressures:

$$\eta_c = 2.2 \cdot \eta_c * ; \lambda_c = 2.6 \cdot \lambda_c * \quad (7.48c)$$

Reduction to Critical Features

Relations (7.48) eventually allow λ_c (or η_c, D_c) elimination, from relation:

$$\forall G \in \{\eta, \lambda, D \cdot P\} : G_1 = G_2 \cdot F_G(P_{r1}, T_{r1}) / F_G(P_{r2}, T_{r2}) \quad (7.48d)$$

when the transport coefficient from point no. 2 is known.

The described method is more indirectly usable since, due to the strong fluctuations in the critical area, the gravitational effects become very pronounced. This makes emergence of strong convection currents unavoidable, currents which lead to errors higher than a magnitude order in molecular transport coefficients measurement.

If transport coefficients full dependence of temperature T at the low pressure ($\eta°$, $\lambda°$, G°), is applied—see Fig. 7.10—another corresponding state principle procedure is obtained, with the reduced coefficients:

$$\eta_r' = \eta/\eta^\circ; \quad \lambda_r' = \lambda/\lambda^\circ; \quad G_r' = G/G^\circ \quad (7.49)$$

where η and $\eta°$ (λ and $\lambda°$, G, and G°) refer to the same temperature.

Using *differences* $\eta - \eta°$ (or $\lambda - \lambda°$, G – G°) instead of *ratios* from relations (7.49) is advantageous when the state equation is known, since in this case the difference depends on *one single* variable (in contrast to the ratio, which depends on *two* variables), namely on the reduced volume V_r:

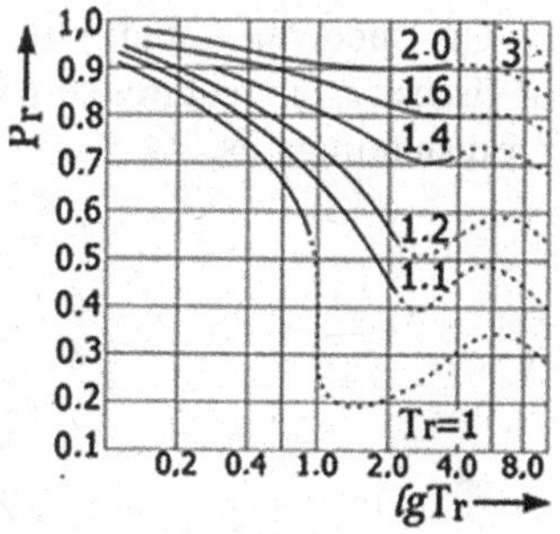

Fig. 7.10 The G_r' (T_r, P_r) dependency

$$\eta = \eta^o + F'_\eta(V_r); \lambda = \lambda^o + F'_\lambda(V_r); G = G^o + F'_G(V_r) \tag{7.50}$$

Similar functions can be also defined for argument V, but in this case, material constants are contained by the dependency function.

For example, in functions having the following form:

$$F''_\eta = A_\eta \cdot V^{-n'}; F''_\lambda = A_\lambda \cdot V^{-n''}; F''_G = A_G \cdot V^{-n'''} \tag{7.51a}$$

A_η, A_λ, and A_G are the material constants, while n′, n″, n‴—the universal constants, generally having values close to 4/3.

In certain cases, more complex functions are used. For example, for the viscosity of the first relation (7.50) Filippov expression is used:

$$F'_\eta = \left(T_C^7 \cdot M^3 \cdot P_C^2\right)^{1/6} \cdot 2730 \cdot V^{-1/3} \cdot (P_C \cdot V - 71.7 \cdot T_C)^{-1} \tag{7.51b}$$

while according to Golubev, viscosities are well correlated by:

$$(\eta - \eta^{\circ})/(\eta_c - \eta_c*) = F'''(V_r) \tag{7.51c}$$

where $\eta^\circ \equiv \eta(P = 0, T)$, $\eta_c \equiv \eta(P = P_c, T = T_c)$, and $\eta_c* \equiv \eta(P = 0, T = T_c)$, with small deviations from an universal function F''' of reduced volume.

Transport in Gas Mixtures

Gases, being given by the formula (7.46a). Constituents and the mixture viscosities are considered at the same reduced temperature and reduced pressure. For a binary mixture, formula (7.46b) can be used.

For gas mixtures at moderate or high pressures, the material constants combination is used, as in Wilke's relations (7.28).

The Buddenberg and Wilke empirical formula has a mean error of 2% in the case of viscosity, but requires corrections for polar or too elongated molecules, as well as for quantum gases.

7.9 Seven Worked Examples

Ex. 7.1

A narrow-graduated cylinder, with a total height L = 0.25 m, initially contains a layer of ethyl ether (liquid) of height H_0 = 0.05 m. The time required for ether complete evaporation at 20 °C is $\tau = 7.45 \cdot 10^5$ s.

It is required to calculate the diffusion coefficient D of the ether-air binary gaseous mixture, knowing density $\rho_l = 716$ kg/m^3 and vapor pressure Pv = 59000 Pa of liquid ether at 20 °C and of molecular mass M = 0.0501 kg/mol.

Solution

Let x be the (variable) distance between cylinder superior part and liquid level. Initially, $x_0 = L - H_0 = 0.25 - 0.05 = 0.2$ and in the end $x_f = L = 0.25$ m. Ether concentration is c = 0 at the superior edge and $c_s = p_v/(R \cdot T)$ at the liquid level.

According to relation (7.6) the diffusion flow density is: $j_D = -D \cdot dc/dx$ or $j_D = -D \cdot c_s/x$, and according to flow definition: $dn/d\tau = s \cdot j_D$ or $dn = -s \cdot D \cdot c_s \cdot x^{-1} \cdot d\tau$, where s is cross section area, while n is the evaporated liquid quantity, expressed in moles.

When denoting by v_l the liquid volume, $v_l = n \cdot M/\rho_l$ it results that: $(\rho_l/M) \cdot dv_l = -s \cdot (D \cdot p_v/(R \cdot T \cdot x) \cdot d\tau$ or, since $v_l = (L - x) \cdot s$ and $dv_l = -s \cdot dx$: $x \cdot dx = [M \cdot D \cdot p_v/(\rho_l \cdot R \cdot T)] \cdot d\tau$. When integrating between values x_0 and x_f the following is obtained: $\rho_l \cdot R \cdot T \cdot \int_{L-H_0}^{L} x \cdot dx = M \cdot D \cdot p_v \cdot \int_0^{\tau} d\tau$.

The first integral being $x^2/2$ and the second being τ, it results that:

$D = \rho_l \cdot R \cdot T \cdot [L^2 - (L - H_0)^2]/(2 \cdot M \cdot p_v \cdot \tau) = 716 \cdot 8.31 \cdot (273 + 20)\ [0.25^2 - (0.25 - 0.05)^2]/(2 \cdot 0.0501 \cdot 59000 \cdot 745000) \rightarrow \mathbf{D = 8.92 \cdot 10^{-6}\ m^2/s}$

Ex. 7.2

Using transport coefficients values of CO_2 at P = 1 atm and T = 273.15 K from Table 7.2, it is required to calculate:

1) the *apparent* diameter δ for CO_2 under these physical conditions, knowing M = 0.044 kg/mol and $C_p(0\ °C) = 36.0$ J/(mol·K).
2) the *limit* diameter δ_0 of the CO_2 molecule, using Sutherland dependency function $\delta(T)$, since at 973 K its viscosity is of $3.91 \cdot 10^{-5}$ kg/(m·s).
3) the *real* diameter σ of the CO_2 molecule, obtained from gas viscosity at 0 °C and using the Lennard Jones energetic parameter, $\varepsilon = 2.61 \cdot 10^{-21}$ J.

Solution

1. By specifying relations (7.13c, 7.16, 7.17) in relation to square apparent diameter, the following is obtained: $\delta^2 = (R \cdot T \cdot M/\pi)^{0.5}/(3.2 \cdot N_A \cdot \eta)$;

$$\delta^2 = [R \cdot T/(\pi\, M)]^{0.5} \cdot (C_p + 5 \cdot R/4)/(3.2 \cdot N_A \cdot \lambda)$$

and respectively

$$\delta^2 = 3 \cdot \left[(R \cdot T)^3/(\pi\, M)\right]^{0.5} /(8 \cdot N_A \cdot P \cdot D)$$

from where, when denoting by indices η, λ, D the corresponding δ values of the three transport coefficients at 0 °C, $\eta_{273} = 1.375 \cdot 10^{-5}$ kg/(m·s), $\lambda_{273} = 0.0142$ J/(m·s·K), and $D_{273} = 0.0000106$ m²/s, it results (at P = 101325 N/m²):

$$(\delta_\eta)^2 = (8.31 \cdot 273 \cdot 0.044/3.142)^{0.5}/(3.2 \cdot 6.02 \cdot 10^{23} \cdot 1.375 \cdot 10^{-5});$$

$$(\delta_\lambda)^2 = [8.31 \cdot 273/(0.044 \cdot 3.142)]^{0.5} \cdot (36 + 1.25 \cdot 8.31)/(3.2 \cdot 6.02 \cdot 10^{23} \cdot 0.0142)$$

$$(\delta_D)^2 = 3 \cdot [8.31 \cdot 273/(0.044 \cdot 3.142)]^{0.5}/(8 \cdot 6.02 \cdot 10^{23} \cdot 101325 \cdot 0.000016)$$

or, by performing calculations and putting 10^{10} Å for 1 m:→$\boldsymbol{\delta_\eta = 4.70}$ **Å**, $\boldsymbol{\delta_\lambda = 4.75}$ **Å**, and $\boldsymbol{\delta_D = 4.18}$ **Å**

2. By applying Eq. (7.18c) at two different temperatures (T_1, T_2), it results that:

$$\eta_1 = \left[M \cdot R \cdot (T_1)^3\right]^{1/2} / \left[3.2 \cdot \pi^{1/2} \cdot N_A \cdot (\delta_0)^2 \cdot (C_S + T_1)\right] \quad (7.8a3)$$

$$\eta_2 = \left[M \cdot R \cdot (T_2)^3\right]^{1/2} / \left[3.2 \cdot \pi^{1/2} \cdot N_A \cdot (\delta_0)^2 \cdot (C_S + T_2)\right] \quad (7.8b3)$$

and when eliminating C_S between relations (7.8a3) and (7.8b3) the following expression is obtained:

$$\begin{aligned} C_S &= \left[\eta_2/(T_2)^{1/2} - \eta_1/(T_1)^{1/2}\right]/\left[\eta_1/(T_1)^{3/2} - \eta_2/(T_2)^{3/2}\right] \\ &= \left[3.91/(973)^{1/2} - 1.375/(273)^{1/2}\right]/\left[1.375/(273)^{3/2} - 3.91/(973)^{3/2}\right] \\ &= 236\ \text{K}. \end{aligned}$$

By introducing the Cs value into one of relations (7.8a3, 7.8b3), its molecule limit-diameter is found, for example:

$$\begin{aligned} (\delta_0/m)^2 &= [M \cdot R/\pi]^{\frac{1}{2}} \cdot \left[(T_2)^{\frac{3}{2}}/\eta_2 - (T_1)^{\frac{3}{2}}/\eta_1\right]/[3.2 \cdot N_A \cdot (T_2 - T_1)] \\ &= [0.044 \cdot 8.31/\pi]^{0.5} \cdot \left[(973)^{1.5}/3.91 - (273)^{1.5}/1.374\right]/ \\ &\quad [3.2 \cdot 6.02 \cdot 10^{23} + 5 \cdot (973 - 273)] \end{aligned}$$

or, expressed in Å:→$\boldsymbol{\delta_0 = 3.38}$ **Å**

3. When using relation (7.20a), the reduced Lennard Jones temperature is obtained, $\tau \equiv k_B \cdot T/\varepsilon = 1.381 \cdot 10^{-23} \cdot 273/(2.61 \cdot 10^{-21}) = 1.445$.

From τ the Curtiss collisions integral is calculated for viscosity, using formula (7.20b), $\Omega_\eta(\tau) = A/\tau^{B} + C/e^{D \cdot \tau} + E/e^{F \cdot \tau}$, with values A-F from Table 7.1:

$$\Omega_\eta = 1.161/1.445^{0.142} + 0.525/e^{0.773 \cdot 1.445} + 2.162/e^{2.438 \cdot 1.445} = 1.336$$

From relation (7.19c), $\eta = (M \cdot R \cdot T/\pi)^{0.5}/(3.2 \cdot N_A \cdot \sigma^2 \cdot \Omega_\eta)$, now can be calculated:

$$\sigma^2 = (M \cdot R \cdot T/\pi)^{1/2}/(3.2 \cdot N_A \cdot \eta \cdot \Omega_\eta) = \ldots$$

$[0.044 \cdot 8.31 \cdot 273/\pi]^{0.5}/(3.2 \cdot 6.02 \cdot 10^{23} \cdot 1.375 \cdot 10^{-5} \cdot 1.336) = 1.592 \cdot 10^{-19} m^2$ from where, by expressing the Lennard Jones (L.J.) diameter in Å:→**σ = 3.99 Å**

Ex. 7.3

At 15 °C in the ideal mixture formed by gases H_2 (1) and SO_2 (2) of molecula masses, $M_1 = 2$ and $M_2 = 64$ g/mol, pure components viscosities are $\eta_1 = 84 \cdot 10^{-7}$ and $\eta_2 = 116 \cdot 10^{-7}$ kg/(m·s) and their thermal conductivities – $\lambda_1 = 168$ and $\lambda_2 = 9.2$ g·m/(s³·K).

It is required to calculate thermal conductivity, at 15 °C, for the mixture having the hydrogen molar fraction $X_1 = 2/3$.

Solution

From $r_{i,j} = [1 + (\eta_i/\eta_j)^{1/2}/(M_i/M_j)^{1/4}]^2/[8 \cdot (1 + M_i/M_j)]^{1/2}$ – definition (7.23c) – r_{12} and r_{21} coefficients values are calculated:

$$r_{12} = \left[1 + (84/116)^{1/2}/(2/64)^{1/4}\right]^2/[8 \cdot (1 + 2/64)]^{1/2} = 3.21$$

$$r_{21} = \left[1 + (116/84)^{1/2}/(64/2)^{1/4}\right]^2/[8 \cdot (1 + 64/2)]^{1/2} = 0.137$$

Through Wilke procedure, when introducing the above r_{12} and r_{21} values in relation (7.23b), $X_1 = 2/3$ from the statement and $X_2 = 1 - X_1 = 1/3$:

$\lambda = \lambda_1 \cdot X_1/(X_1 + X_2 \cdot r_{12}) + \lambda_2 \cdot X_2/(X_2 + X_1 \cdot r_{21})$ or numerically:
$\lambda = 168(2/3)/[2/3 + (1/3) \cdot 3.21] + 9.2 \cdot (1/3)/[1/3 + (2/3) \cdot 0.137]$ therefore
$\lambda = 71.7$ g·m/(s³·K)
→**λ = 0.072 kg·m/(s³·K)**

Ex. 7.4
For the quaternary ideal gaseous mixture having the following composition:

i	1	2	3	4
substance	CO_2	O_2	N_2	H_2
x_i %	10	20	30	40

It is required to determine the CO_2 diffusivity towards the mixture, D'_1, at t = 20 °C and P = 1 atm (760 torr). The Sutherland constants for the 4 gases are known: $C_{S1} = 254$ K, $C_{S2} = 125$ K, $C_{S3} = 104$ K, $C_{S4} = 89$ K, as well as the $D_{1,j}$* values of interdiffusion coefficients $D_{1,j}$ of CO_2 with the j-th component; calculated under somehow different conditions – (p_j*, t_j*):

j	1	2	3	4
p_j*, torr	/	760	755	749
t_j*, ^{0}C		0	25	20
$10^7 \cdot D_{1,j}$*, m^2/s		139	159	614

Solution
From relation (7.24) the Sutherland constants for binary mixtures formed by CO_2 with the other gases, are calculated:

$$C_{S12} = (C_{S1} \cdot C_{S2})^{0.5} = (254 \cdot 125)^{0.5} = 178.2 \text{ K};$$

$$C_{S13} = (C_{S1} \cdot C_{S3})^{0.5} = (254 \cdot 104)^{0.5} = 161.7 \text{ K and}$$

$$C_{S14} = (C_{S1} \cdot C_{S4})^{0.5} = (254 \cdot 89)^{0.5} = 150.4 \text{ K}.$$

According to relation (7.18e), interdiffusion coefficient dependence on temperature and pressure for a given gas pair, has the following form: $D_{ij} = K_{ij} \cdot T^{2.5}/[P \cdot (T + C_{Sij})]$ where K_{ij} (as well as C_{Sij}) do not depend on T and P.

For conditions (T*, P*), relation (7.18e) becomes:

$$D_{ij} = K_{ij} \cdot (T*)^{2.5} / \left[P * \cdot (T * + C_{Sij})\right].$$

Constant K_{ij} elimination and index i replacement with 1 (CO_2) lead to:

$$D_{1j} = D_{1j} * \cdot (P_j * /P) \cdot (T/T_j*)^{2.5} \cdot (T_j * + C_{S1j}) / \cdot (T + C_{S1j}) \quad (7.8a4)$$

From relation (a), with T ≅ 20 + 273 = 293 K and P = 760 torr, it results that:

$$D_{12} = 139 \cdot 10^{-7} \cdot (293/273)^{2.5} \cdot (273 + 178)/(293 + 178) = 159 \cdot 10^{-7} \; m^2/s$$

$$D_{13} = 139 \cdot 10^{-7} \cdot (755/760) \cdot (293/298)^{2.5} \cdot (293 + 162)/(273 + 162)$$
$$= 153 \cdot 10^{-7}\, m^2/s \text{ and}$$
$$D_{14} = 614 \cdot 10^{-7} \cdot (749/760) = 605 \cdot 10^{-7} m^2/s.$$

From relation (7.28a) with C = 4, i = 1 it results that CO_2 diffusivity toward the mixture is: $D'_1 = (1 - x_1)/(x_2/D_{12} + x_3/D_{13} + x_4/D_{14})$ or:

$$D'_1 = (1 - x_1)/(x_2/D_{12} + x_3/D_{13} + x_4/D_{14})$$
$$= 10^{-7} \cdot (1 - 0.1)/(0.2/159 + 0.3/153 + 0.4/605)$$
$$\rightarrow D'_1 = 232.10^{-7} m^2/s$$

Ex. 7.5
A layer of N_2—nitrogen, with M = 0.028 kg/mol is comprised at pressure P = 3500 Pa between two parallel alumina walls, with thermal accommodation coefficients $\alpha_\lambda = 0.8$, at temperature T = 293 K, where isobaric molar heat capacity of gas is $C_p = 28.5$ J/(mol·K), thermal conductivity is $\lambda = 0.0125$ J/(m·s·K) and its molecules collision diameter (from Table 3.2) is: $\delta = 3.71 \cdot 10^{-10}$ m. It is required to calculate distance a between two plates, by considering both transport mechanisms—molecular and Knudsen.

Solution
Relation (7.39) of "combined" transport becomes, for thermal conductivity: $1/\lambda = 1/\lambda_n + 1/\lambda_K$ or, when introducing the *normal* value λ_n (through intermolecular collisions) of conductivity from relation (7.16) and its *Knudsen* value λ_K (through molecule/wall collisions) given by relation (7.36a), respectively:

$\lambda_n = [R \cdot T/(\pi \cdot M)]^{0.5} \cdot (C_p + 1.25 \cdot R)/(3.2 \cdot N_A \cdot \delta^2)$ and
$\lambda_K = P \cdot L_g \cdot (C_p - R/2)/[(8 \cdot \pi \cdot M \cdot R \cdot T)^{0.5} \cdot (2/\alpha_K - 1)]$, it is found that:

$$1/\lambda = 1/A + B/L_g \tag{7.8a5}$$

where A and B are:

$$A \equiv [R \cdot T/(\pi \cdot M)]^{0.5} \cdot (C_p + 1.25 \cdot R)/(3.2 \cdot N_A \cdot \delta^2)$$
$$= [8.31 \cdot 293/(\pi \cdot 0.028)]^{0.5} \cdot (28.5 + 1.25 \cdot 8.31)/\left[3.2 \cdot 6.02 \cdot 10^{23} \cdot (3.87 \cdot 10^{-10})^2\right]$$
$$= 0.02242 J/(m \cdot s \cdot K)$$

$$B \equiv (8 \cdot \pi \cdot M \cdot R \cdot T)^{1/2} \cdot (2/\alpha_K - 1)/\left[P(C_p - R/2)\right]$$
$$= (8 \cdot \pi \cdot 0.028 \cdot 8.31 \cdot 293)^{1/2} \cdot (2/0.8 - 1)/[3500 \cdot (28.5 - 8.31/2)]$$
$$= 7.288 \cdot 10^{-4} s^3 \cdot K/kg.$$

When introducing A, B, and value λ from the statement into relation (a) it results that:

$$L_g = 7.288 \cdot 10^{-4}/(1/0.0125 - 0.0244)$$
$$= 1.868 \cdot 10^{-5}\text{m}.$$

Since $L_g = 2 \cdot a$ for infinite plates at distance a (Sect. 7.5):

$$a = L_g/2$$
$$\rightarrow \mathbf{a = 9.34 \cdot 10^{-6}m}$$

Ex. 7.6
The following parameters are known for Neon: molar mass M = 0.0202 kg/mol, critical parameters $T_c = 44.5$ K, $P_c = 2.62 \cdot 10^6$ N/m^2 and thermal conductivity value, $\lambda' = 0.0456$ J/(m·s·K) at temperature T′ = 273 K.

It is required to calculate, using the corresponding states theorem, the transport phenomenological coefficients at $P_1 = 5 \cdot 10^4$ N/m^2 and $T_1 = 450$ K, namely (1) Thermal conductivity, λ_1; (2) Viscosity η_1, graphically; (3) Viscosity, through an analytical procedure; (4) Self-diffusion coefficient D_1.

Solution

1. Since value λ_c* (at T_c and moderate temperatures) it is unknown, then value λ′ is used in calculations. From T′ and T_1 the corresponding reduced temperatures are found: $T'_r = T'/T_c = 273/44.5 = 6.14$ and $T_{r1} = T_1/T_c = 450/44.5 = 10.1$. On the thermal conductivity curve (reduced by λ_c*) from Fig. 7.6 the function F_λ values are read, values which correspond to the two reduced temperatures: $F'_\lambda = F_\lambda(T'_r) = F_\lambda(6.14) = 1.41$ and $F_{\lambda 1} = F_\lambda(T_{r1}) = F_\lambda(10.11) = 1.97$. Relation (7.42) for λc applied to points T′ and T_1 becomes:

$$\lambda_1 = \lambda' \cdot F_{\lambda 1}/F'_\lambda = 0.0456 \cdot 1.97/1.41$$
$$\rightarrow \boldsymbol{\lambda_1 = 0.0637\ J/(m \cdot s \cdot K)}$$

2. Viscosity η_c* at critical temperature and moderate pressuresis obtained from Trautz relation (7.43):

$$\eta_C^* = 0.631 \cdot \left(\frac{M^3 \cdot P_C^4}{N_A^2 \cdot R \cdot T_C}\right)^{\frac{1}{6}} = 0.631 \cdot \left(\frac{0.0202^3 \cdot 2.62^4 \cdot 10^{6.4}}{6.02^2 \cdot 10^{23 \cdot 2} \cdot 8.31 \cdot 44.5}\right)^{\frac{1}{6}} \text{ or}$$

$$\eta_c* = 7.54 \cdot 10^{-6}\text{kg}/(\text{m} \cdot \text{s}).$$

From Fig. 7.6 it results that: $F_\eta = F_\eta(T_{r1}) = F_\eta(10.1) = 5.55$ and from relation (7.42) for η: $\eta_1 = \eta_c^* \cdot F_\eta = 7.54 \cdot 10^{-6} \cdot 5.55$ or:

$$\eta_1(\mathbf{graphically}) = \mathbf{4.18 \cdot 10^{-5} kg/(m \cdot s)}$$

3. From η_c^* value of point (2) and using Golubev relation (7.41) at $T_r > 1$ the following is obtained:

$F_\eta = T_r^{0.71+0.29/T_r} = 10.1^{0.71+0.29/10.1} = 5.52$, further resulting, from relation (7.40b): $\eta = \eta_c^* \cdot F_\eta = 7.54 \cdot 10^{-6} \cdot 5.52$ or:

$$\eta_1(\mathbf{analytical}) = \mathbf{4.16 \cdot 10^{-5} kg/(m \cdot s)}$$

4. From Slatery relation:

$(D \cdot P)_c^* = 0.826 \cdot [(R \cdot T_c)^5 \cdot (P_c)^4]^{1/6} / [(N_A)^2 \cdot M^3]^{1/6}$, (7.44):

$$(D \cdot P)_c^* = 0.826 \cdot \left(\frac{2.62^4 \cdot 10^{6.4} \cdot 8.31^5 \cdot 44.5^5}{6.02^2 \cdot 10^{23.2} \cdot 0.0202^3} \right)^{1/6} \quad \text{or}$$

$$(D \cdot P)_c^* = 0.181\ kg \cdot m/s^3.$$

From relation (7.45) with $b = 1.823$, it results that: $F_{D \cdot P} = (T_r)^b = 10.1^{1.823} = 67.9$. From relation (7.40b) it is obtained that $D \cdot P = (D \cdot P)_c^* \cdot F_{D \cdot P} = 0.181 \cdot 67.9 = 12.3$ kg·m/s^3, from where $D_1 = (D \cdot P)_1/P_1 = 12.3/(5 \cdot 10^4)$ or → $\mathbf{D_1 = 2.46 \cdot 10^{-4}\ m^2/s}$

Ex. 7.7

Neon molar critical volume is $V_c = 4.17 \cdot 10^{-5}$ m^3/mol. It is required to calculate using Hirschfelder and Curtiss theory the transport coefficients λ_1 and η_1 at $P_1 = 5 \cdot 10^4$ N/m^2 and $T_1 = 450$ K, with diameters from Ex. 7.6.

Solution

From relation (5.4) the compressibility critical factor results having the following form:

$$\begin{aligned} z_C &= P_C \cdot V_C/(R \cdot T_C) \\ &= 2.62 \cdot 10^6 \cdot 41.7 \cdot 10^{-6}/(8.31 \cdot 44.5) \\ &= 0.296 \end{aligned}$$

The maximum energy ε and molecular diameter σ—Lennard-Jones intermolecular potential parameters—are obtained from Stiel and Thodos relations (5.58) and (5.59):

$$\begin{aligned}\varepsilon/k_B =&65.3 \cdot T_C \cdot (z_C)^{3.6}\\ =&65.3 \cdot 44.5 \cdot 0.296^{3.6}\\ =&36.3\ K\end{aligned}$$

$$\begin{aligned}\sigma =&0.1576 \cdot (V_c/N_A)^{1/3}/(z_c)^{1.2}\\ =&0.1576 \cdot \left(4.17 \cdot 10^{-28}/6.02\right)^{1/3}/(0.296)^{1.2}\\ =&2.79 \cdot 10^{-10}\text{m}.\end{aligned}$$

Constant volume heat capacity is given—Neon being monoatomic—by relation (2.29e), Cv = 1.5 R from where, since the gas is supposed to be perfect, thus $C_p = C_v + R$: $C_p = 2.5{\cdot}R = 2.5{\cdot}8.31 = 20.8$ J/(mol·K).

When using relation (7.20a) the Lennard Jones reduced temperature is obtained, $\tau \equiv T_1{\cdot}(\varepsilon/k_B)^{-1} = 450/36.3 = 12.4$, from where Curtiss collisions integral is calculated for viscosity using relation (7.20b), $\Omega_\eta(\tau) = A/\tau^{\ B} + C/e^{D\cdot\tau} + E/e^{F\cdot\tau}$, with values A-F from Table 7.1:

$$\Omega_\eta = 1.161/12.4^{0.142} + 0.525/e^{0.773\cdot 12.4} + 2.162/e^{2.438\cdot 12.4} = 0.798.$$

From relation (7.19c):

$\eta_1 = (M{\cdot}R{\cdot}T_1/\pi)^{0.5}/(3.2{\cdot}N_A{\cdot}\sigma^2{\cdot}\Omega_\eta) = (0.0202{\cdot}8.31{\cdot}450/\pi)^{0.5}/[3.2{\cdot}6.02{\cdot}10^{23}{\cdot}(2.79{\cdot}10^{-10})^2{\cdot}0.798]$

or:

$\rightarrow \eta_1 = 4.10{\cdot}10^{-5}$ kg·m/s

The Curtiss integral for thermal conductivity is, according to relation (7.20b), equal to the viscosity one, namely $\Omega_\lambda = \Omega_\eta = 0.798$.

From relation (7.19d):

$\lambda = [R{\cdot}T_1/(\pi{\cdot}M)]^{1/2}{\cdot}(C_p + 1.25{\cdot}R)/[3.2{\cdot}N_A{\cdot}\sigma^2{\cdot}\Omega_\lambda(\tau)] = [8.31{\cdot}450/(\pi{\cdot}0.0202)]^{1/2}{\cdot}(20.8 + 1.25{\cdot}8.31)/[3.2{\cdot}6.02{\cdot}10^{23}{\cdot}(2.79{\cdot}10^{-10})^2{\cdot}0.798]$

$\rightarrow \boldsymbol{\lambda_1 = 0.0633}$ **kg·m/(s³·K)**

Chapter 8
Transport in Liquids and Solids

8.1 Transport and State of Aggregation

Thermal Conduction

From heat transmission and electric charge transmission point of view, solids and liquids are highly differentiated from one substance to another, distinguishing thus insulators (thermal or electric) and conductors, among which conductivity can reach 20 orders of magnitude in differences.

All gases are bad conductors of heat and electricity, with the exception of highly ionized gases, designated as *plasma*, which is a good conductor of electricity and heat; however, plasma existence requires special conditions and its properties are so much differentiated from nonionized usual gases that plasma is often considered as matter's fourth distinct state of aggregation.

Diffusion

From mass transport point of view, the three states of aggregation greatly differ from one another through their diffusivity values: typical diffusivity values (m^2/s) of gases, liquids, and solids can be 10^{-4}, 10^{-10}, and 10^{-14}, respectively. If from D values point of view, the liquids occupy an intermediate position between the other two states of aggregation, there are other criteria that allow a differentiation of gases, on the one hand, and condensed states of aggregation on the other.

F. E. Daneş et al., *Molecular Physical Chemistry for Engineering Applications*,
https://doi.org/10.1007/978-3-030-63896-2_8

Therefore,

- *D* has a low dependence on substance nature (of one order of magnitude) in gases, while in condensed states this dependency is much more pronounced, reaching up to 10 orders of magnitude.
- *D* slowly increases with temperature (power function form, $D = A{\cdot}T^{B}$) in gases, while diffusivity increase with temperature in liquids and solids is much rapid (according to the exponential expression $D = A{\cdot}e^{-B/T}$).
- *D* depends on pressure ($D = A/P$, except for the Knudsen domain) in gases, while in condensed states of aggregation D does not depend on pressure, when below 100 atmospheres.

These changes depending on the state of aggregation are due to a totally different mechanism of diffusion phenomenon: in the case of gases, diffusion occurs mainly by rectilinear displacement of molecules, which are free between two successive collisions, while in liquids and solids the mobile molecule moves among the other fixed molecules and this displacement can only take place in the case of a "hot" molecule, characterized by a sufficiently high energy to overcome the activation barrier.

Rheology

Rheology is the study of matter deformation in a mechanical force field. From the point of view of momentum transport, the three states of aggregation are grouped as follows: liquids and gases *flow*—they have the property of being *fluids*—whereas crystalline solids cannot flow (in common expressing manner, only solids have their own form, in contrast to gases and liquids). When subjected to a *shear stress* f (the ratio between shear force and its application surface), the typical solid has an *elastic* behavior—respects *Hooke's* law of shear stress f proportionality with the angle of rotation α:

$$f = -G \cdot \alpha, \tag{8.1}$$

while the gas, as well as the typical liquid has a *viscous* behavior, characterized by shear stress proportionality with temporal velocity $d\alpha/dt$ of the angle of rotation α, which measures the shear deformation:

$$f = -\eta \cdot d\alpha/dt. \tag{8.2}$$

Usually, the shear modulus G as well as viscosity η are independent of values (f, α, $d\alpha/dt$), while $d\alpha/dt$ from relation (8.2) is equal to the flow velocity gradient

Fig. 8.1 Shear angle

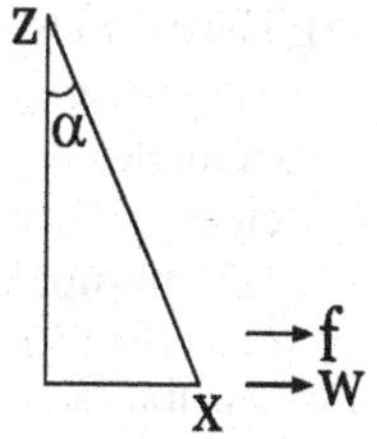

v along a perpendicular direction to the shear stress direction, since $v = dx/dt$ and $dx = z \cdot d\alpha$, while $v/z = grad\ v$, where x and z are the distances of flow direction and perpendicular to it, respectively (Fig. 8.1). Relation (8.2) is thus equivalent to Newton's law (7.1). The practical importance of distinguishing between behaviors reflected by the relations (8.1) and (8.2) is so big that justifies grouping the liquid and gaseous states of aggregation under the general term of *fluid states of aggregation*, opposed to the solid state of aggregation.

From a rheological point of view, the *usual* gases and liquids therefore form together a category—that of *fluid* bodies, characterized by *viscosity*. However, there are many differences between the viscous behavior of gases and liquids, as presented in Sect. 6.2. Liquids rheological characteristics will be discussed in Sect. 8.5.

The actual (crystalline) solids do not show at all this phenomenon (f does not depend on $d\alpha/dt$), being either elastic or plastic. Amorphous solids, however, have a fluidity, although their deformation becomes measurable only after long time periods due to their enormous viscosity, $\eta > 10^{13}$ kg/(ms).

Amorphous solids are considered to be subcooled liquids, a conclusion confirmed by their structure study on a molecular scale: as in liquids case, in the amorphous solids only the *near* order is manifested, while the *distance* order which is characteristic to crystals (Sect. 6.1) is absent.

As a viscous type rheological behavior consequence, transverse oscillations (crossing the solids) cannot cross the fluids, while longitudinal ones can. Their velocity of propagation (velocity of sound) differs according to the state of aggregation, namely, it is of the order of hundreds of m/s for gases and of thousands of m/s for liquids and solids.

Liquids Viscosity

Most liquids, like gases, respect Newton's law (8.2) $f = -\eta \cdot dv/dz$; but between the two categories of fluids viscosity, essential differences are found, namely,

(a) Liquids viscosity is much higher than the one of gas.

(b) The various liquid viscosities may differ with several orders of magnitude, whereas gas viscosity has a relative low dependence on composition. For example, at 20 °C the viscosities expressed in g/(ms) are as follows: ethyl ether 0.23, acetone 0.33, chloroform 0.56, benzene 0.65, water 1, methanol 1.2, nitrobenzene 2, glycol 20, glycerin with 10% water 200, and pure glycerin 1500.

(c) The liquid viscosity *decreases* with temperature increase, that is, rapid, while for gases the temperature rising leads to a slow *increase* of viscosity, as shown in Table 8.1.

Table 8.1 Water viscosity on the vaporization curve, expressed in kg/(10^5 ms)

t (°C)	0	25	50	75	100	125
η (liquid)	179	92	55	38	28	22
η (steam)	0.86	0.96	1.06	1.16	1.25	1.35

These differences are due to the different momentum transfer mechanism of the two fluids types: in gases, the transfer takes place mainly through molecule free translation between two successive collisions. In liquids, momentum transfer takes place when the molecule is moved on a fixed distance (equal to a "cell" dimension) from an equilibrium position to another, displacement which is driven by molecule vibration.

8.2 Variation of Viscosity with State Quantities

Variation of Viscosity with Temperature

Liquid viscosity depends on temperature according to Guzman's law:

$$\eta = A_\eta \cdot e^{B/T}, \tag{8.3}$$

where A_η and B are positive material constants, relatively independent of temperature; A_η is the pre-exponential viscosity factor, η_0 (i.e., viscosity extrapolated to $1/T \to 0$), and B is the viscosity activation energy, E_η, relative to the gas constant, R. Guzman's law (8.3) thus becomes: $\eta = \eta_0 \cdot exp[E_\eta/(R \cdot T)]$, so that *fluidity* φ—viscosity inverse, $\varphi \equiv 1/\eta$—depends on temperature according to relation $\varphi = \varphi_0 \cdot \exp[-E_\eta/(R \cdot T)]$, with $\varphi_0 = 1/\eta_0$, relation which is similar to temperature dependence of the reaction rate constant, $k = k_0 \cdot \exp[-E/(R \cdot T)]$—Arrhenius law of chemical kinetics. This resemblance results from the fact that reaction's degree of advancement is a process *with activation*, similar to momentum transmission within liquids: in both cases it is necessary that molecules which either change their structure (during the chemical reaction) or their position (during the momentum transfer) have been previously accumulated a certain energy amount, exceeding thus an energy barrier height: $\varepsilon^* = E/N_A$ (or E_η/N_A) required to trigger the process; or the molecule proportion having their energy higher than ε* is $[-\varepsilon^*/(k_B \cdot T)]$—the Boltzmann distribution (3.19). Relations (8.3), implying an activation energy, are valid for all transport phenomena in condensed states, proving thus that the transfer mechanism is the same.

Voids Theories to Explain the Temperature Effect on Viscosity

Temperature viscosity exponential dependence (8.3) is explained by Bachinsky through the fluidity $1/\eta$ proportionality with voids free volume:

$$\eta = A_f / V_f, \tag{8.4}$$

where A_f is a material constant independent of temperature and V_f is the total volume of void from one mole of liquid.

The V_f value is N_A times greater than the one of free volume for one molecule of Eyring, v_f, defined by relation (6.9) and calculated as in Sect. 6.3:

$$V_f = N_A \cdot v_f. \tag{8.5}$$

In fact, V_f is obtained as the difference $V_f = V - V_{pr}$ between the molar volume V and the *proper* volume V_{pr}, the volume which is not available for molecules displacement.

Usually, V_{pr} is considered equal to covolume b from the van der Waals ES (4.6). Bachinsky's relation (8.4) thus becomes the following:

$$\eta(T) = A_f / \left[V(T) - V_{pr})\right]. \tag{8.6}$$

If, however, it is admitted that the volume, v_f, of a single void does not depend on T, while void forming is an activated phenomenon (with energy $E\eta$ for one mole) so that the void number of N_f in one mole increases exponentially with temperature, $N_f = N_{f,0} \cdot exp[-E_\eta/(R \cdot T)]$, it results in the void's total volume as given below:

$$V_f = N_{f,0} \cdot v_f \ \exp\left[-E_\eta/(R \cdot T)\right],$$

so that, by replacing V_f in relation (8.4), Guzman dependence (8.3) is found for $\eta(T)$, where $\eta_0 = A_f / (N_{f,0} \cdot v_f)$:

$$\eta = [A_f / (N_{f,0} \cdot v_f)] \exp\left[E_\eta/(R \cdot T)\right]. \tag{8.7}$$

Mobility of Voids

Eyring procedure represents another manner to obtain Guzman theoretical dependence of η (T), which starts from considerations regarding the *mobility of voids*.

By considering the void positions as periodic potential minimums, as in Fig. 8.2a, which correspond to the absence of an external force field, Eyring introduces a shear force F, which modifies the potential in a linear manner:

$$u = u_0 - F \cdot x, \tag{8.8}$$

$u_0(x)$ being the potential in the absence of shear stress, and $u(x)$—the potential in its presence.

In an interval of length δ between two successive voids, interval which is centered at the maximum point of the potential barrier, the aspect of potential dependence u on distance x is according to Fig. 8.2b, where A and C correspond to two successive void positions, while B—the top of potential barrier.

The values of x at A, B, and C are $-\delta/2$, 0, and $\delta/2$, respectively, and potential u_o values in the *absence* of shear at the same points are 0, u_b, and 0, respectively.

According to relation (8.8), the u potential value in the *presence* of shear is $u = F \cdot \delta/2$ at A, $u = u_b$ at B, and $u = -F\delta/2$ at C.

Frequency of Jumps between Voids

The medium frequency of jumps becomes: $\nu_+ = (\nu_0/2) \cdot \exp[(u_A - u_B)/(k_B \cdot T)]$ for A to C jumps and $\nu_- = (\nu_0/2) \cdot exp[(u_C - u_B)/(k_B \cdot T)]$ for C to A jumps, where $(u_A - u_B)$ and $(u_C - u_B)$ are the potential barriers heights in both directions, and ν_0 is molecule vibration frequency.

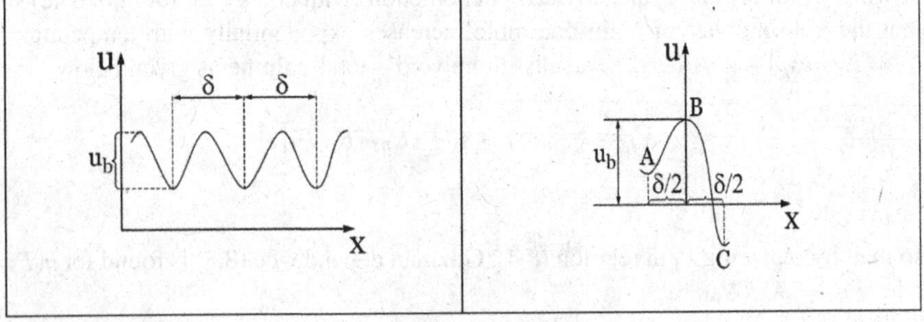

Fig. 8.2 Intermolecular potential and shear stress

The net frequency of passing from void A to C is: $\nu = \nu_+ - \nu_-$, where

$$\nu = (\nu_0/2) \cdot \{\exp[(u_A - u_B)/(k_B \cdot T)] - \exp[(u_C - u_B)/(k_B \cdot T)]\}$$

and when replacing $u_A - u_B = -u_b + F \cdot \delta/2$ and $(u_C - u_B) = -u_b - F \cdot \delta/2$ and with the notation $y \equiv F \cdot \delta/(2 \cdot k_B \cdot T)$, the following expression is obtained:

$$\nu = (\nu_0/2) \cdot (e^y - e^{-y}) \cdot \exp[-u_b/(k_B \cdot T)]. \tag{8.9}$$

For low shear force values, $F \ll 2 \cdot k_B \cdot T/\delta$, it results that $y \ll 1$, and series expansion leads to

$$(e^y - e^{-y}) = 2 \cdot y + y^3/3 + \ldots \cong 2 \cdot y.$$

Therefore,

$$\nu = [\delta \cdot \nu_0 F/(2 \cdot k_B \cdot T)] \cdot \exp[-u_b/(k_B \cdot T)]. \tag{8.2a2}$$

According to Newton's law (8.2), $\eta = -f \cdot (dv/dz)$. But shear stress $f = F/\delta^2$, and velocity gradient $dv/dz = -\nu$, so that Newton's law becomes: $\eta = F/(\delta^2 \cdot \nu)$ or, by replacing ν from (8.2a2):

$$\eta = \left[2 \cdot k_B \cdot T/(\delta^3 \cdot \nu_0)\right] \cdot \exp[u_b/(k_B \cdot T)], \tag{8.10}$$

which corresponds to Guzman's law, where $E_\eta = N_A \cdot u_b$ and the pre-exponential factor $\eta_0 = 2\ k_B \cdot T/(\delta^3 \cdot \nu_o)$ depends relatively slower on temperature.

In relation (8.10), δ^3 is the void volume (approximately equal to the V/N_A volume of a particle), and ν_o is determined from the liquid vibrational spectrum maximum. For common liquids, $V = 10^{-5}$ m^3/mole and $\nu_0 = 10^{13}$ s^{-1}, so that at room temperature (300 K) the pre-exponential factor is obtained $\eta_o = 2 \cdot 1013 \cdot 1.38 \cdot 10^{-23} \cdot 300/[10^{-5}/(6.02 \cdot 10^{23}) \cdot 10^{13}] \cong 10^{-4}$kg/(ms), in accordance with the experimental date order of magnitude.

For polymeric or oxide melts (melted bottles), Guzman's relation (8.3) is no longer valid, being replaced by

$$\eta = A_\eta \cdot e^{B/(T-C)}, \tag{8.11}$$

where C represents the *vitrification* temperature at which η becomes infinite.

Variation of Liquid Viscosity with Pressure

Unlike for ideal gas, where viscosity does not depend on pressure, liquids viscosity increases with pressure increase. This increase is generally exponential, namely,

$$\eta = A_P \cdot \exp(B_P \cdot P), \tag{8.12}$$

where A_P (liquid viscosity at pressure tending toward 0) and B_P are positive constants, which depend on liquid nature and on temperature.

Generally, the B_P constant is relatively small—below 10^{-4} bar^{-1}—so that relation (8.12) can be approximated by the following linear dependence:

$$\eta = A_P \cdot (1 + B_P \cdot P)$$

at moderate pressures (below 1000 atm), while at low pressures (below 20 atm) the viscosity dependence on pressure can be neglected.

It is found that B_P depends on the liquid nature, being small for small molecule substances such as metals and high for large molecule substances (mineral oils and other organic substances with M > 200 g/mol, silicates, etc.).

Viscosity variation with pressure is explained in Eyring's theory by the pressure effect on void formation.

Indeed, during the void formation process, a mechanical work is done not only against the cohesion forces but also against pressure, because a void formation (of volume v_g) increases with v_g the liquid volume. If n_g is the void number from one mole, and E_o is the energy required to overcome the cohesion forces from one mole, the viscosity activation energy results the following expression:

$$E_\eta = E_o + P \cdot n_g \cdot V_g,$$

where $P{\cdot}n_g{\cdot}v_g$ is the mechanical work against the external pressure during voids formation process. Guzman's relation (8.3) becomes the following:

$$\eta = \eta_o \cdot \exp\left[(E_o + P \cdot n_g \cdot V_g)/(R \cdot T)\right] \tag{8.13}$$

and coincides with Eq. (8.12), if expressions $A_P = \eta_o{\cdot}\exp[E_o/(R{\cdot}T)]$, $B_P = n_g{\cdot}V_g/(R{\cdot}T)$ are adopted. This explains the reason for B_P increase with molecule dimension increase: $v_g \cong V/N_A$, increasing when molecular mass increases.

The B_P coefficient from relation (8.12) is not strictly constant and decreases when pressure increases. It is also found that B_P increases rapidly with temperature (especially near the critical temperature) since the void number of n_g from relation (8.13) increases exponentially with T.

Calculation of Viscosity Dependence on Pressure

I. At high reduced temperatures ($T_r > 0.7$, or $V_r > 0.55$), the real (compressed) gas appropriate methods are used—see Sect. 7.7.—for example, the "reduced viscosity" calculation, η/η_c:

$$\eta(T,P) = \eta_c \cdot F(T_r, P_r), \tag{8.14}$$

where η_c is the critical point viscosity, and function F value is read from a diagram, according to the reduced temperature $T_r = T/T_c$ and the reduced pressure $P_r = P/P_c$.

II. For other physical states, $\eta(P)$ is generally calculated from $V(P)$ dependence, using different viscosity functions (and other physical quantities, especially the molar volume V) that do not depend on P, as rheochor and to a lesser extent the orthochor, defined in Sect. 8.3, by relations (8.19) and (8.18). *Andrade's* relation is also usable

$$\eta = V^{1/6} \cdot (\chi_S)^{1/2} \cdot \exp\left[A + B/(V \cdot T)\right], \tag{8.15}$$

with A, B—constants independent of P and T. Thus, if η_o (at $P = 0$) and η' (at $P = P'$) are known, as well as the volumes and the corresponding adiabatic compressibilities, v_o, v', χ_{So}, χ'_S, then viscosity value η can be obtained for any pressure P, having as correspondents v and χ_S (all measurements are at the same temperature), by calculating A and B coefficients.

Variation of viscosity with pressure is important in *tribology*—the technical study of friction, wear, and lubrication of machine parts—for best lubrication conditions choice. The lubrication phenomenon consists of introducing a fluid between two machine parts found into a relative motion, in order to reduce friction; this results in both a wear reduction and a decrease of gear energy consumption. Lubricating fluids are usually mineral oils and more rarely gases, or graphite suspensions.

Lubricants are subject to a number of conditions regarding the most reduced viscosity variation with pressure and temperature: at too high η values, the energy consumption required for movement increases, but for the case of too low viscosities, the oil-to-metal adhesion is not sufficient to hold the oil within the gear, the fluid is therefore gradually expelled and the gear wear is accelerating (Sect. 8.3).

8.3 Viscosity Variation with the Nature of the Liquid

The pre-exponential factor η_0 increases with molecule dimensions. The viscosity activation energy E_η can be correlated with both latent heat of fusion λ_m and of vaporization λ_v, as can be seen in Table 8.2.

Table 8.2 shows for several liquids in comparable physical conditions, the viscosity η and its pre-exponential factor η_0, expressed in g/(ms), then—in

Table 8.2 Viscosity of liquids at melting temperature T_m/K

Liquid	T_m (K)	η (g/(ms))	η_0	E_η	λ_m	λ_v	E_η/λ_v
Ar	84	0.28	1.2	2.2	1.1	6.2	0.35
O_2	55	8.8	2.0	1.8	0.46	7.1	0.24
C_2H_5Br	154	2,5	6.2	4.6	5.9	28	0.17
C_2H_5OH	152	40	2.1	9.6	4.6	39	0.25
H_2O	273	1.8	0.6	12.5	31	41	0.32
CH_3COOH	290	1.2	4.2	7.9	11	24	0.33
NaCl	1077	1.5	2.1	38	22	18.7	0.20
Hg	234	2	56	2.5	2.5	59	0.043
Ag	1233	4.1	57	21	15	140	0.082

kJ/mol—its activation energy E_η and the respective latent heats of fusion, λ_m and of vaporization λ_v.

The first correlation (η, T_m) leads to confirmation of the "through voids" mechanism for viscous flow, since energies E_η and λ_m have the same order of magnitude, while melting follows such a mechanism (Sect. 6.3).

The correlation between E_η and λ_v is, however, more complex. The λ/E_η ratio depends on molecule dimensions and on chemical bonds nature: 3–4 for molecular liquids (higher values for more asymmetric or for longer elongated molecules), 5–15 for ionic liquids, and 8–25 for metallic liquids. Viscosity increases when hydrogen bonds are possible to occur: at an almost identical melting temperature, ethanol has a viscosity 20 times higher than ethyl bromide.

Viscosity increase is even higher in polyols, whose molecules can bind via hydrogen bonds into long chains (diols) or even into spatial structures (substances with at least 3 OH groups).

Water viscosity is small although hydrogen bonds are formed, due to the water structure which is loose, with large voids (at least close to the melting temperature) so that δ is high, leading according to Eyring's relation (8.10) to low values of the pre-exponential factor.

On the contrary, η_o takes high values for metals, so that the value δ is lower, which shows that the metals' mobile particles are not the neutral atoms' atomic groups but the positive ions, which are much smaller; this conclusion is confirmed by the high electrical conductivity of molten metals.

Comparison of Liquid Viscosities

In some cases, the $\ln \eta = f(1/T)$ dependence is not linear as in Guzman's law. In such situations, however, it is possible to define activation energy and pre-exponential factor *apparent* values, varying with temperature:

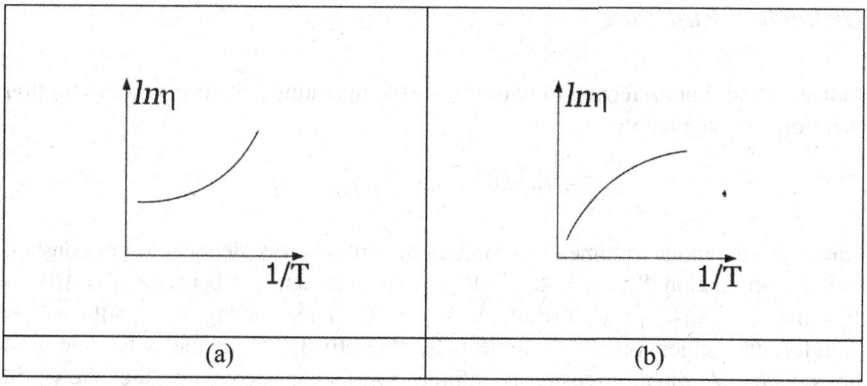

Fig. 8.3 Apparent activation energies dependent on temperature. (**a**) Hydrogen bonds (**b**) Elongated molecules

$$E_{\eta,\mathrm{ap}} \equiv R \cdot [d(\ln \eta)/d(1/T)];\ \ln \eta_{0,\mathrm{ap}} = d(T \cdot \ln \eta)/dT. \quad (8.16\mathrm{a, b})$$

In the case of hydrogen bonding liquids, temperature increase leads to breaking of these bonds, which reduces not only the actual activation energy but also the apparent one, $dE_{\eta,\mathrm{ap}}/d\,T < 0$ (Fig. 8.3a).

On the contrary, for elongated molecule liquid case (mineral oils), temperature increase affects molecule arrangement disorder increase, leading to void dimension increase and thus a pre-exponential factor η_0 decreases. In this case, $d\,E_{\eta,\mathrm{ap}}/d\,T > 0$ as seen in Fig. 8.3b.

Systems of Increments

For the case of molecular organic liquids, η can be expressed directly according to the liquid structure, if other macroscopic quantities Y_1, Y_2, …, are known, like density or surface tension; F is a universal function, and k – atomic increments, of group and of bond. Therefore:

$$F(\eta, Y_1, Y_2, \ldots) = \sum k_\eta. \quad (8.17)$$

The viscosity containing additive-constitutive functions which are mostly used are Orthochor, Rheochor, and Souders function.

Orthochor Function

Indices 1 and 2 are referring to two different temperatures, the orthochor definition therefore is given below:

$$O_{ch} \equiv (\eta_1 \cdot V_1 - \eta_2 \cdot V_2)/(\eta_1 - \eta_2), \tag{8.18}$$

where *V* is the molar volume; it shows that the orthochor is identical to *V*pr constant in Bačinski relation (8.6), $\eta = A/[V - V_{pr}]$; when eliminating A between $\eta_1 = A/[V_1 - V_{pr}]$ and $\eta_2 = A/[V_2 - V_{pr}]$ it results in $V_{pr} = O_{ch}$. The meaning of the orthochor is therefore the "eigenvolume" of molecules, respectively, of the molar volume fictitious value V when viscosity is infinite. Orthochor increments are shown in Table 8.3.

Rheochor Function

When expressing viscosity in mP (1 mP = 10^{-4} kg m^{-1} s^{-1}), the molar mass M—in g/mol and ρ_L, $\rho_{G,sat}$ —the liquid and saturated vapor density in g/cm^3, the R_{ch} rheochor is defined by

$$R_{ch} \equiv \eta^{1/8} \cdot M/\left(\rho_L + 2 \cdot \rho_{G,\ sat}\right). \tag{8.19}$$

The increments for the calculation of rheochor are presented in Table 8.4.

The rheochor is virtually independent of *P* (up to 6000 atm) and of *T*. Thus, knowing the density variation with pressure and temperature, the dependence

Table 8.3 Orthochor function increments (m^3/mol)·10^9

• from a cycle: 6 atoms 6101; 5 atoms – 1258
• from a bond: C-H 6722, C-C 3872, C=C 10621, C-O 5722, C=O 13483, C-N 3725, C=N 14525, C≡N 19755, C-F 14074 , C-Cl 21461, C-Br 26018, C-I 33773, C-S 10835, C=S 23184 , O-H 10677, O-S 17070, N-H 7375, N-O 13186, N=O 11563 , N-N 8641, N-Cl 29075, N-Br 45999

Table 8.4 Components of the rheochor function

Atomic increments
H: 5.5 (in C-H) and 10 (in O-H) or 6 (in N-H)
O: 10 (in C-O-C) and 13 (in C-O-H) or 13.2 (in C=O)
other: C 12.8; N 6.6; Cl 27.3; Br 35.8
Group increments
C_6H_5 101.7; COO (esters, acids) 36; CN 38.9; NO_2 38.9; ONO 39.3

$\eta(P, T)$ can be obtained from (8.19). At lower temperatures, ρ_G is negligible and relation (8.19) becomes (by replacing $\rho_L = M/V_L$):

$$\eta = (R_{ch}/V_L)^8,$$

where V_L of liquid is expressed in cm^3/mole and η in mP.

The Souders function is defined by

$$F_{So} \equiv V \cdot [2.9 + lg\,(lg\;\eta)], \tag{8.20}$$

with V in cm^3/mole and η in mP. Similar to orthocor, F_{So} depends on pressure and it can be used only for the calculation of η at moderate pressures.

All three additive-constitutive functions make possible the precise calculation of viscosity—or its variation with temperature—by means of liquid density. Changes in the heating fluid structure (e.g., dissociation, hydrogen bonds breaking) can be detected by the occurrence of variations with temperature of functions R_{ch}, O_{ch}, or F_{So}.

Viscosity of Liquid Mixtures

It is generally difficult to evaluate the liquid (homogeneous) mixture viscosity starting from the components' viscosities. This is due to the interactions leading to solution structure changes as compared to pure liquids structure.

In the absence of strong interactions, empirical relations are used, such as

$$\eta^{1/3} = \sum\nolimits_{\rm i} X_{\rm i} \cdot \eta_{\rm i}^{1/3};\, lg\,\eta = \sum\nolimits_{\rm i} X_{\rm i} \cdot lg\,\eta_{\rm i};\, R_{\rm Ch} = \sum\nolimits_{\rm i} X_{\rm i} \cdot R_{\rm Ch,i}, \qquad (8.21\text{a, b, c})$$

$$F_{\rm So} = \sum\nolimits_{\rm i} X_{\rm i} \cdot F_{\rm So,i};\, \eta = \sum\nolimits_{\rm i} \left[X_{\rm i} \cdot \eta_{\rm i}/(\gamma_{\rm i})^{1/3}\right], \qquad (8.21\text{d, e})$$

elaborated by Kendall and Monroe, Arrhenius and Kendall, Shukla, Souders, Lima, and Reik; i is the component index, x is the molar function, γ is the "molar fraction"-type activation coefficient, and R_{ch} and F_{So} are the rheochor and Sounder's functions, respectively, from definitions (8.19) and (8.20).

The dependence of composition for mixture viscosity, $\eta(x1, x2, \ldots)$ at given temperature can be obtained from the following relations:

$$lg\,\eta = \sum\nolimits_{\rm i} x_{\rm i} \cdot lg\,\eta_{\rm i} + \sum\nolimits_{\rm i}\sum\nolimits_{\rm j} x_{\rm i} \cdot x_{\rm j} \cdot C_{\rm ij}\eta, j > i, \tag{8.22a}$$

$$lg\,\eta = \sum\nolimits_{\rm i} x_{\rm i}^2 \cdot lg\,\eta_{\rm i} + \sum\nolimits_{\rm i}\sum\nolimits_{\rm j} x_{\rm i} \cdot x_{\rm j} \cdot C_{\rm ij}, j > i, \tag{8.22b}$$

$$V \cdot lg\eta = \sum_i x_i^2 \cdot V_i \cdot lg\eta_i + \sum_i \sum_j x_i \cdot x_j \cdot C_{ij}, j > i, \tag{8.22c}$$

elaborated by Grunberg and Nissan, van der Wyk, respectively, Tamura and Kurata, where V is the molar volumes and C_{ij} are the constants which depend on temperature and on the nature of the two components, but not on their proportion.

Relations (8.21) and (8.22a, 8.22b, and 8.22c) have no sense for solutions with distinct solvent, if the dissolved substances are in their pure state either solids or gases. By denoting with η_0 the solvent viscosity and with η the solution viscosity at a certain concentration C, the behavior of such solutions can be characterized by the relative viscosity η_{rel}, the specific viscosity η_{sp}, or the reduced viscosity η_{red}, which are defined as follows:

$$\eta_{rel} \equiv \eta/\eta_0; \eta_{sp} \equiv \eta/\eta_0 - 1; \eta_{red} \equiv (\eta/\eta_0 - 1)/C. \tag{8.23a, b, c}$$

The η_{rel} and η_{sp} concentration functions are usually increasing. In the diluted solutions field, their increase with C is linear, that is η_{red} depends on components' nature and on temperature, but not on C.

Intrinsic Viscosity

For concentrated solutions, the reduced viscosity depends on C and the solvent–solute system is defined by the intrinsic viscosity, denoted by $[\eta]$, representing the reduced viscosity value at infinite dilution:

$$[\eta] \equiv \lim_{C \to 0} \eta_{red}. \tag{8.23d}$$

For polymer solutions, $[\eta]$ is useful when determining the polymer mean molecular mass M_p, as well as when determining polymer conformation in solution:

$$[\eta] = A(M_p)^B, \tag{8.23e}$$

where A, B do not depend on the molecular mass. Constant B, however, depends on the shape of polymer molecules, ranging from 0 for the spherical shape up to 1 for the elongated shape (as for example in the substituted cellulose aqueous solutions).

In heterogeneous systems of immiscible liquids, Taylor's relation is used:

$$\eta = \eta' \cdot [1 + \varnothing \cdot (\eta' + 2.5 \cdot \eta'')/(\eta' + \eta'')]. \tag{8.24a}$$

Here, η', η'' are the continuous phase and the dispersed phase viscosities, and Ø is the volume fraction of the dispersed phase.

For suspensions, one of the most used relations is Einstein's relation:

$$\eta = \eta_0 \cdot (1 + Ø/2)/(1 - Ø)^2, \qquad (8.24b)$$

where η_o and Ø are the pure liquid viscosity and the volume fraction of the solid phase in suspension.

By Ø powers series expansion of $(1 - Ø)^{-2}$, relation (8.24b) is reduced for diluted suspensions ($Ø \ll 1$) to

$$\eta = \eta_0 \cdot (1 + 2.5 \cdot \phi). \qquad (8.24c)$$

Relations (8.24a, 8.24b and 8.24c) also applies to emulsions and colloidal solutions; if the dispersed phase is lyophilic, the Ø fraction has to be replaced by $Ø \cdot (r'/r)^3$, where r is the *actual* particle radius and r' is the *solvated* particle radius.

8.4 The Flow Process

Flowing Regimes

The flow process—fluid displacement within a force field—can be either laminar or turbulent (Fig. 8.4).

In the *laminar* regime, each particle moves in the applied force direction, or in the maximum potential gradient opposite direction, so that in a homogeneous potential domain all the fluid particles move along parallel trajectories (Fig. 8.4a).

In the *turbulent* regime, in addition to the velocity component which is parallel to the applied force, there is a perpendicular to force velocity component, thus occurring vortices (turbines) as can be seen in Fig. 8.4b.

In the industrial installations, the flow process is usually turbulent, excepting those processes whose rates depend on phase contact surface, processes such as heterogeneous catalysis, adsorption, some distillation or absorption operations achieved within packed columns.

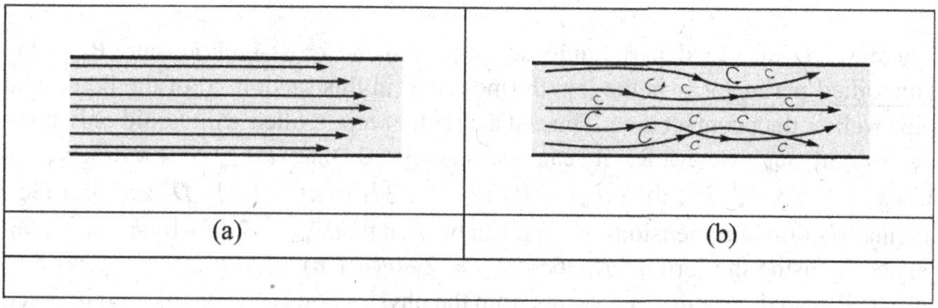

Fig. 8.4 Flowing regimes. (**a**) Laminar flow (**b**) Turbulent flow

For laboratory operations (e.g., in the viscosity measuring devices) a laminar regime is almost always carried out.

In the turbulent regime, friction (and energy consumption for fluid displacement) is much higher than laminar, and the mass or heat transport intensity is greatly increased.

Transport laws are more complex in turbulent regime, for example, friction depends on the solid nature with which the fluid comes into contact and on its roughness, parameters that do not interfere in the laminar regime.

The two flow regimes are differentiated by the different expressions obtained for the friction coefficient f_{fr}, defined by Fanning as below:

$$f_{\mathrm{fr}} \equiv R \cdot (\Delta P) / \left[L \cdot \rho \cdot \overline{v}^2\right], \tag{8.25}$$

where $R = D/2$, D is the D diameter of the pipe and L ($\gg D$) is its length.

Reynolds Criterion

For flow regime determination, the decisive dimension is the value taken by the Reynolds criterion—dimensionless number is defined by the following:

$$[\mathrm{Re}] = \overline{v} \cdot d_{\mathrm{hid}} \cdot \rho / \eta, \tag{8.26a}$$

where $\overline{v}$ is fluid displacement mean velocity (relative to a solid body), ρ and η are the density (volume mass, kg/m^3) and fluid viscosity, and d_{hid} is section's *hydraulic* diameter. When fluids flow through circular cross-section pipelines filled with fluid, d_{hid} is the 2·R pipe diameter, and

$$[\mathrm{Re}] = 2 \cdot \overline{v} \cdot R \cdot \rho / \eta. \tag{8.26b}$$

For noncircular cross-section pipelines, as well as for liquid flow through channels or pipelines only partially filled with liquid, the d_{hid} is given as

$$d_{\mathrm{hid}} \equiv 4 \cdot s_{\mathrm{lic}} / P_{\mathrm{um}}. \tag{8.26c}$$

where s_{lic} is the filled part liquid surface from the cross-section, and P_{um}—the humidified perimeter —is the length (measured in this section, a) of the liquid and pipe wall or duct contact area. Thus, at a circular section filled with liquid only up to the height $h_{\mathrm{lic}} = 1.5 \cdot R$, it can be calculated that $P_{\mathrm{um}} = 4 \cdot \pi \cdot R/3$, $s_{\mathrm{lic}} = (2 \cdot \pi/3 + \sqrt{3}/4) \cdot R^2$, then $d_{\mathrm{hid}} = R \cdot [2 + 3\sqrt{3}/(4 \cdot \pi) \cong 1.21 \cdot D$, and at a rectangular section of dimensions a × b it can be seen that $P_{\mathrm{um}} = 2 \cdot (a + b)$ and $s_{\mathrm{lic}} = ab$, where, by using the formula (8.26c), $d_{\mathrm{hid}} = 2 \cdot ab/(a + b)$.

The Reynolds criterion represents from the physics point of view the ratio between inertial forces (proportional to the unit volume kinetic energy: $\rho \cdot g \cdot \overline{w}^2/2$), and fluid internal friction forces, proportional to $\eta \cdot grad\,\overline{v} = \eta \cdot \overline{v}/a$. For long cylindrical pipes, the limit value is given as

$$[\mathrm{Re}]_{\mathrm{lim}} = 2300. \tag{8.26d}$$

At [Re] < 2300, the flow is laminar, and turbulences appear at values more than 2300. The turbulent regime is unavoidable at [Re] > 10^5.

In the intermediate T, laminar flow is unstable and can only be maintained until a first shock or vibrations occur.

Laminar Flow within a Circular Section Tube

A cylindrical radius R tube filled with fluid is taken into account (Fig. 8.5), which is considered to be a coaxial volume element with the tube and having the radius r and the length L.

It is denoted by v, the fluid velocity local value. It can be seen that v does not depend on time τ (due to a stationary flow), it is parallel to the tube axis and does not depend on the longitudinal position coordinate, x. With the system having a cylindrical symmetry, the velocity v will depend only on the distance r between the considered point and the cylinder axis

There are three forces acting on the hatched section element, namely,

- The F_{fr} friction force, actually distributed over the entire lateral surface (but with a resultant force acting on cylinder axis, due to symmetry).
- The two compression forces F_1 and F_2, acting on cylinder bases and which are due to static pressures P_1 and P_2. At equilibrium,

$$F_{fr} = F_2 - F_1, \tag{a}$$

where $F_{fr} = f \cdot S_a$; $F_2 = P_2 \cdot S_b$; $F_1 = P_1 \cdot S_b$, where S_a and S_b are the cylinder lateral and frontal area, respectively, $S_a = 2 \cdot \pi \cdot r \cdot L$; $S_b = \pi \cdot r^2$.

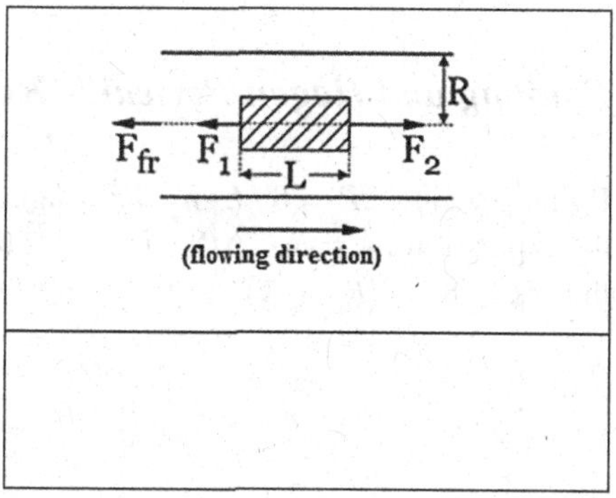

Fig. 8.5 Forces balance at flowing into a cylindrical tube

f from equality (a) represents the specific friction force from the law (8.2): $f = -\eta \cdot dv/dr$.

By replacing S_a, S_b, and f in relation (a) it results into the following: $-2 \cdot \pi \cdot r \cdot L \cdot \eta \cdot dv/dr = \pi \cdot r^2 \cdot (P_2 - P_1)$. By denoting with $\Delta P = P_2 - P_1$ the pressure drop and by separating the variables, it is found that $2 \cdot L \cdot \eta \cdot dv = -(\Delta P) \cdot r \cdot dr$, where, through undefined integration,

$$4 \cdot L \cdot \eta \cdot v = -(\Delta P) \cdot r^2 + C_1. \tag{b}$$

The integration constant C_1 is obtained by imposing the condition that at tube periphery—in contact with the wall—the fluid velocity is null: $v_{r = R} = 0$, where $C_1 = (\Delta P) \cdot R^2$, or after replacing the constant C_1 in (b):

$$v = \left(R^2 - r^2\right) \cdot (\Delta P)/(4 \cdot L \cdot \eta). \tag{8.27}$$

Relation (8.27) represents the $v(r)$ distribution of local velocities in the tube cross-section, depending on distance r from the tube center; the velocity gradually increases from tube's periphery to its center. The maximum velocity will be reached on the cylinder axis, $v_{max} = v_{r = 0}$, namely,

$$v_{max} = R^2 \cdot (\Delta P)/(4 \cdot L \cdot \eta). \tag{c}$$

The mean velocity is obtained by integrating on the tube cross-section $s : \overline{v} = \int v \cdot ds / \int ds$, where the infinitesimal element ds is the ring area, centered on the tube axis, between the circles of radius r and $r + dr$. As $s = \pi \cdot r^2$, therefore, $ds = 2 \cdot \pi \cdot r \cdot dr$. Thus, by calculating the integral from the denominator (equal with $\pi \cdot R^2$) and by simplifying: $\overline{v} = \left(2/R^2\right) \cdot \int_{r=0}^{r=R} v \cdot r \cdot dr$. By replacing $v(r)$ from (8.27): $\overline{v} = \left[(\Delta P)/(2 \cdot L \cdot \eta) \cdot \int_{r=0}^{r=R} \left(r - r^3/R^2\right) \cdot dr\right.$ and after a definite integration:

$$\overline{v} = R^2 \cdot (\Delta P)/(8 \cdot L \cdot \eta). \tag{8.28}$$

Thus, the mean velocity is half of the maximum velocity given by relation (c).

Fanning and Hagen-Poiseuille Relations

By introducing $\Delta P = 8 \cdot L \cdot \eta \overline{v}/R^2$ obtained for laminar flow from relation (8.28) in the expression $f_{fr} = R \cdot (\Delta P)/\left[L \cdot \rho \cdot \overline{v}^2\right]$ (8.25) of the friction coefficient, it is found that $f_{fr} = 8 \cdot \eta/(R \cdot \rho \cdot \overline{v})$.

By comparing with expression (8.26b),

$$[\mathrm{Re}] = 2 \cdot \overline{v} \cdot R \cdot \rho/\eta$$

of Reynolds number for filled pipes of circular cross section, it results in the following:

$$f_{\mathrm{fr}} = 16/[\mathrm{Re}]. \tag{8.29}$$

Fanning's relation (8.29) shows that in the laminar regime, the friction coefficient depends only on [Re] and is inversely proportional to it.

The mean velocity is also related to flow rate. The mass flow $\dot{m}$, defined as the mass of liquid flown in the unit of time, namely,

$$\dot{m} = m/\tau,$$

where m being the mass and τ the flow duration, is obtained by the following equation:

$$\dot{m} = \overline{v} \cdot S_{\mathrm{tr}} \cdot \rho, \tag{8.30a}$$

where $S_{\mathrm{tr}} = \pi{\cdot}R^2$ is the total cross-section area, ρ is the density, while the mean velocity is $\overline{v} = R^2 \cdot (\Delta P)/[8 \cdot L \cdot \eta]$ according to relation (8.28), where

$$\dot{m} = \pi \cdot R^4 \cdot \rho \cdot (\Delta P)/(8 \cdot L \cdot \eta). \tag{8.30b}$$

which represents the *Hagen-Poiseuille* relation.

Turbulent Flow

In the turbulent regime, velocity distribution is—Fig. 8.6b—a much more flattened dependence than the second-degree parabola in Fig. 8.6a of the laminar flow. Instead of the $v(r)$ dependence from relation (8.27), for the turbulent regime, the following expression is found:

$$v = K \cdot \left[1 - (r/R)^{\mathrm{A}}\right], \tag{d}$$

where the K factor is a dependency function of density, viscosity and ΔP L(pressure gradient), while the A exponent depends on liquid adhesion at the wall and on the wall smoothness.

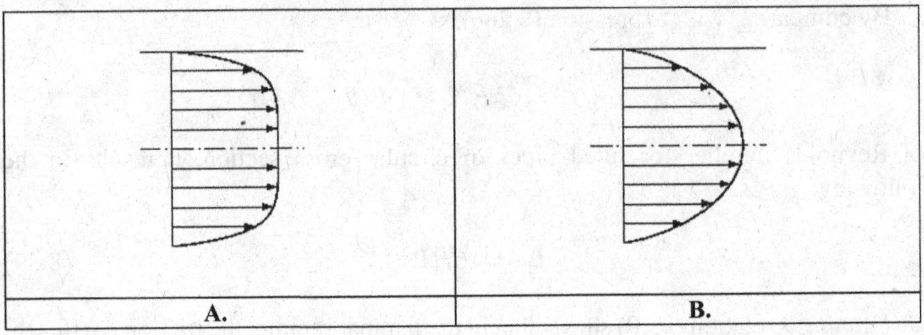

Fig. 8.6 Profile of flow velocity through a circular section pipe (**a** Laminar (**b**) Turbulent

By introducing the velocity profile (d) in relation (8.28) of the medium velocity, it results in the following:

$$\overline{v} = K \cdot \int_{r=0}^{r=R} 2 \cdot \pi \cdot \left[1 - (r/R)^{A}\right] \cdot dr \Big/ \int_{r=0}^{r=R} 2 \cdot \pi \cdot r \cdot dr,$$

where after integration, the following relation is obtained $\overline{v} = K \cdot A/(A+2)$, therefore, since—according to relation (d)—K is the maximum velocity:

$$\overline{v} = v_{\max} \cdot A/(A+2). \tag{8.31a}$$

For the friction coefficient in the turbulent regime, it results in the following:

$$f_{\mathrm{fr}} = B/[\mathrm{Re}]^{2/A}, \tag{e}$$

in which B is obtained from the following condition: at $[\mathrm{Re}] = [\mathrm{Re}]_{\mathrm{lim}} = 2300$, the value f_{fr} is identical both in laminar and turbulent regime; according to relations (8.29) and (e):

$B/\{[\mathrm{Re}]_{\mathrm{lim}}\}^{2/A} = 16/\{[\mathrm{Re}]_{\mathrm{lim}}$, that is, $B = 16{\cdot}2300^{2/A-1}$, where

$$f_{\mathrm{fr}} = (16/2300)/\{[\mathrm{Re}]/2300\}^{2/A}. \tag{8.31b}$$

In smooth pipes $A = 7 \pm 1$ while in the rough ones (with asperities) $A = 11 \pm 1$.

Energy Consumption in Different Flow Regimes

From the Hagen and Poiseuille relation, the P_{fr} power necessary to overcome the friction forces $P_{\mathrm{fr}} = L_{\mathrm{fr}}/\tau$ is also obtained, where L_{fr} is the work of friction forces, consumed during time τ.

But $L_{fr} = \Delta P \cdot v_{ext}$, where v_{ext} is the total volume flown during the same period of time, which is expressed as the ratio of total leaked mass and fluid density: $v_{ext} = m/\rho$ so that $P_{fr} = \Delta P \cdot m/(\rho \cdot \tau)$ or, by replacing the m/τ ratio with the mass flow $\dot{m}$:

$$P_{fr} = \Delta P \cdot \dot{m}/\rho. \tag{f}$$

Laminar Regime By introducing $\dot{m} = \pi \cdot R^4 \cdot \rho \cdot (\Delta P)/(8 \cdot \pi \cdot L)$ from relation (8.30b) in the (f) expression, it results in the following:

$$P_{fr} = P_{tr} = 8 \cdot L \cdot \eta \cdot \dot{m}/\left(\pi \cdot R^4 \cdot \rho^2\right). \tag{g}$$

Therefore, for a certain given pipe and fluid (L, R, η, ρ constants), the transport necessary power increases in the *laminar* regime with flow's square value.

Turbulent Regime By replacing $[\mathrm{Re}] = 2 \cdot R \cdot \overline{v} \cdot \rho/\eta$ from definition (8.26b) in relation (8.31b), $f_{fr} = (16/2300)/\{[\mathrm{Re}]/2300\}^{2/A}$, valid for turbulent flow regime, the expression $f_{fr} = (16/2300)/\left[2 \cdot R \cdot \overline{v} \cdot \rho/(2300 \cdot \eta\right\}^{2/A}$ is found, which is introduced in relation (8.25) $f_{fr} \equiv R \cdot (\Delta P)/\left[L \cdot \rho \cdot \overline{v}^2\right]$, leading to the following: $\Delta P = \left[16 \cdot L \cdot \rho \cdot \overline{v}^2/(2300 \cdot R)\right]/\left[2 \cdot R \cdot \overline{v} \cdot \rho/(2300 \cdot \eta\right\}^{2/A}$, which finally leads, for the power consumption from relation (f), at the following dependence:

$$P_{fr} = \left(\frac{16 \cdot L}{2300 \cdot \pi^2 \cdot R^5 \cdot \rho^2}\right) \cdot \left(\frac{2300 \cdot \pi \cdot \eta \cdot R}{2}\right)^{2/A} \cdot (\dot{m})^{3-(2/A)}, \tag{8.32}$$

which is reduced to expression (g) in case of laminar flow ($A = 2$).

Thus, in turbulent mode—where A value is well above 2—the consumed energy increases almost as the *cube* of the flow rate, while at laminar regime it was proportional to the *square* of the flow rate.

8.5 Rheology of Liquids

Rheology is the study of the deformation and flow of bodies under external forces action. In rheology, the body surface tangential stress is particularly important, being characterized by the tangentially applied *shear* stress f and defined as force F applied to the unit of area, S denoting the area:

$$f \equiv F/S. \tag{8.33a}$$

Tangential forces produce an angular *shear* deformation, where the body volume remains constant, but the angles between a given volume faces change. The

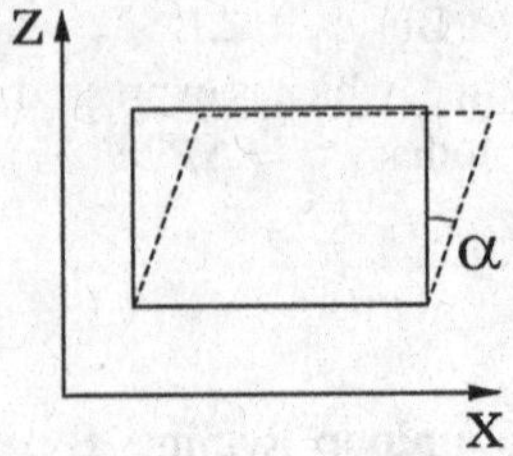

Fig. 8.7 Shear deformation

magnitude of these deformations is characterized, as presented in Fig. 8.7, by the angle α between the considered sides initial and final position. The main feature of the flow process is the strain rate $d\alpha/d\tau$ – the variation rate of the rotation angle velocity, where τ is time, the velocity being equal to the gradient dv/dz of linear flow velocity v along a direction z, perpendicular to the flow direction x:

$$d\alpha/d\tau = -dv/dz, \tag{8.33b}$$

The most general rheological law has the following form:

$$\varphi(f, \alpha, df/d\tau, d\alpha/d\tau) = 0, \tag{8.34a}$$

wherein the φ function is considered, besides the *four* rheological listed variables, also different material constants, depending on the substance nature, pressure, and temperature. Three particular rheological law limits—cases are often encountered, that is, cases where two variables occur, namely,

$$\text{viscous behavior}: \ \varphi(f, d\alpha/d\tau) = 0 \tag{8.34b}$$

$$\text{elastic behavior}: \ \varphi(f, \ \alpha) = 0 \tag{8.34c}$$

$$\text{plastic behavior}: \ \varphi(\alpha, df/d\tau) = 0 \tag{8.34d}$$

Sometimes there are behaviors—intermediaries between the precedent dependencies—where the phenomenological description contains *three* rheological variables:

$$\text{viscoelastic behavior}: \ \varphi(f, \alpha, d\alpha/d\tau) = 0 \tag{8.34e}$$

$$\text{elastic-plastic behavior}: \ \varphi(f, \alpha, d\,f/d\tau) = 0 \tag{8.34f}$$

The elastic, plastic, and elastic–plastic behaviors—the last two being described by integral or integral–differential equations—are characteristic of solids, forming the study object of material technology. Gases can only have a "viscous" behavior, and liquids can be viscous (usually) or viscoelastic. Rheological laws are analytical relations in the case of fluids.

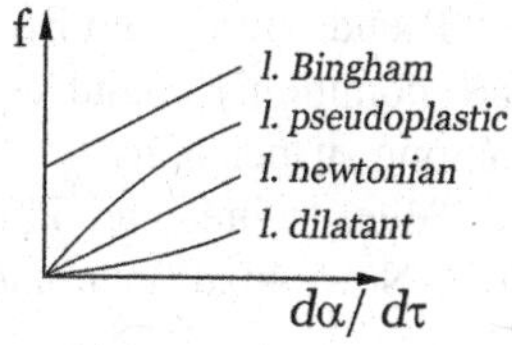

Fig. 8.8 Non-Newtonian liquids

Non-Newtonian Liquids

Those viscous liquids for which there is a proportionality between deformation velocity $d\alpha/d\tau$ and tension f, are referred to as Newtonian liquids, since $f = -\eta \cdot dv/dz$ and $\eta \cdot d\alpha/d\tau$ are equivalent, when considering identity (8.33b), with Newton's law, $f = \eta \cdot d\alpha/d\tau$ from relation (8.2).

Non-Newtonian liquids are liquids with viscous behavior (f depending only on $d\,\alpha/d\,\tau$) but at which tension f is not proportional with shear velocity, as in three of Fig. 8.8 examples. For non-Newtonian liquids, two viscosities can be distinguished, namely, an *integral* viscosity η' and a *differential* one η," at which Newtonian liquids coincide and do not depend on f (or on $d\alpha/d\tau$), while for non-Newtonian liquids η' and η'' are generally different, that is, they vary with tension and are calculated from the following definitions:

$$\eta' \equiv \mathrm{f}/(d\alpha/d\tau); \eta'' \equiv df/d(d\alpha/d\tau), \tag{8.35}$$

while shear velocity dependence of tension $f\ d\alpha/d\tau$ is no longer a straight line passing through the origin.

Types of Non-Newtonian Rheology Liquids

The non-Newtonian fluids from Fig. 8.8 are characterized as follows:

(a) *Dilatant* liquids: the curve $\mathrm{f}(d\alpha/d\,\tau)$ passes through the origin and has a positive curvature so that both η' and η'' increase with f (or with $d\alpha/d\tau$) and $\eta' < \eta''$.
(b) *Pseudoplastic* liquids: the curve $\mathrm{f}(d\alpha/d\tau)$ also passes through the origin, but has a negative curvature, therefore: $\eta' > \eta''$, and both of them decrease when f increases.
(c) *Bingham*-type liquids, at which, for tensions of values below f_0, the liquid does not flow nor deforms, and at $f > f_o$, the f dependence on $(d\alpha/d\tau)$ is linear, so that η'' is independent of tension; η' decreases when f increases, is always higher than η'' and tends to infinity when $d\alpha/d\tau$ tends toward 0 (i.e., at tensions no greater than $\mathrm{f_o}$):

$$f \le f_0 \to d\alpha/d\tau = 0; f \ge f_0 \to f = f_0 + \eta' \cdot d\alpha/d\tau. \tag{8.36}$$

Pseudoplastic and Bingham liquids $f \leq f_0 \rightarrow d\alpha/d\tau = 0; f \geq f_0 \rightarrow f = f_0 + \eta' \cdot d\alpha/d\tau$ are commonly found among colloidal and macromolecular solutions, as well as among suspensions.

Ellis, Reiner and Philippoff, and Shulman equations have been proposed for non-Newtonian flow, namely,

$$g = A \cdot f + R \cdot f^{C}, \tag{8.37a}$$

$$g = f/\left[A + B/\left(C + f^2\right)\right], \tag{8.37b}$$

$$g = C \cdot \left(f^{A} - B\right)^{D}, \tag{8.37c}$$

with *A*, *B*, *C*, *D* are non-negative constants. Note that Eyring's viscosity model involves a flow that is Newtonian only at tensions tending toward 0; otherwise, the relation (8.9) series expansion is no longer truncated, and a pseudoplastic behavior is obtained at high tensions, namely:

$$g = A \cdot sh(f/B). \tag{8.37d}$$

Viscoelastic Behavior

The stress depends not only on the shear velocity $d\alpha/d\tau$ but also on the shear deformation α, according to relation (8.34e), for viscoelastic bodies (polymers, amorphous solids, gels, concentrated suspensions, rocks during geological periods) case. However, since α can be in turn considered a dependency function of the shear velocity $d\alpha/d\tau$ and of duration τ (measured from the shear process beginning), $\alpha = \alpha(\tau, d\alpha/d\tau)$, it can be assumed that tension f itself depends on the same parameters:

$$f = F(\tau, d\alpha/d\tau). \tag{8.38}$$

At a sufficiently long duration shear, f does not depend on time τ anymore, the body presenting a viscous behavior, $lim_{\tau \rightarrow \infty} (df/d\tau)_g = 0$.

This limit behavior at long periods of time can be described using coordinates (f, g) through the curve in Fig. 8.9, which brings together the Bingham fluid and pseudoplastic fluid features, that is, by denoting $d\alpha/d\tau$ with g:$f < f_o \rightarrow g = 0$; but

$$f > f_o \rightarrow \left(df/dg = 0 \cap d^2 f/dg^2 < 0\right) \tag{8.39}$$

Fig. 8.9 Viscoplasticity

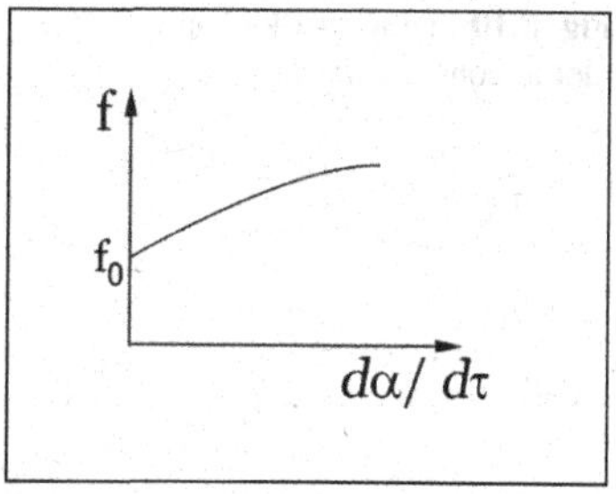

Time as State Variable in Rheology

Often the "viscoelastic" adjective is improperly attributed to the non-Newtonian pure viscous fluid with the dependence of the f value on $d\alpha/d\tau$, as can be seen in Fig. 8.9, but which does not have any prior shear stresses.

The confusion is due to the fact that time (as well as the spatial position) is *not* a state quantity in thermodynamics or kinetics, matter properties at one certain point depending only on the other physical variables (pressure, temperature, composition, etc.) at the same time values.

However, the situation is totally different in rheology, where (excepting the limit cases of purely viscous or purely elastic behaviors) matter properties also depend on the respective body history, which complicates the analysis of phenomena.

The understanding of rheological behavior is sometimes facilitated by the use of temporal plots, which show the evolution of one or more physical quantities (on the ordinate) as functions of the time represented on abscissa.

Thus, as can be seen in Fig. 8.10, in relation to the temporal evolution $f = f(\tau)$ of tangential tension at finite flow times (τ), the viscoelastic liquids are either *thixotropic* if, at constant shear rate (denoted by g), tension decreases in time $(df/d\tau)_g < 0$, or *rheopectic* if tension increases in time, $(df/d\tau)_g > 0$.

Structural Explanations of Viscoelasticity

If the dissolved substance dispersed particles or its (large) molecules present an elongated (water-soluble viscose) or lamellar (kaolin suspensions in water) shape, during the flow process, a particle axis parallel alignment to the flow direction occurs, reducing thus the tension.

Sometimes this thixotropic tension drop is so strong that an apparently solid material suddenly begins to flow when the applied force exceeds a certain limit (quicksand case).

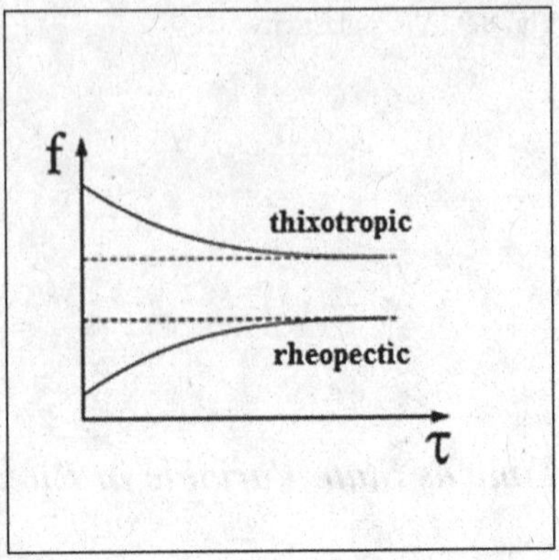

Fig. 8.10 Time rheological plot at constant $d\alpha/d\tau$

In rare cases (polymethacrylic acid aqueous solutions, gypsum suspensions), the flow leads to structuring—increasing the contact points number between particles—producing "hardening" of solution or suspension (rheopexia). Concentrated bentonite aqueous suspensions are thixotropic, but by dilution they become rheopectic.

Maxwell Viscoelastic Model

In *Maxwell's* model, the total time variation of deformation is the sum of a viscous component $d\alpha_v/d\tau = f/\eta$, given by Newton's law (8.2), and an elastic component, obtained by deriving Hooke's law (8.1) deformation $\alpha_e = f/G$ in relation to time $d\alpha_e/d\tau = (d\,f/d\tau)/G$:

$$d\alpha/d\tau = f/\eta + (df/d\tau)/G, \tag{8.40a}$$

For the case of zero shear-rate the solution of the differential equation (8.40a) is:

$$f = f_o \cdot exp\,(-\tau/\tau_r) \tag{8.40b}$$

where f_o is tension's initial value and τ_r is the time required for the e times tension drop, a time duration referred to as *relaxation* time of the tangential mechanical stresses and defined by viscosity to elastic modulus G ratio:

$$\tau_r = \eta / \cdot G. \tag{8.40c}$$

The relaxation time is of milliseconds order for common liquids and of millions of years for crystals, while pastes and amorphous solids (polymers, glasses) present intermediate values.

Equations (8.40) have two limit cases: $\tau_r \to 0$ for the purely viscous body and $\tau_r \to \infty$ for the elastic body.

8.6 Heat Conduction

Mass and Heat Transfer in Condensed States of Aggregation

Similar to gases, liquids and solids respect Fourier and Fick laws, but the phenomenological coefficients have completely different values. D, η, and λ variations with different physical factors are totally different — both from each other and from these quantities' variations for the gases case. The only common traits are the very pronounced dependence on substance nature (from one liquid or solid to another, the transport coefficients could vary with several orders of magnitude, while in gas they vary at most as from 1 to 10), as well as the fact that pressure effect is relatively weak and seems to depend mainly on the volume reduction (pressure variation is even more pronounced as temperature is higher): for a decrease by half (D, η) or a doubling (λ) of these quantities with respect to their low pressure values, it is generally necessary to raise the pressure to several thousands of atmospheres if the temperature is sufficiently distanced from the critical temperature.

Thermal Conductivity of Liquids

Table 8.5 shows a relatively pronounced variation of thermal conductivity λ with the liquid nature, namely of two orders of magnitude—a stronger variation than for gas case, but much less pronounced than viscosity variation from one liquid to another.

Thermal conductivity order of magnitude in liquids depends especially on liquid chemical bonds nature: λ is small for molecular fluids, slightly higher when hydrogen bonds are formed within the molecular bonds and very high for ionic and especially metallic liquids.

Ultraviscous liquids (amorphous solids) have the same order of magnitude, related to the shares of ionic and covalent bonds numbers, or the polar bond degree of ionicity.

Table 8.5 Thermal conductivity of different liquids

		λ (J/(ms K))		
Substance	t (°C)	Liquid	Gas	Remarks
C_6H_6	20	0.145	0.011	Molecular
n-C_5H_{12}	20	0.134	0.013	Molecular
CH_3OH	20	0.215	0.017	H bounds
H_2O	20	0.60	0.017	Intense H bounds
NaCl	820	4.5	/	Ionic
Hg	20	8.3	0.002	Metallic
Na	100	85	/	Metallic
Glass	20	0.8	/	Amorphous
Celluloid	20	0.4	/	Amorphous

Thermal conductivity is higher for liquids (while ratio is inverse for viscosity) toward vapors, under comparable conditions. The λ_L/λ_G ratio ranges from 10 to 4000 in the examples from the table, the molecular fluids corresponding to lower ratio values.

Temperature increase leads to liquids thermal conductivity decrease, unlike gases where λ increases with temperature; therefore, when moving toward the critical point on the liquid–vapor equilibrium curve, the λ_L/λ_G ratio decreases, tending toward 1. The λ(T) variation is, however, less pronounced unlike the analogous exponential viscosity variation: from the melting point to the critical one, λ decreases 2–3 times.

Pressure increase leads to liquid conductivity increase, the variation being of the same order of magnitude, as in the viscosity case (doubling the phenomenological transport coefficient to 2000–10,000 bar).

Models of Energy Transfer into Liquids

The thermal conductivity theoretical models in liquids have led to the conclusion that λ increases with velocity of propagation of sound u_s, but decreases when liquid molar volume increases, as in the Bridgman's relation where $(V/N_A)^{1/3}$ is the mean distance between two molecules, namely,

$$\lambda = 2,8 \cdot k_B \cdot u_s/(V/N_A)^{2/3}, \tag{8.41a}$$

in Kardos model λ is proportional to the liquid (isobaric) heat capacity C_p and to the distance of molecule free displacement, $L_f = 95 \cdot 10^{-12}$ m:

$$\lambda = L_f \cdot C_P \cdot u_s/V. \tag{8.41b}$$

The sound velocity is computable from thermodynamic data: $u_s = (\rho \cdot \chi_S)^{-1/2}$, the adiabatic compressibility χ_S being obtained from the isotherm (χ_T) by $\chi_S = \chi_T \cdot C_V / C_P$, so that

$$u_s = [C_P/(C_V \cdot \rho \cdot \chi_T)]^{1/2}, \tag{8.42a}$$

where the heat capacity C_v can be obtained when knowing the coefficient of expansion α, through the following:

$C_V = C_P - \alpha^2 \cdot V \cdot T/\chi_T$, where

$$u_s = \left(\rho \cdot \chi_T - \alpha^2 \cdot M \cdot T/C_P\right)^{-1/2}. \tag{8.42b}$$

As an alternative, $C_V \cong C_P$ and relation (8.42a) is approximated by the following:

$$u_s = (\rho \cdot \chi_T)^{-1/2}. \quad (8.42c)$$

Relations (8.41a) and (8.41b) are relatively well verified for molecular fluids but cannot explain thermal conductivity high values of ionic and metallic liquids. At these, λ is proportional to electrical conductivity λ_{el} (which is more readily to be accurately measured) as in relation (8.43):

$$\lambda = K_{Lo} \cdot \lambda_{el} \quad (8.43)$$

of Wiedemann and Franz, where Lorenz coefficient K_{Lo} slightly depends on temperature (increases typically by 10% for a 600 ± 100 K heating), on pressure or on substance chemical formula, but is different for ionic liquids (saline melts) when compared to the metallic ones, where free electrons are the energy and electric charge carriers, not the ions.

Heat Conduction in Solids

For solids, thermal conductivity greatly depends on the presence of macropores and of empty reticular nodes. Thus, while compact glass has $\lambda = 1.2$ J/(ms K), the glass wool has a 30 times lower thermal conductivity due to the high space proportion occupied by air, which is a poor heat conductor. Conductivity varies roughly proportional to the apparent density for such porous materials. The solid texture also has an important effect on its thermal conductivity.

Compact solids have a conductivity that depends on the lattice nature: λ is higher for pure ionic solids and especially for metals, where λ expressed in J/(ms K) is: Fe 60, Al 230, Cu 370, Ag 410. Conductivity has reduced values in molecular crystals or polarized covalent ones (organic polymers, glass). Correlation with lattice nature is not absolute—metal alloys with low λ, such as invar (10) or kanthal (13), can be obtained.

The Wiedemann and Franz law (8.43) applies equally to solids and liquids, but the factor of proportionality between electrical and thermal conductivity differs according to the state of aggregation.

8.7 Diffusion

Diffusion in Solids

Similar to heat conduction, diffusion depends great lv on solid's nature and texture.

The presence of macropores, two-dimentional defects (contact surfaces between granules) or of open miripores (with a few Å dimensions) greatly faciliate the impurity diffusion in solids.

Diffusion in Liquids

The hydrodynamic theories developed by Sutherland and Einstein consider the liquid as a continuous medium (solvent, index 1) in which molecules of the dissolved substance (also called solute, index 2) move. The transfer intensity is given by the compensation of the driving force $F_m = (N_A)^{-1} \cdot grad\ \mu_2$ applied to a molecule (μ_2 is the chemical potential of solute) by the resistant force $F_r = f_S \cdot v$ – the product of frictional resistance f_S and velocity v given to the molecule:

$$(N_A)^{-1} \cdot \text{grad}\ \mu_2 = f_S \cdot v. \tag{8.44a}$$

The mean displacement velocity v of solute molecules through the solvent molecules is given by mass flow density J_2 (amount of substance in mol, transported in the unit of time through the unit of area): $v = - J_2/c_2$, where c_2 is the molar concentration (i.e., expressed in mol/m^3) of species 2. But according to Fick's first law (7.6): $J_2 = D_{21} \cdot grad\ c_2$, hence, since $c^{-1} \cdot dc = d(lnc)$:

$$v = D_{21} \cdot d(\ln c_2)/dy, \tag{8.44b}$$

where y is the spatial coordinate (length) in the diffusional flow direction. A diffusion coefficient is found from relations (8.44a) and (8.44b):

$$D_{21} = (N_A \cdot f_s)^{-1} d\mu_2/d(\ln c_2). \tag{8.45a}$$

For ideal solutions case, $\mu_2 = \mu_2^{o} + R \cdot T \cdot ln\ c_2$, where it results into the following: $d\mu_2/d(ln\ c_2) = R \cdot T$. Thus it is found that $D_{21} = R \cdot T/(N_A \cdot f_S)$ or, by replacing R with k_B/N_A:

$$D_{21} = k_B \cdot T/f_S, \tag{8.45b}$$

which represents the Nernst and Einstein equation of diffusion.

Diffusivity/Viscosity Correlation for Liquids

With the aid of classical hydrodynamics, Stokes calculates the frictional resistance f_s through that of a rigid sphere in laminar motion within an incompressible continuous liquid:

$$f_s = 6 \cdot \pi \cdot \eta_1 \cdot r_2 \cdot (\beta \cdot r_2 + 2 \cdot \eta_1)/(\beta \cdot r_2 + 2 \cdot \eta_1). \qquad (8.46a)$$

In relation (8.46a), η_1 is the liquid viscosity, r_2 is the spherical radius of the solute molecule, and β is a sphere/liquid coefficient of friction. The limiting case β values are as follows: $\beta = \infty$ if the fluid completely wets the sphere and $\beta = 0$ for the liquid that does not wet at all the sphere. For these limit cases, $fs = 6 \cdot \pi \cdot \eta_1 \cdot r_2$, respectively, $fs = 4 \cdot \pi \cdot \eta_1 \cdot r_2$. Sutherland admits that moistening is total, that is, it can be considered that $\beta = \infty$; the *Stokes and Einstein* equation is obtained as follows:

$$D_{21} = k_B \cdot T/(6 \cdot \pi \cdot \eta_1 \cdot r_2). \qquad (8.46b)$$

Relation (8.46b) is accurate only for large solute molecules (as compared to solvent molecules) and of quasi-spherical shape. From this equation, it follows that:

- Diffusivity D varies with the solute molecule dimension. Thus, macromolecules slowly diffuse in any solvent (r_2 radius is high), for certain impurity diffusion in molten metals it is found that there is a more rapid diffusivity in the case of small impurities; for example, Li atoms diffuse faster than those of Na or K in molten Al.

  On the other hand, solvent change can modify the diffusivity order of various ions: the diffusion of the Na^+ ion in water is faster than the one of Li^+ ion, since the Li^+ ion, with its lower radius before solvation, can polarize more strongly the water molecules, which leads to the formation of a much larger hydration shell; the hydrated ion radius is therefore considered in calculations.
- Walden's law: "for a solvent and a certain solute, the $D \cdot \eta/T$ product does not depend on temperature." Since viscosity presents an exponential decrease with T, it is deduced that diffusivity increases exponentially when temperature rises.

Diffusion activation energy in liquids is practically equal to the viscous flow activation energy and is the same for all dissolved solutes into a certain solvent.

Crystalline Lattice Defects

Any crystal encountered in reality has a different structure from the ideal one, by the presence of a certain number of *lattice defects*—deviations from the regular particles arrangement in the nodes.

Defects presence is inherent even under thermodynamic equilibrium conditions, except the 0 K temperature. But frequently the defects occur as a result of the crystal formation irreversible process, as in the case of three-dimensional "gas bubbles" defects.

Gas Bubbles from the real crystal are formed when a certain melt crystallization takes place by a cooling so rapid that the melt dissolved gases but insoluble in the crystal does not have time to leave the melt.

Defect Classification According to their Dimension Number

Crystal defects are classified according to dimension number as follows: three-dimensional, two-dimensional, one-dimensional, and zero dimensional.

Three-Dimensional Defects

The three-dimensional defects extend in all three spatial directions with a distance of at least several hundred atoms, being thus a solid *textural feature*, rather than a structural one. Besides the above-mentioned gas bubbles, three-dimensional defects also include also solid *inclusions* from other materials, as well as *amorphous areas* (occurring in crystalline polymers).

Three-dimensional defects form, in particular, the object of study for material technology, metallography, and petrography. Three-dimensional defects are nonequilibrium formations, and their occurrence can be avoided if measures are taken to grow crystals at a sufficiently low rate.

Two-Dimensional Defects

The two-dimensional defects are inherent to any crystal: the entire free surface of a crystal, even ideal (crystal and vacuum interface, or crystal and another phase interface), represents a two-dimensional defect, since the coordination number is different—lower for particles on the crystal surface than for the analogous particles arranged in the crystal depth.

The crystal surface is thus an area of increased energy and special properties, which is studied by surface chemistry and which will be described in a future book of a physical chemistry course.

Another type of two-dimensional defects occurs on same composition microcrystals contact surfaces, in the case of polycrystalline solid masses (composed of microscopic or submicroscopic dimension crystals). Due to crystalline lattice different orientation, the connection between a certain microcrystal and another cannot be made directly, but through *superficial contact areas* of thickness equivalent to several atomic diameters.

These areas are characterized by a continuous flexion of crystalline lattice planes, which ensures a gradual passage between the two neighboring microcrystal orientations.

Strong lattice deformations equate to the contact area *amorphization*, resulting in reduced density, increased energy, and a higher degree of disorder (equivalent to the area entropy growth).

Impurities are easily concentrated in the superficial contact area, so that the polycrystalline mass mean purity is always lower than that of the monocrystal obtained from the same raw material, even when the isolated microcrystals do not have a high content of impurities.

The low density of superficial contact areas facilitates diffusion and self-diffusion through solid material: the superficial diffusion flow can be comparable or even higher to the normal (volume) diffusion flow.

One-Dimensional Defects

The one-dimensional defects, with a length of at least several hundred of atoms, have a width in the range of one to several atomic diameters. The number and mobility of these defects—referred to as *dislocations*—determine the solid mechanical properties: the ideal crystal tear resistance is several hundred times higher than that of the real crystal with dislocations.

Zero-Dimensional Defects

The zero-dimensional or *punctiform* defects, ineluctable in any real crystal, are located in a single lattice point, as in Fig. 8.11.

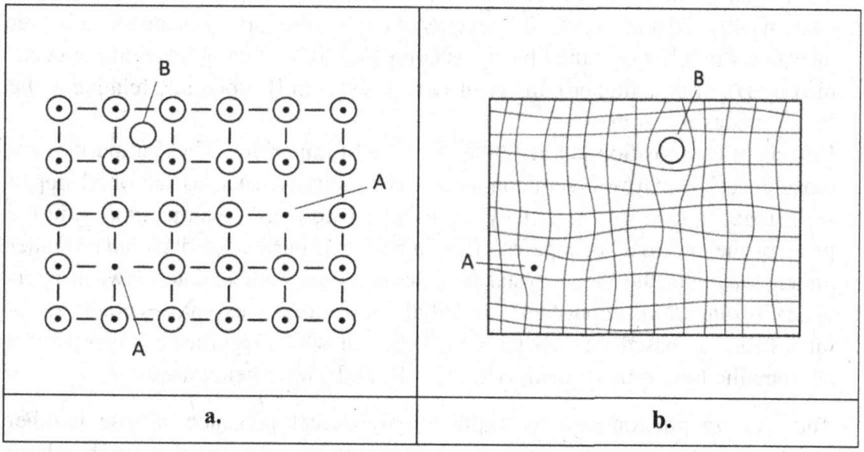

Fig. 8.11 Punctiform defects. (**a**) Type of defect (**b**) Lattice deformation

Punctiform Defects in Simple Lattices

In simple crystals (metals, B, diamond, Si), which have only equivalent monatomic particles in their nodes, there can be only two types of punctiform defects: a vacant *node* or an occupied *internode* ("*internode*"—the space between neighboring nodes), denoted by A and B, respectively, in Fig. 8.11.

The black points from Fig. 8.11a represent the lattice nodes while the circles—the atoms. Both the vacant nodes and the occupied internodes modify somehow the structure of the lattice neighboring area, phenomenon illustrated in Fig. 8.11b; the crystalline planes are slightly close to each other in a free node area and slightly scattered in an internode area, as compared to the normal interplane distances from an ideal crystal, or from the real crystal area without defects. Similar to all the other defects, vacant nodes and occupied internodes present an increased energy relative to the ideal crystal, which favors chemical reaction and diffusion.

Punctiform Defects in Nonequivalent Node Lattices

In crystals with more complex lattices, other punctiform defects types may also appear, for example,

- Lattice location inversions: Type A particles occupy the nodes normally intended for type-B particles and vice versa.
- Stoichiometry defects: AB crystal composition may deviate from the stoichiometric composition of 50% A and 50% B, for example, the NiO crystal seldom contains 49% Ni and 51% O. The excess of component A may be due to: a higher internodes number occupied by A (as compared to the internodes number occupied by B), or by a higher number of vacant nodes in B sublattice (relative to the vacant nodes A number).
- Defects of composition facilitate the presence of impurities. These impurities are those ensuring extrinsic semiconductors conductivity, such as Ge or Si doped with impurities; in this case, they are usually placed in the lattice nodes, in the place of the "normal" occupants of these nodes. If these impurities have smaller dimensions than the ideal crystal particles, they may intercalate between internodes as *interstitial* impurities. Interstitial impurification explains metallic structures forming, based on a metal ideal lattice in which significant amounts of a nonmetallic low-ionic element (H, C, Si, B, N, P) have penetrated.

The *melting* phenomenon is explained by defect presence, whose number increases with temperature rise until their proportion becomes so high (about 1 per 1000 nodes), that the particle cohesion forces can no longer balance the thermal fluctuations of the equilibrium position and thus, *melting* occurs—a reorganization consisting of sudden disappearance of the scattered order present in solids and absent in fluids (Sect. 6.2).

Diffusion in Solids

Similar to heat conduction, diffusion depends greatly on solid's nature and texture.

The presence of macropores, two-dimensional defects (contact surfaces between granules) or of open micropores (with a few Å dimensions) greatly facilitate the impurity diffusion in solids. However, it is observed that diffusion and self-diffusion occur even in compact macrocrystalline solids lacking micropores, where the only defects are zero-dimensional.

Diffusion Mechanisms in Solids

Self-diffusion in the compact solid can be conceived in the ideal solid as a simultaneous position exchange of two or more particles in the nodes, but such a mechanism involves very high activation energy. In most cases, diffusion characteristics in compact solids can be interpreted by admitting the punctiform defects participation in the elemental diffusion, since this type of defects is present in any real solid.

Self-diffusion through defects can occur in the simple crystalline solid through two paths: vacant nodes or interstices (internodes).

(I) At diffusion through vacant nodes, atoms' position changing is conditioned by the simultaneous fulfilment of following two conditions:

(a) The presence of a free node, adjacent to the atom that will migrate.
(b) The accumulation by thermal fluctuations of a sufficient amount of energy to this atom, so that it can overcome the energy barrier ε_{bn} required for the displacement from an occupied node to a neighboring free node.

The following expression is obtained for the diffusion coefficient:

$$D = \nu \cdot d^2 \cdot \exp \quad \left[- \left(\varepsilon_{bn} + \varepsilon_{gl}\right)/(k_B \cdot T)\right], \tag{8.47a}$$

where ν is atoms' oscillation frequency of lattice atoms, d is the distance between neighboring nodes, the factor $exp[-\varepsilon_{bn}/(k_B \cdot T)]$ is the fraction of atoms of sufficient energy to overcome the energy barrier between the node, and the $exp[-\varepsilon_{gl}/(k_B \cdot T)]$ factor represents the free nodes share.

(II) By the diffusion through interstices, there will also be an energy barrier ε_{bi} of passage from an internode into the unoccupied neighboring internode, and the probability of passing from one interstitial position to another is proportional to the number of occupied internodes through relation (6.48) of Frenkel.

For self-diffusion, the following expression results:

$$D = \nu \cdot d^2 \cdot \exp \quad \left[-\left(\varepsilon_{bi} + \varepsilon_{gl} + \varepsilon_{io}\right)/(k_B \cdot T)\right]. \tag{8.47b}$$

At impurities diffusion it results in the following:

$$D = \nu' \cdot d^2 \cdot \exp\left[-\varepsilon'_{bi}/(k_B \cdot T)\right], \quad (8.47c)$$

where ν' and ε'_{bi} are the oscillation frequency and the energy barrier height for impurities, respectively. It can be seen that in both cases, D increases exponentially with temperature, following a law of form:

$$D = A \cdot \exp\left[-E_D/(R \cdot T)\right] \quad (8.48)$$

with A (pre-exponential factor) and E_D (diffusion activation molar energy, $E_D = N_A \cdot \varepsilon_D$) slightly modified by temperature variations.

The law (8.48) for the variation of solids diffusivity with temperature is identical to that of liquids, but the diffusion activation energy is much higher in solids, so that at equal temperature in liquids case, D exceeds by some order of magnitude the diffusivity of a similar solid.

The dependence A(T) has the form of relations (8.47a, 8.47b and 8.47c), that is, since the oscillation frequency is independent of temperature:

$$D(T) = \nu \cdot [d(T)]^2 \cdot exp\left[-E(T)/(R \cdot T)\right]. \quad (8.49)$$

8.8 Nine Worked Examples

Ex. 8.1

Calculate the viscosity of η_3 of the n-dodecane liquid at $t_3 = 373$ K where this substance has a molar volume $V_3 = 247 \cdot 10^{-6}$ m³/mole. It is known, at temperatures $t_1 = 273$ K and $t_2 = 323$ K, expressed in kg/(ms): $\eta_1 = 227 \cdot 10^{-5}$, as well as the corresponding molar volumes (m³/mole) $V_1 = 223 \cdot 10^{-6}$, $V_2 = 234 \cdot 10^{-6}$.

Solution

Three methods are available: Guzman, Bačinski, and Eyring.

1. According to Guzman's relation (8.3): $\eta_1 = A \cdot \exp(B/T_1)$, $\eta_2 = A \cdot \exp(B/T_2)$ and $\eta_3 = A \cdot \exp(B/T_3)$.

   After eliminating constants (A, B) between the three relations it results into the following: $\eta_3 = \exp\left\{\frac{[-T_1 \cdot (T_3 - T_2) \cdot \ln \eta_1 + T_2 \cdot (T_3 - T_1) \cdot \ln \eta_2]}{T_3 \cdot (T_2 - T_1)}\right\}$ or, with values (η_1, η_2, T_1, T_2, T_3) from the statement:

$$\eta_3 = \exp \frac{\left[-273 \cdot (100 - 50) \cdot ln\left(227 \cdot 10^{-6}\right) + 323 \cdot (100 - 0) \cdot ln\left(922 \cdot 10^{-6}\right)\right]}{3735 \cdot (50 - 0)}$$

$$\rightarrow \quad \eta_3 = 477 \times 10^{-6} \text{kg}/(\text{ms}) \quad (8.8a2)$$

2. According to Bačinski's relation $A_f = \eta_1 \cdot (V_1 - V_{pr}) = \eta_2 \cdot (V_2 - V_{pr}) = \eta_3 \cdot (V_3 - V_{pr})$. After elimination of constants A_f, V_{pr}, for η_3 it results into the following: $\eta_3 = \eta_1 \cdot \eta_2(V_2 - V_1)/[\eta_1 \cdot (V_3 - V_1) - \eta_2 \cdot (V_3 - V_2)]$, therefore,

$$\eta = \frac{227 \cdot 10^{-6} \cdot 922 \cdot 10^{-6} \cdot (234 - 223) \cdot 10^{-6}}{\left[227 \cdot 10^{-6} \cdot (247 - 223) \cdot 10^{-6} - 922 \cdot 10^{-6} \cdot (247 - 234) \cdot 10^{-6}\right]} \quad (8.8b2)$$

$$\rightarrow \quad \eta_3 = 541 \cdot 10^{-6} \mathrm{kg/(ms)}.$$

3. In Eyring's relation (8.10), $\eta = [2 \cdot k_B \cdot T/(\delta^3 \cdot \nu_c)] \cdot \exp[u_B/(k_B \cdot T)]$, are introduced the independent temperature constants $A = 2 \cdot k_B/(\delta^3 \cdot \nu_c)$ and $B = u_B/k_B$, the relation thus becomes: $\eta = A \cdot T \cdot \exp(B/T)$, which becomes, for temperatures $T_1 = 273$ K, $T_2 = 323$ K, and $T_3 = 373$ K: $\eta_1 = A \cdot T_1 \cdot exp\,(B/T_1)$, $\eta_2 = A \cdot T_2 \cdot exp\,(B/T_2)$, and $\eta_3 = A \cdot T_3 \cdot \exp\,(B/T_3)$. By eliminating the constants A and B: $\eta = [T_2 \cdot (T_3 - T_1) \cdot \ln(\eta_2/T_2)(T_3 - T_2) \cdot \ln(\eta_1/T_1)]/[T_3 \cdot (T_2 - T_1)]$. By putting $\eta_1 = 227 \cdot 10^{-5}$, $\eta_2 = 922 \cdot 10^{-5}$, and the values of temperatures T_1, T_2, T_3 it results into the following:

$$\eta_3 = 487 \cdot 10^{-6} \mathrm{kg/(ms)}. \quad (8.8c2)$$

4. The arithmetic mean of methods resulted from points (1)–(3) is obtained from values (8.8a2), (8.8b2) and (8.8c2), namely,

$$\eta_{3,\mathrm{calc}} = 10^{-6} \cdot (477 + 541 + 487)/3$$
$$\Rightarrow \eta_3(\mathbf{calc}) = \mathbf{502 \cdot 10^{-6} kg/(ms)}.$$

The experimental value is: η_3 (exp) $= 511 \cdot 10^{-6}$ kg/(ms) and differs with 2% from the calculated one.

Ex. 8.2

At $t_1 = -20$ °C, the mercury viscosity is $\eta_1 = 18.6$ mP. The following parameters' calculation is demanded:

1. Mercury viscosity η_2 at temperature $t_2 = 100$ °C, knowing that mercury volume varies with temperature according to relation:

$$v = v_0 \cdot \left(1 + a \cdot t + b \cdot t^2\right). \quad (8.8a3)$$

where v_0 is the volume at 0 °C, $a = 1.82 \cdot 10^{-4}$ K^{-1}, $b = 7.8 \cdot 10^{-10}$ K^{-2}.
2. The mean activation energy of mercury on the given temperature range.

Solution

1. The relation (8.20) of Souders customized at t_1 and t_2, states that:

$F_{So} = V_1 \cdot (2.9 + lg\, lg\, \eta_1)$; $F_{So} = V_2 \cdot (2.9 + lg\, lg\, \eta_2)$. By equalizing, it results that:

$$lglg\eta_2 = (V_1/V_2) \cdot (2.9 + lglg\eta_1) - 2.9. \tag{8.8b3}$$

But by using relation (8.8a3) and equality $V_1/V_2 = v_1/v_2$, it results:

$$\frac{V_1}{V_2} = \frac{v_0 \cdot \left(1 + a \cdot t_1 + b \cdot t_1^2\right)}{v_0 \cdot \left(1 + a \cdot t_2 + b \cdot t_2^2\right)}$$
$$= \frac{\left(1 - 20 \cdot 1.82 \cdot 10^{-4} + 20^2 \cdot 7.8 \cdot 10^{-10}\right)}{\left(1 + 100 \cdot 1.82 \cdot 10^{-4} + 100^2 \cdot 7.8 \cdot 10^{-10}\right)} \text{ or } (V_1/V_2) = 0.9786.$$

By replacing (V_1/V_2) and η_1 in (8.8b3), it results into the following:

$$lglg\eta_2 = 0.9786 \cdot (2.9 + lglg\, 18.6) - 2.9 = 0.03888 \text{ from where} \Rightarrow \boldsymbol{\eta_2} = \mathbf{12.4\ mP}.$$

The activation energy is obtained from relation (6.9a), logarithm at t_1 and t_2: $ln\ \eta_1 = ln\ \eta_0 + E\eta/(R{\cdot}T_1)$ and $ln\eta_2 = ln\eta_0 + E\eta/(R{\cdot}T_2)$. By eliminating $(ln\ \eta_0)$ between the two equations, the following is obtained:

$$E_\eta = [R \cdot T_1 \cdot T_2)/(T_2 - T_1)] \cdot ln\,(\eta_1/\eta_2) = \ldots$$
$$\ldots = [8.31 \cdot 253 \cdot 373)/(100 - 20)] \cdot ln\,(18.6/12.4) = 2631$$
$$\Rightarrow \quad \mathbf{E\eta = 2.63\ kJ/mole}.$$

Ex. 8.3

A silicate has at temperatures t_1, t_2, t_3 (°C) 1350, 1400, and 1450, the following corresponding viscosities, expressed in kg/(ms): η_1, η_2, and η_3 of 11.4, 4.3, and 3.05, respectively. It is required to determine the vitrification temperature.

Solution

By applying logarithm to relation (8.11), $\eta = A\,e^{B/(T-C)}$, the following is obtained: $T{\cdot}ln\ \eta - C{\cdot}ln\ \eta = T{\cdot}ln\ A - C{\cdot}ln\ A + B$. By denoting with $A = e^E$ and $B = D - C{\cdot}E$, the relation becomes (with the new unknown constants C, D, E)

$$C \cdot ln\ \eta + D \cdot T + E = T \cdot ln\ \eta. \tag{8.8a4}$$

By customizing relation (a) for t_1, t_2 and t_3 the following system is obtained: $C{\cdot}ln\ \eta_1 + D{\cdot}T_1 + E = T_1{\cdot}ln\ \eta_1$; $C{\cdot}ln\ \eta_2 + D{\cdot}T_2 + E = T_2{\cdot}ln\ \eta_2$; $C{\cdot}ln\ \eta_3 + D{\cdot}T_3 + E = T_3{\cdot}ln\ \eta_3$. By eliminating D and E between the 3 equations, C is obtained:

$$C = \frac{[(T_3 - T_2) \cdot T_1 \cdot ln\,\eta_1 + (T_1 - T_3) \cdot T_2 \cdot ln\,\eta_2 + (T_2 - T_1) \cdot T_3 \cdot ln\,\eta_3]}{[(T_3 - T_2) \cdot ln\,\eta_1 + (T_1 - T_3) \cdot ln\,\eta_2 + (T_2 - T_1) \cdot ln\,\eta_3]}$$

or

$$C = \frac{[50 \cdot 1623 \cdot ln\,11.4 - 100 \cdot 1673 \cdot ln\,4.3 + 50 \cdot 1723 \cdot ln\,3.05]}{[50 \cdot ln\,11.4 - 100 \cdot ln\,4.3 + 50 \cdot ln\,3.05]}$$

$\Rightarrow$ **vitrification temperature = 1569 K**

Ex. 8.4

The viscosity at 1550 °C of slags of composition $5CaO{\cdot}3Al_2O_3{\cdot}YSiO_2$ depends of Y as following:

k	1	2	3
Y	4.34	6.62	10.0
η, kg/(m·s)	1.12	2.43	5.92

It is required to establish, based on the Arrhenius and Kendall relationship, whether the calculations of the molar fractions by the number of cations (Ca, Al, Si) in the oxides or by the number of O atoms of the oxides are more correct.

Solution

The molar fraction of SiO_2 is in the first—cationic—variant is given as

$$X_c = n_{SiO_2}/(n_{SiO_2} + n_{CaO} + 2 \cdot n_{Al_2O_3}) = Y/(Y + 5 + 2 \cdot 3)$$

or

$$X_c = Y/(Y + 11) \qquad (8.8a5)$$

and in the second, or anionic, variant:

$$X_a = 2 \cdot n_{SiO_2}/(2 \cdot n_{SiO_2} + n_{CaO} + 3 \cdot n_{Al_2O_3}) = 2 \cdot Y/(2 \cdot Y + 5 + 3 \cdot 3)$$

or

$$X_a = Y/(Y + 7) \qquad (8.8b5)$$

The "cationic" type molar fractions, $X_{j,c}$, in the melts with index $j \in \{1,2,3\}$, are—by relation (8.8a5): $X_{1,c} = Y_1/(Y_1 + 11) = 4.34/15.34 = 0.283$, respectively, as well as $X_{2,c} = Y_2/(Y_2 + 11) = 6.62/17.62 = 0.377$, and $X_{3,c} = 10/(10 + 11) = 0.476$.

The "anionic" type molar fractions, $X_{j,a}$, are obtained similarly, with relation (8.8b5):

$$X_{1,a} = Y/(Y_1 + 7) = 4.34/(4.34 + 7) = 0.383,$$
$$X_{2,a} = Y/(Y_2 + 7) = 6.62/13.62 = 0.486 \text{ and } X_{3,a} = 10/17 = 0.588.$$

Relation (8.21b), $lg\eta = \Sigma_j X_j \cdot lg\,\eta_j$ becomes, for a homogeneous mixture of two components with molar fractions X, (1 – X) and viscosities A, B: $lg\,\eta = X \cdot lg\,A + (1 - X) \cdot lg\,B$. Or, by denoting with $lg\,B = C$ and $lg(A/B) = D$:

$$lg\eta = C + D \cdot X. \tag{8.8c5}$$

From relation (8.8c5) for the points (1) and (3): $lg\eta_1 = C + D \cdot X_1$ and $lg\eta_3 = C + D \cdot X_3$, we obtained $C = (X_3 \cdot lg\eta_1 - X_1 \cdot lg\eta_3/(X_3 - X_1)$ and $D = (lg\eta_1 - lg\eta_3)/(X_1 - X_3)$.

It results in the following for the cationic model:

$$C_c = (0.476 \cdot lg1.12 - 0.283 \cdot lg5.92)/(0.476 - 0.283) = -1.012 \text{ and}$$
$$D_c = (lg1.12 - lg5.92)/(0.283 - 0.476) = 3.747.$$

Similarly, the "anionic" model parameters are obtained:

$$C_a = (0.588 \cdot lg1.12 - 0.383 \cdot lg5.92)/(0.588 - 0.383) = -1.299 \text{ and}$$
$$D_a = (lg1.12 - lg5.92)/(0.383 - 0.588) = 3.522.$$

From equation (8.8c5) for point 2, the values η expressed in kg/(m·s) are obtained: $lg\,\eta_{2,c} = C_c + D_c \cdot x_{2,c} = -1.012 + 3.747 \cdot 0.377 = 0.402$, from where $\eta_{2,c} = 2.52$ $lg\,\eta_{2,a} = C_a + D_a \cdot x_{2,a} = -1.299 + 3.522 \cdot 0.486 = 0.412$ from where $\eta_{2,a} = 2.58$.

The cationic model error is therefore lower (4%) than that of the anionic model (6%) when the experimental value is 2.43 kg/(ms).

Ex. 8.5

Calculate the Tamura and Kurata viscosity at 10 °C of a liquid with composition $X_1 = 0.1$; $X_2 = 0.2$; $X_3 = 0.7$ where components 1, 2, 3 are $C_6H_5NH_2$ aniline, C_2H_5OH ethanol, and $(C_2H_5)_2O$ diethyl ether, respectively.

Molar volumes (cm^3/mole) V_1, V_2, V_3 are at this temperature of 90.5, 57.9, and 101.7, respectively, and the viscosities (mP) are $\eta_1 = 65$, $\eta_2 = 14.6$, and $\eta_3 = 2.38$.

The molar volume of the solution, V, depends on the composition according to the following relation:

$$V\left(\frac{cm^3}{mole}\right) = \sum_{i=1}^{3} V_i \cdot X_i - 0.24 \cdot X_1 \cdot X_2 + 0.15 \cdot X_2 \cdot X_3 - 0.18 \cdot X_3 \cdot X_1. \tag{8.8a6}$$

The binary interaction constants Cij of relation (8.22c) for the viscosity, $V \cdot lg\eta = \sum_i x_i^2 \cdot V_i \cdot lg\eta_i + \sum_i \sum_j x_i \cdot x_j \cdot C_{ij}$, are as follows:

$$C_{12} = 1.08 \cdot C_{12}^0; C_{23} = 0.97 \cdot C_{23}^0; C_{31} = 1.02 \cdot C_{31}^0, \tag{8.8b6}$$

where C_{ij}^0 are the "ideal" values of constants:

$$C_{ij}^0 = 0.5 \cdot (V_i \cdot V_j)^{1/2} \cdot lg\left(\eta_i \cdot \eta_j\right). \quad (8.8c6)$$

Solution

Using relation (8.8a5) it results into the following: $V = 90.5{\cdot}0.1 + 57.9{\cdot}0.2 + 101.7{\cdot}0.7 – 0.24{\cdot}0.2{\cdot}0.1 + 0.15{\cdot}0.2{\cdot}0.7 – 0.18{\cdot}0.7{\cdot}0.1 = 92.2\ cm^3/mole$.

From relation (8.8c6) the following expression is obtained: $C_{12}^0 = 0.5{\cdot}(V_1{\cdot}V_2)^{1/2}{\cdot}lg(\eta_1{\cdot}\eta_2) = 0.5{\cdot}(90.5{\cdot}57.9)^{0.5}{\cdot}lg\ (65{\cdot}14.6) = 107.7$ and similarly:

$$C_{23}^0 = 0.5 \cdot (57.9 \cdot 101.7)^{0.5} \cdot lg(14.6 \cdot 2.38) = 59.10 \text{ and}$$
$$C_{31}^0 = 0.5 \cdot (101.7 \cdot 90.5)^{1/2} \cdot lg(2.38 \cdot 90.5) = 111.91.$$

Using relations (8.8b5) it is found: $C_{12} = 1.08{\cdot}107.7 = 116.3$; $C_{23} = 0.97{\cdot}59.1 = 57.3$, and $C_{31} = 1.02{\cdot}111.9 = 114.1$.

Tamura and Kurata relation (8.22c) becomes, for three components:

$$V \cdot \lg \eta = V_1 \cdot (X_1)^2 \cdot \lg \eta_1 + V_2 \cdot (X_2)^2 \cdot \lg \eta_2 + V_3 \cdot (X_3)^2 \cdot \lg \eta_3 + X_1 \cdot X_2 \cdot C_{12} + X_2 \cdot X_3 \cdot C_{23} + X_3 \cdot X_1 \cdot C_{13}, or\ 92.2 \cdot \lg \eta = (90.5 \cdot 0.1^2 \cdot lg65 + 57.9 \cdot 0.2^2 \cdot lg14.6 + 101.7 \cdot 0.7^2 \cdot lg2.38. + 0.1 \cdot 0.2 \cdot 116.3 + 0.2 \cdot 0.7 \cdot 57.3 + 0.7 \cdot 0.1 \cdot 114.1) = 49.43$$

from where
$lg\eta = 0.536 \quad \Rightarrow \quad$ **$\eta = 3.44$ mP**

Ex. 8.6

In a viscous liquid of viscosity $\eta = 8$ mP and density $\rho_L = 900\ kg/m^3$, a density lipophilic solid $\rho_S = 3600\ kg/m^3$ is dispersed in $g_S = 40\%$ of the slurry mass. Knowing that the solid particles are spherical, of radius $r = 25$ Å, it is required to determine the thickness **a** of the solvate coating if the suspension viscosity is $\eta = 21$ mP.

Solution

Einstein's relation (8.24b) can be used, whereby the solid volume fraction Ø is replaced by the solvated particles volume fraction:

$$\eta = \eta_0 \cdot (1 + \varnothing'/2)/(1 + \varnothing')^2 \text{ or } 21 = 8 \cdot (1 + \varnothing'/2)/(1 + \varnothing')^2.$$

After rearrangement, the following equation results: $21 \cdot (\varnothing')^2 - 46 \cdot \varnothing' + 13 = 0$.

Its roots are 1/3 and 13/7, but the second is unacceptable ($\varnothing' > 1$) so that finally: $\varnothing = 1/3$.

The actual solid volume fraction $\varnothing'$, is obtained from densities, denoting by m_S, m_L the masses of solid and liquid and by V_S, V_L—the corresponding volumes.

For 1 kg of solution: $m_S = g_S = 0.4$ kg; $m_L = 1{-}0.4 = 0.6$ kg; $v_S = m_S/\rho_S = 0.4/3600 = 1.11{\cdot}10^{-4}$ m^3; $v_L = m_L/\rho_L = 0.6/900 = 6.67{\cdot}10^{-4}$ m^3.

$$\O = v_S/(v_S + v_L) = 1.11 \cdot 10^{-4}/(1.11 \cdot 10^{-4} + 6.67 \cdot 10^{-4})\, or\, \O = 1/7.$$

According to Sect. 8.4., $\O' = \O\ (r/r')^3$, where r′ is the solvated particle radius; is obtained that:

$$r' = r \cdot (\O'/\O)^{1/3} = 25 \cdot [(1/3)/(1/7)]^{1/3} = 33.2\ \text{Å}.$$

The solvation layer thickness therefore becomes: $a = r' - r = 33.2 - 25 \Rightarrow$ **a = 8.2 Å**

Ex. 8.7
It is required to calculate: (1) the local velocity radial distribution v(r) and (2) the mean velocity $\overline{v}$ of the laminar flow through a cylindrical tube of radius R under the effect of a longitudinal pressure gradient ΔP/L constant (ΔP being the pressure drop, and L is the length) in the case of a Bingham liquid of parameters (f_0, η″).

Solution
1. For Bingham fluid from relations (8.36)

$$f \le f_0 \rightarrow d\alpha/d\tau = 0;\ \ f \ge f_0 \rightarrow f = f_0 + \eta' \cdot d\alpha/d\tau, \qquad (8.8a7)$$

it equals, similar to deduction of relation (8.27), the frictional force on the lateral side of the L-shaped cylinder and the variable radius r, with compression force difference on the cylinder base, $F_{fr} = F_2 - F_1$ or $f{\cdot}2{\cdot}\pi{\cdot}r{\cdot}L = \Delta P{\cdot}\pi{\cdot}r^2$:

$$f = 0.5 \cdot r \cdot \Delta P/L. \qquad (8.8b7)$$

By replacing f from (8.8b7) to (8.8a7), the following is obtained:

$$r \le R_1 \rightarrow dv/dr = 0,\quad r \ge R_1 \rightarrow dv/dr = \Delta P/(2 \cdot L \cdot \eta'') \qquad (8.8c7)$$

$$\text{where}: R_1 = 2 \cdot L \cdot f_0/\Delta P. \qquad (8.8d7)$$

There are two possible cases, A and B, according to relations between R and R_1:

A. $R \le R_1$; $dv/dr = 0$ across the entire tube cross section. Since, at the wall (r = R) v = 0, it results that in the whole tube v = 0. In conclusion,

$$v(r) = 0,\ \Delta P/L \le 2 \cdot f_0/R.$$

B. $R > R_1$, that is, the pressure gradient is sufficiently high:

$$\Delta P/L > 2 \cdot f_0/R. \tag{8.8e7}$$

In this case, the first equation (8.8c7) is obtained by integration

$$v = C_1, \quad r \le R_1 \tag{8.8f7}$$

and the second results, by denoting with (C_1, C_2) the integration constants

$$v = \Delta P \cdot \left(R_1 \cdot r - r^2\right)/(4 \cdot L \cdot \eta'') + C_2, r \in [R_1, R]. \tag{8.8g7}$$

When introducing in (8.8g7) the condition $v = 0$ at $r = R$, it results into the following: $C_2 = -\Delta P \cdot (R_1 \cdot R - R^2)/(4 \cdot L \cdot \eta'')$, from where

$$v = F_1(r) = \left[\left(R_1 \cdot r - r^2\right)/2 - R_1 \cdot (R - r)\right] \cdot \Delta P/(2 \cdot L \cdot \eta''), r \in [R_1, R]. \tag{8.8h7}$$

The constant C_1 is obtained by introducing the condition that at $r = R_1$, v satisfies both relations (8.8f7) and (h): $C_1 = (R - R_1{}^2) \cdot \Delta P/(4 \cdot L \cdot \eta'')$, therefore (8.8f7) becomes the following:

$$v = F_2(r) = (R - R_1)^2 \cdot \Delta P/(4 \cdot L \cdot \eta''), r \in [0, R_1]. \tag{8.8i7}$$

The radial velocity distribution is therefore given in case **B** by the intersection of relations (8.8g7) ∩ (8.8i7).

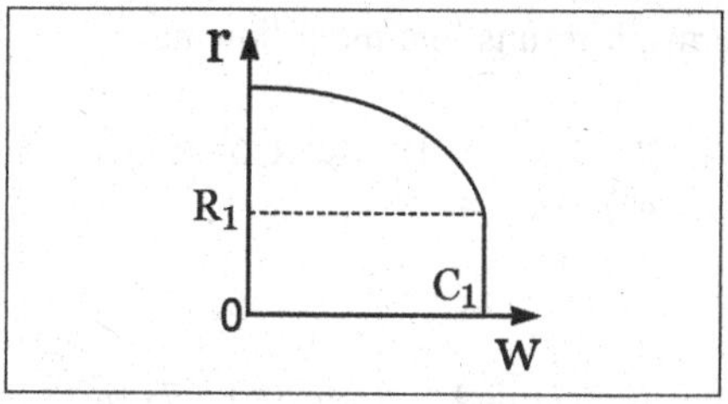

It is found in the adjacent figure, the existence of a peripheral area, located at $r \in [R_1, R]$, parabolically lowered velocity, and a central zone (with r between 0 and R_1), where the velocity is constant.

2. In case **A**, the velocity being null in the whole tube, a null flow results.

In case B, the mediation average is used: $\overline{v} = 2 \cdot R^{-2} \cdot \int_{r=0}^{R} v \cdot r \cdot dr$ or, since v (r) is expressed by two different relations, (8.8j7) and (8.8k7):

$$\overline{v} = 2 \cdot R^{-2} \cdot (I_1 + I_2), \tag{8.8j7}$$

$$\text{where } I_1 \equiv \int_{r=0}^{R_1} F_2(r) \cdot r \cdot dr; \qquad I_2 = \int_{r=R_1}^{R} F_1(r) \cdot r \cdot dr. \tag{8.8k7}$$

By introducing $F_2(r)$ from (8.8i7) in the second definition (8.8k7) I_1 results:

$$I_1 = \int_{r=0}^{R_1} \left(\frac{\Delta P}{4 \cdot L \cdot \eta''} \right) \cdot (R - R_1)^2 \cdot r \cdot dr$$
$$= \left(\frac{\Delta P}{4 \cdot L \cdot \eta''} \right) \cdot (R - R_1)^2 \cdot 0,5 \cdot r^2 \Big|_{r=0}^{r=R_1},$$

from where

$$I_1 = \Delta P \cdot (R_1)^2 \cdot (R - R_1)^2 / (8 \cdot L \cdot \eta''). \tag{8.8l7}$$

By replacing in the second definition (8.8k7), the function $F_1(r)$ from (8.8h7) the following expression is obtained:

$$I_2 = \int_{r=R_1}^{r=R} \left(\frac{\Delta P}{2 \cdot L \cdot \eta''} \right) \cdot \left[\left(\frac{R^2}{2} - \frac{R_1}{R} \right) + R_1 \cdot r - \frac{r^2}{2} \right] \cdot r \cdot dr,$$

from where

$$I_2 = \left(\frac{\Delta P}{4 \cdot L \cdot \eta''} \right) \cdot \left[\frac{(R^2 - 2 \cdot R_1 \cdot R) \cdot (R^2 - R_1^2)}{2} + \frac{2 \cdot R_1 \cdot (R^3 - R_1^3)}{3} - \frac{(R^4 - R_1^4)}{4} \right].$$

After regrouping terms, it results into the following:

$$I_2 = (3 \cdot R^4 + 2 \cdot R^3 \cdot R_1 - 6 \cdot R^2 \cdot R_1^2 + 6 \cdot R \cdot R_1^3 - 5 \cdot R_1^4) \cdot \Delta P / (8 \cdot L \cdot \eta''). \tag{8.8m7}$$

By replacing I_1 from (8.8l7) and I_2 from (8.8m7) in (8.8j7), the following is obtained:

$$\overline{v} = (R - R_1) \cdot \left[\left(3 \cdot R^3 + 5 \cdot R^2 \cdot R_1 + 5 \cdot R \cdot (R_1)^2 - (R_1)^3 \right) \cdot \Delta P / (48 \cdot L \cdot \eta''), \right.$$

or by replacing R_1 from (8.8d7) and by denoting with Y the ratio $Y \equiv R \cdot \Delta P / (2 \cdot L \cdot f_c)$:

$$\Rightarrow \quad \mathbf{\overline{v} = R^2 \cdot \Delta P \cdot (1\text{-}Y) \cdot (3 + 5 \cdot Y + 5 \cdot Y^2\text{-}Y^3) / (24 \cdot L \cdot \eta'')}.$$

Ex. 8.8
For acetone (CH_3COCH_3, the molecular mass M = 0.06009 kg/mol), liquid viscosity and compressibility, $\eta = 0.000903$ J/(ms) and $\chi_T = 1.23 \cdot 10^{-9}$ ms^2/kg, are known in the liquid state at 300 K and below 1 atm. Vapor pressure at 300 K is $P_{sat} = 31$ kPa.

It is required for liquid: 1° the density; 2° thermal conductivity; 3° caloric capacity isobar.

Solution

1. From the molecular formula, the R_{ch} rheochor is calculated by means of the increments in Table 8.4, namely, 12.8 (C); 5.5 (H from C–H) and 13.2 (O from C=O): $R_{ch} = 3 \cdot 12.8 + 6 \cdot 5.5 + 1 \cdot 13.2 = 84.6$.

From the rheochor definition (8.19), $R_{ch} \equiv \eta^{1/8} \cdot M/(\rho + 2 \cdot \rho_{G,sat})$ with the viscosity expressed in mP (1 mP = 10^{-4} kg m^{-1} s^{-1}), M—in g/mol, and ρ, $\rho_{G,sat}$ —liquid density and of saturated vapors—in g/cm^3, the sum of densities results to be (in S.I.): $(\rho_L + 2 \cdot \rho_{G,\,sat}) = 10^3 \cdot (\eta/10^{-4})^{1/8} \cdot (M/10^{-3})/R_{ch} = 1000 \cdot 9.03^{1/8} \cdot 60.09/84.6$ or $\rho + 2 \cdot \rho_{G,\,sat} = 935.17$ kg/m^3 (8.8a8)

Saturated vapors density is obtained from the vapor pressure, which is small enough to justify the perfect gas laws use:

$$\rho_{G,sat} = M \cdot P_{sat}/(R \cdot T) = 0.06009 \cdot 31000/(8.31 \cdot 300) = 0.747\ \text{kg/m}^3.$$

By introducing in equality (a): $\rho = 935.17 - 2 \cdot 0.747 \Rightarrow$ **$\rho = 933.7$ kg/m^3**

2. From relation (8.42c), $u_s = (\rho \cdot \chi_T)^{-1/2}$, with χ_T from the statement and p from the first point, the sound velocity in the liquid is $u_s = (933.7 \cdot 1.23 \cdot 10^{-9})^{-1/2} = 933$ m/s, and its molar volume – $V = M/\rho = 0.06009/933.7 = 6.44 \cdot 10^{-5}$ m^3/ mol. The thermal conductivity is given by Bridgman's formula (8.41a): $\lambda = 2.8 \cdot k_B \cdot u_s/(V/N_A)^{2/3} = 2.8 \cdot 1.38 \cdot 10^{-23} \cdot 933/[6.44 \cdot 10^{-5}/(6.02 \cdot 10^{23})]^{2/3} \Rightarrow$ **$\lambda = 0.160$ W/(m·K)**
3. From relation (8.41b), $\lambda = L_f \cdot C_P \cdot u_s/V$, where the value proposed by Kardos is admitted for the molecule free displacement $L_f = 95 \cdot 10^{-12}$ m and are introduced C_P, u_s and V calculated at point 2: $C_P = (\lambda \cdot V)/(L_f \cdot u_s) = (0.16 \cdot 6.44 \cdot 10^{-5})/(95 \cdot 10^{-12} \cdot 933) \Rightarrow$ **$C_P = 116$ J/(mole · K)**

The experimental value is 8% higher: 126 J/(mol K).

Ex. 8.9

Coefficients of gold self-diffusion at $T_1 = 1175$ and $T_2 = 1225$ K are, respectively (in m^2/s), $D_1 = 1.53 \cdot 10^{-15}$ and $D_2 = 3.45 \cdot 10^{-15}$. It is required to determine the following:

1. The diffusion activation energy E, and its pre-exponential factor A, in a first approximation.
2. The mean linear thermal relaxation coefficient, in the temperature range 0–1200 K, if it is admitted that the activation energy is proportional to the repulsive factor d^{-n} of the intermolecular potential Mie (4.21) at the same temperature, therefore,

$$E/E_0 = (d_0/d)^n, \tag{8.8a9}$$

where E(T) and d(T) are the activation energy and the distance d between the lattice nodes and thus, E_0 and d_0—corresponding values extrapolated to 0 K. The following physical quantities are known: the interatomic distance at 0 K, $d_0 = 2.88 \cdot 10^{-10}$ m, the exponent n = 12.5 of repulsive potential and frequency of oscillations, $\nu = 1.8 \cdot 10^{14}\ s^{-1}$.

Solution

1. From relation (8.48), $lnD_1 = lnA - E/(R \cdot T_1)$, $lnD_2 = lnA - E/(R \cdot T_2)$, where, by eliminating A: $E = [R \cdot ln(D_2/D_1)]/(1/T_1 - 1/T_2) = \ldots$.

$\ldots = [8.31 \cdot ln(3.45/1.53)]/(1/1175 - 1/1225) = 1.946 \cdot 10^5$ J/mole. Thus, $A = D_1 \cdot exp[E/(R \cdot T_1)] = 1.53 \cdot 10^{-15} \cdot exp[194600/(8.31 \cdot 1175)] = 6.86 \cdot 10^{-7}\ m^2/s$. E and A are assigned to the mean measurement temperature (1175 + 1225)/2 = 1200 K

$$\Rightarrow \quad \mathbf{E_{1200} = 195\ kJ/mole; A_{1200} = 6.86 \cdot 10^{-7}\ m^2/s},$$

2. The *linear* dilatation coefficient γ is assumed to be independent of temperature; the distance between nodes will vary with temperature, according to the following law:

$$d(T) = d_0 \cdot (1 + \gamma \cdot T). \tag{8.8b9}$$

By introducing d from (8.8b9) in (8.8a9) the following expression is obtained:

$$E(T) = E_0 \cdot (d_0)^n \cdot (1 + \gamma \cdot T)^{-n}. \tag{8.8c9}$$

With d(T) and E(T) of dependencies (8.8b9) and (8.8c9), the relation (8.49), $D = \nu \cdot d^2 \cdot exp\,[-E/(R \cdot T)]$, becomes after applying the following logarithm:

$$ln\,D(T) = \left\{ ln \left[\nu \cdot (d_0)^2 \right] \right\} + [2 \cdot ln\,(1 + \gamma \cdot T)] - E_0 \cdot (d_0)^n \cdot (1 + \gamma \cdot T)^{-n}/(R \cdot T). \tag{8.8d9}$$

If $\gamma \cdot T \ll 1$ and $n \cdot \gamma \cdot T \ll 1$, the expressions $ln\,(1 + \gamma \cdot T)$ and $(1 + \gamma \cdot T)^{-n}$ can be expressed by $\gamma \cdot T$ and by $1 - n \cdot \gamma \cdot T$. The relation (8.8d9) will be simplified at the following:

$$ln\,D(T) = \left\{ ln \left[\nu \cdot (d_0)^2 \right] \right\} + 2 \cdot \gamma \cdot T - E_0 \cdot (d_0)^n/(R \cdot T) + E_0 \cdot (d_0)^n \cdot n \cdot \gamma/R. \tag{8.8e9}$$

The subtraction of equation (8.8e9) with index 1 (at 1175 K) from the same equation, but with the index 2 (at 1225 K) will give, expressed for the energy at 0 K:

$[E_0 \cdot (d_0)^n/R] = T_2 \cdot T_1 \cdot \{[ln(D_2/D_1)]/(T_2 - T_1) - 2 \cdot \gamma\}$ and with $n = 12.5$ and by introducing the values (T_1, D_1, T_2, D_2) from the statement:

$$[E_0 \cdot (d_0)^n/R] = 23407 - 2.879 \cdot 10^6 \cdot \gamma. \tag{8.8f9}$$

The expression (8.8e9) at 1175 K is given as:

$lnD_1 = \{\ln[\nu \cdot (d_0)^2]\} + 2 \cdot \gamma \cdot T_1 + (E_0 \cdot (d_0)^n \cdot n/R) \cdot (\gamma - 1/T_1)$ or, by replacing $E_0 \cdot (d_0)^n/R$ from the relation (8.8f9) and with (T_1, D_1, ν, d_0) from the statement $ln[1.8 \cdot 10^{14} \cdot (2.88 \cdot 10^{-10})^2/(1.53 \cdot 10^{-15})] = (23407 - 2.879 \cdot 10^6 \cdot \gamma) \cdot (1/1175 - 12.5 \cdot \gamma) - 2 \cdot 1175 \cdot \gamma$, which after terms regrouping becomes the following second degree equation in γ:

$$3.60 \cdot 10^7 \cdot \gamma^2 + 2.95 \cdot 10^4 \cdot \gamma - 3.08 = 0, \tag{8.8g9}$$

whose only acceptable root (the solids dilatability being always positive) is $\gamma = 7.79 \cdot 10^{-6}\ K^{-1}$.

The actual relaxation coefficient (of the volume) is three times higher than the mean (after the space directions) of the linear relaxation, therefore,

$$\overline{\alpha} = 3 \cdot \gamma = 3 \cdot 7.79 \cdot 10^{-6} \quad \Rightarrow \quad \boldsymbol{\alpha = 2.34 \cdot 10^{-5}\ K^{-1}}.$$

Mathematical Annex

A.1 Basic Notions

The following notions, already known in the mathematical knowledge, are shortly related:

- The factorial double
- By parts (p.p. integration)
- Recurrence (relation of r., solution of r.)

The Factorial Double

The factorial double of a nonnegative integer number N is denoted by using the symbol N!!. It represents the product of all the integer positive numbers smaller or equal to N and which have the same parity as N, that is, N is odd $N = 2 \cdot k + 1 \rightarrow N!! \equiv 1 \cdot 3 \cdot 5 \cdot 7 \cdot \ldots \cdot (2 \cdot k - 1) \cdot (2 \cdot k + 1)$; or:

$$(2 \cdot k + 1)!! = \frac{(2 \cdot k + 2)!}{(k + 1)!} \cdot 2^{-(k+1)} \tag{A.1a}$$

and for even N:

$$N = 2 \cdot k \rightarrow N!! \equiv 2 \cdot 4 \cdot 6 \cdot 8 \cdot \ldots \cdot (2 \cdot k - 2) \cdot (2 \cdot k);$$

or:

$$(2 \cdot k)!! = (k!) \cdot 2^k \tag{A.1b}$$

F. E. Daneş et al., *Molecular Physical Chemistry for Engineering Applications*,
https://doi.org/10.1007/978-3-030-63896-2

The Integration by Parts

The integration by parts implies $f(x)$ (from integral $\int f(x) \cdot dx$) decomposition into a two factors product, namely $u(x)$ and $w(x)$. The integral $\int f(x) \cdot dx$ depends only one integration variable, which is x, in a product of two factors, which are $u(x)$ and $w(x)$:.

$$f(x) = v(x) \cdot w(x)$$

so that on the one hand the integration $\int w(x) \cdot dx$ to be immediate – that is to simply result in an expression $u(x)$ so that $u'(x) = w(x)$, where the function u' designates the derivate du/dx – and on the other hand the integral $\int v'(x) \cdot u(x) \cdot dx$ to be easier to calculate (eventually through recurrence) than the starting integral, $\int a(x) \cdot b'(x) \cdot dx$.

The basic equality of the integration through parts method is as follows:

$$\int v \cdot u' \cdot dx = u \cdot v - \int u \cdot v' \cdot dx$$

where u and v are functions for the variable x and the sign « '» shows the derivation reported to x.

The Recurrence

Let X be an ensemble of real variables, k, an integer positive variable, and $F(k, X)$, a function defined by these two variables.

An equation is qualified as "a recurrence of the order n of the function $F(X, k)$," n being an integer positive number, if it is presented under the form of the dependency:

$$\Phi[F(k, X), F(k+1, X), F(k+2, X) \ldots, F(k+n-1, X), F(k+n, X)] = 0.$$

For example, the following relation is a recurrence of the order 3:

$$\Phi[F(k, X), F(k+1, X), F(k+2, X), F(k+3, X)] = 0 \;\; F(k+n, X)] = 0.$$

The recurrence is named a *linear* recurrence if the expression $\Phi[F(k, X), F(k+1, X), F(k+2, X) \ldots, F(k+n-1, X), F(k+n, X)]$ is a linear function for each of the arguments $F(i, X)$ where i is any whole number between 0 and n, that is, if the equation $\Phi = 0$ had the form:

$$a_0 + \sum_{i=l}^{i=n} a_i \cdot F(k+i, X) = 0$$

in which the coefficients a_i (i being all integers between 0 and n) do not depend on the *real* variables of the ensemble X but only on the whole variables (i, k, n).

The solution of the recurrence is the method through which the expressions of the functions *F(k, X)* are established for any value *k*. A necessary (but not sufficient) condition for determining the solution is knowing beforehand a number of n expressions for the functions *F(k, X)*, generally speaking being about the *first* n between these functions, that is, F(1, X), F(2, X), . . ., F(*n*, X).

A.2 Gamma Functions

These are functions of a nonnegative, whole argument n and a real positive argument α functions that are equal to a defined integral depending on another real argument x, which can vary between the limits 0 and infinite; the integrand's number is a squared exponential, and the integral solves by parts, using recurrence. Thus, the function $\Gamma_n(\alpha, n)$ is given by:

$$\Gamma_n(\alpha) = \int_0^{\infty} x^n \cdot e^{-\alpha \cdot X^2} \cdot dx \tag{A.2a}$$

It represents a particular case of integration of a defined real function:

$$\Psi_n(\alpha, x) \equiv \int \left[x^n \cdot \exp\left(-\alpha \cdot x^2\right)\right] \cdot dx$$

where x is a real, variable, nonnegative parameter; α a real, constant positive parameter; and n a nonnegative integer constant. The result of the integration from (A.2a) is obviously dependent on x, because both limits of the integral are numerically defined.

The recurrence relation is of the second (II) degree, linear and homogenous. It is expressed as follows:

$$\Psi_n = [(n-1)/2 \cdot \alpha] \cdot \Psi_{n-2}$$

The starting points of the recurrence – those for the two smallest values of n (that is n = 0 and n = 1) – are obtained through standard methods. These are:

$$\Psi_0(\alpha) \equiv \int_{x=0}^{x=\infty} e^{-\alpha \cdot x^2} \cdot dx = 0.5 \cdot \sqrt{\pi/\alpha}$$

$$\Psi_1(\alpha) \equiv \int_{x=0}^{x=\infty} x \cdot e^{-\alpha \cdot x^2} \cdot dx = \frac{1}{2 \cdot \alpha} \quad \Psi_n(\alpha) \equiv \int_{x=x'}^{x=x''} x^n \cdot e^{-\alpha \cdot x^2} \cdot dx$$

The final formula – adapted to the present notation – is detailed in Gradshteyn (2007) on pg. 364, positions 3.461–2 and 3.461–3, respectively.

The Parity of the Gamma Function Index

The result of the recurrence depends on the x power n's parity before the exponential, namely:

The next subdivisions are cases of this statement : A) when n is even; B) when n is uneven

When *n* Is Even

$$\Psi_{2\cdot k}(\alpha) = \int_0^{\infty} x^{2\cdot k} \cdot e^{-\alpha \cdot x^2} \cdot dx = \frac{(2\cdot k-1)!!}{2\cdot(2\cdot\alpha)^k} \cdot \sqrt{\pi/\alpha}.$$

or, by expressing the factorial double through two simple factors, with the relation (9.1a):

$$\int_0^{\infty} x^{2\cdot k} \cdot e^{-\alpha \cdot x^2} \cdot dx = \frac{(2\cdot k)!}{k!} \cdot (4\cdot\alpha)^{-(k+1/2)} \cdot \sqrt{\pi} \tag{A.2b}$$

With the particular cases:

$$k=0 : \Psi_0(\alpha) \equiv \int_{x=0}^{x=\infty} e^{-\alpha \cdot x^2} \cdot dx = (1/2) \cdot \sqrt{\pi/\alpha} \tag{a}$$

$$k=1 : \Psi_2(\alpha) \equiv \int_{x=0}^{x=\infty} x^2 \cdot e^{-\alpha \cdot x^2} \cdot dx = (1/4) \cdot \sqrt{\pi/\alpha^3} \tag{b}$$

$$k=2 : \Psi_4(\alpha) \equiv \int_{x=0}^{x=\infty} x^4 \cdot e^{-\alpha \cdot x^2} \cdot dx = (3/8) \cdot \sqrt{\pi/\alpha^5} \tag{c}$$

When *n* Is Uneven

$$\Psi_{2\cdot k+1}(\alpha) \equiv \int_0^{\infty} x^{2\cdot k+1} \cdot e^{-\alpha \cdot x^2} \cdot dx = k!/(2\cdot\alpha^{k+1}) \tag{A.2c}$$

with the cases:

$$k = 0 : \Psi_1(\alpha) \equiv \int_{x=0}^{x=\infty} k \cdot e^{-\alpha \cdot x^2} \cdot dx = 1/(2 \cdot \alpha) \quad \text{(d)}$$

$$k = 1 : \Psi_3(\alpha) \equiv \int_{x=0}^{x=\infty} x^3 \cdot e^{-\alpha \cdot x^2} \cdot dx = 1/(2 \cdot \alpha^2) \quad \text{(e)}$$

$$k = 2 : \Psi_5(\alpha) \equiv \int_{x=0}^{x=\infty} x^5 \cdot e^{-\alpha \cdot x^2} \cdot dx = 3/\alpha^3 \quad \text{(f)}$$

A.3 The Integration of the Exponential/Polynomial Product

It is integrated undefined through parts an integral of a degree n defined by:

$$F_n(x, a) \equiv \int x^n \cdot e^{ax} \cdot dx \quad \text{(A.3a)}$$

with x – real variable, a – real parameter, and n – nonnegative, whole parameter. Choosing the parts as:

$$du = e^{a \cdot x} \cdot dx, \quad v = x^n$$

It results:

$$u = e^{a \cdot x}/x, \quad dv = n \cdot x^{n-1},$$

and the transformation result,

$$\int v \cdot du = v \cdot u - \int u \cdot dv,$$

will be

$$\int x^n \cdot e^{a \cdot x} \cdot dx = n \cdot x^{n-1} \cdot e^{a \cdot x} - (n/a) \cdot \int x^{n-1} \cdot e^{a \cdot x} \cdot dx$$

which represents a nonhomogeneous, linear, first (I)-degree recurrence:

$$F_n(x, a) = x^{n-1} \cdot e^{a \cdot x} - (n/a) \cdot F_{n-1}(x, a) \quad \text{(A.3b)}$$

The solution of the recurrence is from Gradshteyn (2007) on pg. 106, relation 2.321–2, that is:

$$F_n(x,a) = (n!/a) \cdot e^{a \cdot x} \sum_{k=0}^{k=n} \left[(-a)^{-k} \cdot x^{n-k}/(n-k)! \right] \tag{A.3c}$$

A.4 Decompositions According to Series of Integer Powers

This type of decomposition allows for approximating a transcending equation into real numbers through an algebraical equation.

The simplification of the solution consists of replacing an analytical function with its approximative value, given the retaining of a finite number of terms from the series of whole powers. For example, in order to approximately determine the nonzero roots of a transcendental equation:

$$4 \cdot x/(x+1) + 12 \cdot \sin x - 15 \cdot atan\ x = 0,$$

whose solution proves to be difficult – even on the computer – turning to developing in three MacLaurin series for the expressions $1/(x + 1)$, $\sin x$ and respectively arc tg x, leads to the form:

$[4 \cdot x \cdot (1 - x + x^2 - x^3 + x^4 - \ldots)] + [12 \cdot (x - x^3/6 + x^5/120 - \ldots)] - [15 \cdot (x - x^3/3 + x^5/5 - \ldots)] = 0$ from which, through regrouping, simplification through x and giving up to the terms where x's power is ≥ 5, leads to:

$$1.1 \cdot x^4 - 4 \cdot x^3 + 7 \cdot x^2 - 4 \cdot x + 1 = 0$$

an equation whose solution, detailed in § A.5., is a lot simpler.

The Definitions of Taylor's and MacLaurin's Series

The series Taylor $F_{Tay}(x)$ equivalent in the point $x = a$ with the function $f(x)$ is

$$F_{Tay}(x) \equiv f(a) + \sum_{n=1}^{n=\infty} [(x-a)^n/n!] \cdot \lim_{X \to a} d^n f(x)/dx^n \tag{A.4a}$$

and the MacLaurin series $F_{McL}(x)$ is the particularization of the Taylor series for $x = 0$:

$$F_{McL}(x) \equiv f(0) + \sum_{n=1}^{n=\infty} (x^n/n!) \cdot \lim_{X \to 0} d^n f(x)/dx^n \tag{A.4b}$$

That is, by denoting with *f'(0), f''(0), f'''(0)*, etc., the values of the successive derivates of the function *f(x)* at the point *x* = *0*:

$$\mathrm{F_{McL}}(x) \equiv f(0) + f'(0) \cdot x + f''(0) \cdot x^2/2 + f'''(0) \cdot x^3/6 + f^{IV}(0) \cdot x^4/24 + \cdots$$

Usual Decompositions in a MacLaurin Series

$$-1 + \exp x = \sum_{n=1}^{n=\infty} x^n/n! \quad \text{(A.4c)}$$

$$\ln(1+x) = \sum_{n=1}^{n=\infty} (-1)^{n+1} \cdot x^n/n \quad \text{(A.4d)}$$

$$\sin x = \sum_{n=\infty}^{n=\infty} (-1)^{n+1} \cdot x^{2n-1}/(2 \cdot n - 1)! \quad \text{(A.4e)}$$

$$1 - \cos x = \sum_{n=1}^{n=\infty} (-1)^{n+1} \cdot x^{2n}/(2 \cdot n)! \quad \text{(A.4f)}$$

$$1 - (1+x)^{-1} = \sum_{n=1}^{n=\infty} (-1)^{n+1} \cdot x^n \quad \text{(A.4g)}$$

$$b \in (-1,1) \to (1+x)^b - 1 = \sum_{n=1}^{n=\infty} \left[(n!)^{-1} \cdot (x^n) \cdot \prod_{k=0}^{k=n} (b-k) \right] \quad \text{(A.4h)}$$

A.5 The Rapid Solution of Algebraical Equations

An approximate solution of an algebraical or transcendental equation can be rapidly obtained, using the Windows calculator in the scientific or probabilistic mode, through one of two methods: (a) iterative and (b) secant.

The Iterative Method

The algebraical (or even transcendental) equation *F(x)* = *0* is brought to the form:

$$x = \mathrm{F_{iter}}(x) \quad \text{(A.5a)}$$

which derived $d\,F_{\text{iter}}/dx$ with values between -1 and 1. Starting from the well-chosen initial value x_0, the successive values $x_1, x_2, \ldots, x_n$ are calculated:

$$\forall n \geq 0 : x_n = \mathrm{F_{iter}}(x_n - 1) \tag{A.5b}$$

The iteration stops when the difference between two successive approximations of x is inferior to the desired precision for the equation's solution.

If the equation $F(x) = 0$ is algebraical, that is, defined through the equality $P(x) = 0$, where $P(x)$ is a polynomial of integer positive powers of the real variable x, then it proves to be effective to bring the equation to the form (A.5a) by adopting one of the iterative functions:

$$\mathrm{F_{iter}}(x) = \pm\sqrt[N]{\mathrm{C_N} \cdot x^N - \mathrm{P}(x)} \tag{A.5c}$$

where N is the degree of the polynomic and C_N is the coefficient of its superior term (the one with the highest power of x), and the sign « – » before the root sign appears only if N is even and, in this case, serves to determining an eventual negative root of the equation.

For instance, in order to obtain the below equation roots with a 5 digits precision

$$\mathrm{P}(x) = 0, \mathrm{P}(x) = x^4 - 3 \cdot x^2 + 2 \cdot x - 5 = 0,$$

the adequate function for the iteration will be $\mathrm{F_{iter}}(x) = \pm\sqrt[4]{3 \cdot x^2 - 2 \cdot x + 5}$, the determining with « – » before the root sign will serve to determine a negative root and the other one at determining a positive root.

The negative root can be found with $x_{k+1} = \sqrt[4]{3 \cdot (x_k)^2 - 2 \cdot x_k + 5}$ through the following steps:

k	**0**	**1**	**2**	**3**	**4**	**5**	**6**
x_k	−4	−2.7947	−2.4151	−2.2864	−2.2418	−2.2262	−2.2207
k	**7**	**8**	**9**	**10**	**11**		
x_k	−2.2188	−2.2181	−2.2179	−2.2178	−2.2178		

The adopted value for the root is, therefore, −2.2178.

Determining the positive root with $x_{k+1} = \sqrt[4]{3 \cdot (x_k)^2 - 2 \cdot x_k + 5}$ is through the following steps:

k	**0**	**1**	**2**	**3**	**4**	**5**	**6**
x_k	0	1.4953	1.7183	1.7967	1.8249	1.8351	1.8388
k	**7**	**8**	**9**	**10**	**11**		
x_k	1.8402	1.8406	1.8408	1.8409	1.8409		

with the approximate value of the root of 1.8409.

The Secant Method

When the interval (x', x'') is known, in which one (or at least one) of the roots of the equation $F(x) = 0$ can be found, that is, when $F(x') \cdot F(x'') \leq 0$, the interval can be gradually narrowed down until the precision of the desired solution using the secant method as it was done in the first point from example 4.1 on pg. 170.

Complementary Readings

F– Physics; G – general chemical physics; M – molecular chemical physics; L – books of physical chemistry in Romanian; I, II, III – complementary works related to part I, II respectively III of this book; C – calculus and mathematical methods; A- auxiliary materials.

F	**Landau L.**, Lifshitz E. "Course of theoretical physics", publ. Butterworth-Heinemann, Oxford/ UK 1975–1986 in 10 volumes. The last editions are:
	Vol. 1 Mechanics 1976; vol. 2 The classical theory of fields 1975; Quantum mechanics. Non-relativistic theory 1977; vol. 4 Quantum electrodynamics 1982; vol. 5 Statistical physics 1980; vol. 6 Fluid mechanics 1987; vol. 7 Theory of elasticity 1986; vol. 8 Electro-dynamics of continuous media 1984; vol. 9 Theory of condensed state (= Statistical physics part II) 1980; vol. 10, Physical kinetics 1981.
	Vol. 3 and 10 are published at Pergamon, Oxford/ U.K.
	Schiller C. "Motion muntain – the adventure of physics. 26th Ed." Publ. University Science Books, Mill Valey/ CA/ USA 2016 in 6 volumes: Vol. I Fall, flow and heat; Vol. II Relativity; vol. III Light, charges and brains; vol. IV The quantum of change; vol. V Motion inside matter – pleasure, technology and stars; vol. VI The strand model, a speculation on unification" available online
	Wallard A. & al. (editors) "Kaye & Laby tables of physical & chemical constants., online 2th Ed." publ. Natl Physical Lab/ USA 2016
G	**Atkins P.**, "Atkins' physical chemistry. 10th rev. ed. " publ. Oxford Univ., Oxford/ U.K. 2014
	Bahl B., Tuli G., Bahl A. "Essential of physical chemistry " publ. S. Chand, New Delhi/ India 2012
	Murgulescu I. & al. „Introduction in physical chemistry" publ. Edit. Academiei, Bucharest 1976–1986 in 6 volumes: vol.I-1. Atoms, molecules, chemical links; vol.I-2. The structure and properties of molecules; vol. II-1. The molecule-kinetics theory of matter; vol. II-2. Chemical kinetics and catalysis; vol. III – Chemical thermodynamics; vol. IV - Electrochemistry.
	Trapp C. "Student's solution manual to accompany Atkins' physical chemistry. 10th ed." publ. Oxford Univ., Oxford/ U.K. 2014"

(continued)

F. E. Daneş et al., *Molecular Physical Chemistry for Engineering Applications*,
https://doi.org/10.1007/978-3-030-63896-2

M	**Atkins P.**, by Paula J., Friedman R. "Quanta, matter and change – a molecular approach to physical chemistry" publ. Oxford Univ., Oxford/ U.K. 2008
	McQuarrie D. „Physical Chemistry: a Molecular Approach. 3nd Ed." Publ. Viva Edn., Berkeley/ CA/ USA 2011
	Silberberg M. "Chemistry: the molecular nature of matter and change. 7th Ed." publ. McGraw-Hill, New York 2015
	Yates J., Johnson J. "Molecular physical chemistry for engineers" publ. Univ. Science Books, Mill Valley/ CA/USA 2007
L	[1] **Daneş F.**, Daneş S., Petrescu V., Ungureanu M. (vol. 1) „TERMODINAMICA CHIMICĂ - O termodinamică a materiei (The chemical thermodynamics. Thermodynamics of matter)" publ. AGIR, Bucharest 2013
	[2] **Daneş F.**, Daneş S., Petrescu V., Ungureanu M. (vol. 2) „CHIMIE FIZICĂ MOLECU LARĂ (Physical molecular chemistry)" publ. AGIR, Bucharest 2016
	[3] **Daneş F.**, Ungureanu M. (vol. 3) „CINETICA TRANSFORMĂRILOR FIZICO-CHIMICE (The kinetics of the physical-chemical transfomations)" publ. AGIR, Bucharest 2009
I	**Jedrzejewski F.** "Modèles aléatoires et physique probabiliste" publ. Springer, Paris 2009
	Sethna J. "Statistical mechanics: Entropy, order parameters, and complexity" publ. Oxford Univ., Oxford/ U.K. 2006
	Velasco E. "Fundamentals of thermodynamics and statistical mechanics. Second Ed." publ. https://www.createspace.com/ 2010
II	**Gavezzotti A.** "Molecular aggregation– Structure analysis and molecular simulation of crystals and liquids" publ. Oxford Univ., Oxford/ U.K. 2013
	Goodstein D. "States of matter. Collection Phoenix" publ. Dover, Mineola/ NY/USA 2014
	Matthews P. "Gases, liquids and solids" publ. Cambridge Univ., Cambridge/ U.K. 1995
III	**Ghajar A.**, Cengel Y. "Heat and mass transfer – fundamentals and applications. third Ed." publ. McGraw-Hill, New Delhi/ India 2016
	Lightfoot E., **Bird** R., Stewart W. "Transport phenomena. 3nd Ed." publ. Wiley, New Delhi/ India 2014
	Plawsky J. "Transport phenomena fundamentals. third Ed." publ. Taylor & Francis, Boca Raton/ FL/USA 2014
C	**Fink J.** "Physical chemistry in depth" publ. Springer, Heidelberg/ Germany 2009
	Gradshteyn I., Ryzhik I. "Table of integrals, series, and products. Seventh Ed." publ. Elsevier, Oxford/ U.K. 2007
	Story T. "Mathematical modeling: kinetics, thermodynamics and statistical mechanics" publ. iUniverse, Bloomington/ IN/ USA 2011
A	**Bunin B.** & al. "Chemoinformatics: theory, practice and products" publ. Springer, Dordrechts/ Netherlands 2007
	IUPAC (ed.) "Quantities, units and symbols in physical chemistry. third ed." publ. Royal Soc. of Chemistry, London 2007
	Sansonetti J., Martin W. "Handbook of basic atomic spectroscopic data" J. Phys. Chem. Ref. Data 34 (4) 1559–2235 (2005)/ Amer. Inst. Physics, 2005 available online

Index

F. E. Daneş et al., *Molecular Physical Chemistry for Engineering Applications*,
https://doi.org/10.1007/978-3-030-63896-2

F

Printed in the United States
by Baker & Taylor Publisher Services

Printed in the United States
by Baker & Taylor Publisher Services